Specialist Periodical Reports Online

Application for Free Access to Electronic Chapters

Catalysis Volume 25

Customers purchasing a print volume are now entitled to free site-wide access to the electronic version of that title.
(For the definition of a site, please consult **www.rsc.org/subagree**)

To apply for free access, please complete and return this form.

Contact Name: ______________________________

Job Title: ______________________________

Organisation: ______________________________

Address: ______________________________

____

____

Town: ______________________________

Country: ______________________________

Post/Zip Code: ______________________________

E-mail: ______________________________

Telephone: ______________________________

Continued over

Please give IP addresses of the site overleaf in the format w.x.y.z.

Wildcards and ranges are allowed

e.g. for Class B 128.128.*.*

for Class C 192.192.192.*

for a range 123.456.1-99.*

Please put each entry on a new line:

____

____

____

____

(please use a continuation page if necessary)

I have read and agree to the terms of the Electronic Information Licence Agreement at **www.rsc.org/subagree**

Signed: ____________________ **Date:** ________________

☐ Please contact me about access for additional sites outside the site definition in the Electronic Information Licence Agreement

☐ Please arrange username/password access and not IP address control

Please return this form by non-electronic means to the Royal Society of Chemistry at the address below.

Photocopies or facsimiles are not acceptable.

Books Dept

Royal Society of Chemistry · Thomas Graham House

Science Park · Milton Road · Cambridge · CB4 0WF · UK

T +44(0)1223 420066 · **F** +44(0)1223 420247

E SPR@rsc.org

Registered Charity No. 207890

The Royal Society of Chemistry will store the information you supply on its electronic records in order that information about its activities, products and services may be sent to you by mail, telephone, email or fax.

If you DO NOT wish to receive information, please put a tick in the box ☐.

Catalysis

Volume 25

A Specialist Periodical Report

Catalysis

Volume 25

A Review of Recent Literature

Editors
James J. Spivey, *Louisiana State University, USA*
Yi-Fan Han, *East China University of Science and Technology, Shanghai, China*
K. M. Dooley, *Louisiana State University, USA*

Authors
Jun Bao, *University of Science and Technology of China, Hefei, China*
A. Beretta, *Politecnico di Milano, Italy*
Natalia Böhmer, *hte Aktiengesellschaft, Heidelberg, Germany*
Mercedes Boronat, *Instituto de Tecnología Química, Valencia, Spain*
Uwe Burghaus, *North Dakota State University, USA*
De Chen, *Norwegian University of Science and Technology, Norway*
A. Donazzi, *Politecnico di Milano, Italy*
Xuezhi Duan, *East China University of Science and Technology, China*
P. Forzatti, *Politecnico di Milano, Italy*
Karin Föttinger, *Vienna University of Technology, Austria*
Cornelia Futter, *hte Aktiengesellschaft, Heidelberg, Germany*
G. Groppi, *Politecnico di Milano, Italy*
Andreas Kuschel, *hte Aktiengesellschaft, Heidelberg, Germany*
M. Maestri, *Politecnico di Milano, Italy*
Eko Prasetyo, *hte Aktiengesellschaft, Heidelberg, Germany*
Thomas Roussière, *hte Aktiengesellschaft, Heidelberg, Germany*
Stephan A. Schunk, *hte Aktiengesellschaft, Heidelberg, Germany*
E. Tronconi, *Politecnico di Milano, Italy*
Noritatsu Tsubaki, *University of Toyama, Japan*
Xinggui Zhou, *East China University of Science and Technology, China*

RSC Publishing

If you buy this title on standing order, you will be given FREE access to the chapters online. Please contact sales@rsc.org with proof of purchase to arrange access to be set up.

Thank you

ISBN: 978-1-84973-578-0
DOI: 10.1039/9781849737203
ISSN: 0140-0568

A catalogue record for this book is available from the British Library

Published by The Royal Society of Chemistry,
Thomas Graham House, Science Park, Milton Road,
Cambridge CB4 0WF, UK

Registered Charity Number 207890

For further information see our web site at www.rsc.org

Printed and bound in Great Britain by CPI Group (UK) Ltd, Croydon, CR0 4YY

Preface

DOI: 10.1039/9781849737203-FP005

This volume consists of reviews of topics of current interest to catalysis researchers, including computational methods, novel synthesis approaches, high-throughput experimental techniques, and energy-related catalysis.

In Chapter 1, Karin Föttinger (Vienna University of Technology, Austria) reviews the synthesis, characterization and utilization of PdZn and related intermetallic compounds. Applications range from selective hydrogenation to steam reforming. PdZn has similar electronic properties as Cu, but better stability in certain applications. Structure – activity correlations are examined for a broad range of materials from oxide supported nanoparticles to single crystal-based model systems. It is shown that PdZn may be attractive in methanol steam reforming for hydrogen production, especially in microreactor technology, but decreasing the amount of Pd used is probably a necessary next step. For optimal H_2 production a Cu-like electronic structure is needed, which is provided by tetragonal 1 : 1 PdZn. Oxidized Zn species at the PdZn/ZnO interface are probably needed to activate water. For selective hydrogenation catalysis, Pd site isolation is desired.

In Chapter 2, Alessandra Beretta, Alessandro Donazzi, Gianpiero Groppi, Matteo Maestri, Enrico Tronconi and Pio Forzatti (Politecnico di Milano, Italy) review their own and others' work on the partial oxidation of light hydrocarbons to syngas mixtures. They explain how novel experimental techniques and theoretical methods have advanced the fundamentals here, leading to the development of reliable kinetic schemes that can be used to interpret the complex behavior of short contact time partial oxidation (CPO) reformers and improve their engineering. The emphasis is on the development of experimental (microreactor designs, spatially resolved measurements of temperature and compositions) and modeling (DFT, microkinetic, and structured reactor models) tools that have formalized the kinetics of the surface reaction, identified the roles of surface chemistry, gas-phase chemistry, and heat and mass transfer phenomena. They conclude with some recent results on the optimal design of CPO-reformers, aimed at improving thermal management and preserving catalyst stability.

In Chapter 3, Xuezhi Duan and Xinggui Zhou (East China University of Science and Technology, Shanghai, China), and De Chen (Norwegian University of Science and Technology, Trondheim, Norway) review recent advances in structural manipulation of the catalysts for ammonia decomposition. Because ammonia decomposition catalysts are aimed for general application in energy and environmental industries, the cost of the catalyst is crucial. The performances of supported Ru, Ni, Fe and bimetallics are summarized, and the sizes and shapes of metal nanoparticles, promoters, and supports are reviewed and correlated with various preparation techniques. It is shown that B_5 sites in supported Ru are critical in catalyst function, while for supported Ni they are inactive. Electron donating groups, surface defects and unsaturated carbon atoms also greatly affect the reaction. For supported Ni, an appropriate ratio of Ni(111) to Ni(110) surface is crucial, while for supported Fe, the active species is α-Fe(N) formed by the nitriding reaction. Recent work in searching for new, more efficient catalysts involving bimetallics, and in reactor modifications, is also reviewed.

In Chapter 4, Uwe Burghaus (North Dakota State University, Fargo, USA) discusses the applications of electron beam lithography (EBL) in surface science and catalysis. Emphasis is placed on ultra-high vacuum model studies utilizing so-called nano-array catalysts, which consist of a periodic arrangement of clusters on a support. The surface reactions, kinetics, and dynamics of the interactions with gas-phase species of such catalysts in typical applications are reviewed, considering both the advantages and disadvantages of this approach, and some related nanofabrication techniques. Efforts to overcome the sequential nature of EBL techniques by developing high-throughput patterning methods are recently under way, and these are also considered briefly.

In Chapter 5, Mercedes Boronat (Universidad Politécnica de Valencia, Spain) reviews computational tools used to study oxygenation and hydrogenation on gold catalysts. These catalysts have been widely studied for reactions such as water gas shift and selective oxidation, among others. In this chapter, the dependence of catalytic activity of gold on factors such as particle size, particle morphology, surface concentration of coordinatively unsaturated Au atoms, and the oxidation state of gold are explored from a theoretical point of view. Systematic and controlled variation of these properties independently over a wide range of interest is virtually impossible experimentally. Using a computational approach, this chapter carefully compares theoretical results to available experimental results to validate the important conclusions about the catalytic importance of the properties of these materials.

In Chapter 6, Stephan Schunk (hte Aktiengesellschaft, Germany), along with co-authors Natalia Böhmer, Cornelia Futter, Andreas Kuschel, Eko Prasetyo and Thomas Roussière, review high-throughput experimentation (HTE) as an approach to screening of catalysts more efficiently than conventional sequential methods. The importance of catalysis to the chemical industry necessitates a constant search for catalysts that are more active, selective, and stable. HTE allows this search to include catalysts that might not be evaluated otherwise, since most catalyst development focuses on incremental improvements to proven formulations. These methods are applicable in both domains: homogeneous and heterogeneous catalysis. One key element to this approach is a robust software architecture to allow efficient experimental planning and analysis of large amounts of information in a short time. This chapter discusses this issue, along with proper methods to develop libraries and to design experiments.

In Chapter 7, Jun Bao (University of Science and Technology of China) and Noritatsu Tsubaki (University of Toyama, Japan) review recent work in the synthesis and characterization of core-shell and bimodal catalysts for Fischer-Tropsch (FT) synthesis. These catalysts are examples of improvements to these well-studied materials. The core-shell morphology is prepared by coating a spherical FT catalyst with a porous zeolite membrane shell. These materials help increase the yield of iso-paraffins, a valuable product, while minimizing selectivity to waxes and heavy products. Bimodal catalysts are prepared using oxides as building blocks to produce small pores inside large pores. This structure increases FT activity by limiting diffusion resistance in the large pores while increasing the dispersion and area concentrartion of the active metal sites in the smaller pores.

I appreciate the contributions of my co-editors (Kerry Dooley, LSU) and Yi-fan Han (East China University of Science and Technology). As always, comments are welcome.

James J. Spivey and Kerry Dooley
Gordon A. and Mary Cain Dept. Chemical Engineering
Louisiana State University
Baton Rouge, LA 70803
jjspivey@lsu.edu; dooley@lsu.edu

Yi-fan Han
East China University of Science and Technology
Shanghai, China
yifanhan@gmail.com

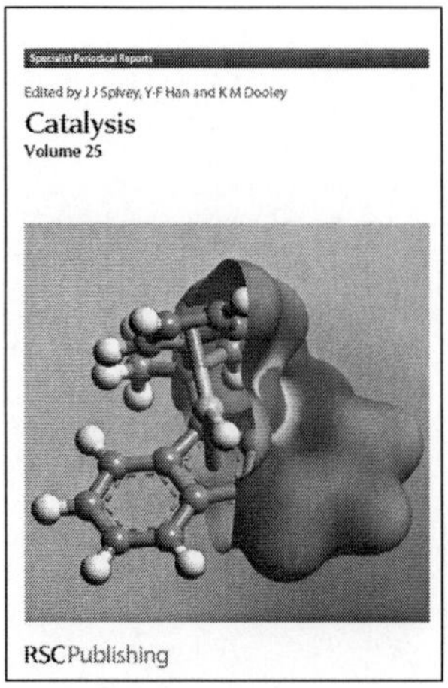

Cover

Image provided courtesy of computational science company Accelrys (www.accelrys.com).
An electron density isosurface mapped with the electrostatic potential for an organometallic molecule. This shows the charge distribution across the surface of the molecule with the red area showing the positive charge associated with the central metal atom. Research carried out using Accelrys Materials Studios®.

Gaining insight into the kinetics of partial oxidation of light hydrocarbons on Rh, through a multiscale methodology based on advanced experimental and modeling techniques

A. Beretta,* A. Donazzi, G. Groppi, M. Maestri, E. Tronconi and P. Forzatti
DOI: 10.1039/9781849737203-00001

This chapter updates previous reviews on the catalytic partial oxidation of methane and light hydrocarbons over noble metals; specifically, it focuses on the development of experimental and modeling tools that in recent years allowed to measure with accuracy and formalize the kinetics of the surface process, thus setting the basis for the engineering of short contact time CPO reformers. Such advanced tools include special micro-reactor designs for the kinetic investigation under isothermal conditions, first-principle microkinetic schemes, techniques for the spatially resolved measurement of temperature and concentration profiles inside working adiabatic reactors, detailed reactor models accounting for the role of transport phenomena in structured catalysts as well as that of homogeneous reactions. These contributions pave a multi-scale path, that runs from the fundamentals of surface kinetics to the reactor optimal design.

1 Introduction

Small scale reformer designs are currently being developed for the distributed and small scale production of H_2.[1–5] Future and futuristic applications are: the fuelling of H_2-vehicles, the use of cogeneration systems based on fuel cells for residential heat and power supply, the valorization of CH_4-rich bio-gas, the on-board generation and utilization of H_2-rich streams as co-fuels, reducing agents and/or feed of Auxiliary Power Unit systems based on SOFC.

Fuel reformers produce hydrogen or hydrogen/CO_X mixtures by exploiting the catalytic conversion of gaseous or liquid fuels in the presence of steam, CO_2 and/or O_2 or air. Among various processes and reactor solutions that have been proposed (externally heated reformers, exo-endo reformers, autothermal reformers), the catalytic partial oxidation (CPO) of hydrocarbon fuels in insulated short contact time reactors has several advantages: flexibility to scale up and down, fast light off, resistance to extinction after load changes, high H_2 yields at millisecond contact times, and the possibility of fully autothermal operation after light off.[6] This is important in systems of a smaller scale than conventional hydrogen or syngas plants, particularly if restrictions on weight and/or volume are tight.

The CPO process consists of the selective conversion of a hydrocarbon fuel and O_2 into CO and H_2 using metals (Ni, Pd, Pt, Rh but also Fe, Co

Laboratory of Catalysis and Catalytic Processes, Dipartimento di Energia, Politecnico di Milano, piazza Leonardo da Vinci 32, 20133 Milano. E-mail: alessandra.beretta@polimi.it

and other transition metals[7–9]), typically deposited in the form of a catalytic wash-coat over high-void fraction structures such as foams or honeycomb monoliths. Lanny Schmidt and his group at Minnesota University have pioneered the process and during the last twenty years largely demonstrated its potential over a wide range of fuels, from gaseous to liquid to solid ones.[1,10–21]

In recent years the development of novel experimental techniques and theoretical methods based on DFT has made possible the advancement of fundamental studies of CH_4 activation on metals, which have led to the development of reliable kinetic schemes that can be used to interpret the complex behavior of short contact time– CPO reformers and address their engineering. The surface kinetics of the partial oxidation of light hydrocarbons have been investigated to a lesser extent, but recent studies have shown important analogies between the paths of activation of hydrocarbon fuels of different nature.

The aim of this chapter is to exemplify such recent advances, by reviewing findings from our and other laboratories that, along a multiscale methodology, run from the surface kinetics to the reactor optimal design. Other previous review papers, in particular the recent paper by Enger, Rune and Holmen,[8] have given a broad view of mechanicistic findings (based also on the development of in-situ or operando characterization techniques) for various catalyst systems. In this work, the emphasis is on the development of experimental and modeling tools that have allowed to recognize and formalize the kinetics of the surface reaction, and also to identify the roles of surface chemistry, gas-phase chemistry, heat and mass transfer phenomena in short contact time CPO reformers.

Kinetic studies on Rh are mainly treated in this chapter, the reason being that experimental, theoretical and modeling studies have presently reached a maturity and a comprehension of the process drivers that is still missing in the case of other metals. However, investigations on other catalysts, mostly Pt, will be also commented because of their important methodological contributions. First (Section III), we illustrate the kinetic findings from isothermal reactor designs, wherein the surface chemistry was isolated and kinetically controlled data were collected on CH_4, C_2H_6 and C_3H_8 activation in the presence of O_2 and steam. Then (Section IV) we present the microkinetic modeling of CH_4 partial oxidation on Rh based on *a priori* estimations of the kinetic parameters; the model was extremely useful in identifying the prevalent reaction paths behind the observed kinetic dependences. Concerning then the more complex and representative scale of adiabatic, short contact time CPO-reformers, a state of the art technique for the experimental investigation is represented by the axially resolved sampling technique; this will be also presented (Section IV). It is an extremely powerful technique that provides detailed information on the axial evolution of the process stoichiometry and of the corresponding temperature evolution. However, a thorough interpretation of the data needs the means of reactor modeling, which in turn needs the best possible description of the transport phenomena characteristic of the specific reactor configuration (Section V).

Finally, the chapter presents (Section VI) some case-studies wherein the spatially resolved measurements of temperature and composition, in combination with predictive modeling, have driven important advances in the comprehension of the CPO process and set the basis for the rational design of improved CPO-reformers. We conclude by presenting recent results on the optimal design of CPO-reformers, aimed at improving the thermal management of the reactor and preserving the catalyst stability (Section VII).

The ending paragraph attempts then to identify issues that still remain open, including some controversial findings between mechanicistic studies (elsewhere treated) and kinetic studies.

2 The severity of the CPO process

Methane partial oxidation, with stoichiometry:

$$CH_4 + 0.5\,O_2 \rightarrow CO + 2\,H_2 \qquad \Delta H^\circ_{298} = -36\,kJ/mol \qquad (1)$$

is a mildly exothermic reaction. However, several reactions may be catalyzed by metals in the presence of CH_4 rich streams, including the total oxidation of CH_4 (2), the steam reforming (3) and CO_2 reforming (4) of CH4, the water gas shift reaction (5).

$$CH_4 + 2\,O_2 \rightarrow CO_2 + 3\,H_2O \qquad \Delta H^\circ_{298} = -\,802\,kJ/mol \qquad (2)$$

$$CH_4 + H_2O \rightarrow CO + 3\,H_2 \qquad \Delta H^\circ_{298} = +206\,kJ/mol \qquad (3)$$

$$CH_4 + CO_2 \rightarrow 2\,CO + 2\,H_2 \qquad \Delta H^\circ_{298} = +247\,kJ/mol \qquad (4)$$

$$CO + H_2O \rightarrow CO_2 + H_2 \qquad \Delta H^\circ_{298} = -\,41\,kJ/mol \qquad (5)$$

The presence of a direct route of partial oxidation has not been excluded by some authors in the literature. But, in practice, the catalytic conversion of CH_4 in the presence of O_2 or air involves simultaneously both strongly exothermic and endothermic reactions; sharp gradients of temperature and concentration will tend to develop within the catalyst volume, when the process is run under adiabatic conditions in insulated short contact time reactors.

Thus, although adiabatic short contact time reactors are extremely flexible and useful for demonstrating the process or testing a new catalytic formulation under representative conditions at the laboratory scale, they are not suitable tools for a rigorous kinetic investigation. The quantitative analysis of the observed performances needs the means of mathematical modeling, but the description of the flow pattern and the quantification of transfer phenomena, if we consider for instance the case of foams, may not be consolidated.[22,23] It is also useful to consider that since the final temperature profile which establishes within an insulated structured catalyst is a result of the operating conditions, it inevitably varies at varying conditions; thus, in an autothermal reactor parameters cannot be varied one at a time as in the traditional kinetic studies.

The understanding of the surface kinetics calls for the development of dedicated lab-scale reactors, wherein the kinetic control of the heterogeneous reaction dominates over other chemical and physical phenomena and the catalyst temperature is uniform in the whole bed. Different approaches have been adopted in the literature and witness an effort of either eliminating any artefact, or to contain them within a measurable form.

3 Experimental tools for the kinetic investigation of the CPO of CH_4 and light hydrocarbons under isothermal conditions: reactor designs

We first discuss two alternative designs, recently applied in the literature for studying the kinetics of CH_4 activation on Rh and other metals: the differential ultra-diluted packed-bed reactor adopted by Iglesia and co-workers to eliminate any transport artifact, and the annular reactor proposed by our group to obtain integral data under reduced and well defined impact of mass transfer. Other reactor designs are also mentioned.

3.1 The gradient-less differential micro-reactor

The ideal reactor for the kinetic investigation is the one where the temperature and the concentration of species (reactants and products) at the active site have the same values as in the gas phase and any transport limitation is ruled out; along this principle, in recent years Iglesia's group has developed an "infinitely" diluted micro-reactor and applied it for the fundamental study of CH_4 activation with O_2, steam and CO_2 over Pt, Rh and other metals.[24–32]

Two levels of dilution are involved: intraparticle dilution and bed dilution. Both levels are pushed up to a point where turnover rates do not depend any longer on them. Preferred values of intra-particle dilution on Pt were found to be higher than 200, while optimal values of bed dilution were found to be higher than 2000.

The intra-particle dilution was obtained through a laborious process: a certain, measurable amount of catalyst was first mixed with a comparable amount of inert powders, then the mixed powders were pelletized, crushed, sieved to a suitable mesh size, and then mixed again with another definite amount of inert powders. This procedure needs to be repeated many, many times before achieving the desired diluent to catalyst ratio (at least 200, as mentioned above). The diluted catalyst particles were then mixed with particles of inert material in order to realize the desired bed dilution.

The preparation procedure is such that the final catalyst density referred to the reactor volume is extremely low, only chemical events become kinetically relevant and the reactor behaves as a gradient-less reactor. However, because of the low amount of catalyst which at the end of the procedure is present in the reactor, contamination becomes a major issue: any part of the reactor (quartz wool, thermocouple sheath, the internal walls) must be carefully cleaned and protected with specific coatings. Coherently, attention must be paid when following the preparation procedures. Finally, from a fundamental viewpoint, this high metal dilution can possibly lead to loss of activity due to metal/support effects (*e.g.* formations of alloys or solid solutions) as pointed out by Jones *et al.*[24]

Concerning the field of application of this reactor design, it must be considered that the influence of bypass in the reacting flow can be neglected only by operating under differential regime.[25] This is in fact the regime within which the kinetic investigations on CH_4 activation with H_2O and CO_2 were studied by Wei and Iglesia[26] over various metals. Similarly, differential regime data were collected to study the dependence of oxygen coverage on the kinetics of CH_4 oxidation on Pt, Rh and Pd surfaces and ethane oxidation on Pt surfaces by Chin, Iglesia *et al.*[27–30]

It was demonstrated that over Ni, Pt and Rh the turnover frequency of the forward reforming of CH_4 with steam or CO_2 (later identified as Regime 4 of CH_4 activation) is independent from the nature and the amount of the co-reactant, while it is proportional to the partial pressure of CH_4, in strict analogy with methane pyrolysis:

$$r_{H2O-ref} = r_{CO2-ref} = r_{pyr} = k\, p_{CH4} \tag{6}$$

In more recent studies, the infinitely diluted fixed bed reactor has been successfully applied to investigate the kinetics of CH_4 oxidation and C_2H_6 oxidation on Pt surfaces; four distinct kinetic regimes have been identified at varying oxygen coverage. It has been shown that over oxygen saturated Pt surfaces (Regime 1) CH_4 oxidation is kinetically limited only by CH_4 activation and the turnover rate is proportional to CH_4 partial pressure and independent from O_2 partial pressure. At decreasing O_2/CH_4 feed ratio, a dramatic enhancement of the turnover rate is observed (up to two orders of magnitude higher than in Regime 1) and an inverse dependence from O_2 partial pressure together with a quadratic dependence from the partial pressure of the fuel come into place (Regime 2). The experimental evidence on the positive effect of decreasing surface oxygen coverage is supported by DFT calculations, according to which CH_4 adsorption is strongly favored on O^*-* site pairs, because of a concerted stabilization of (Pt-CH_3---H–O^*) transition state. At further decreasing the O_2 surface coverage, O_2 activation becomes the limiting reaction step and the turn over rate becomes proportional to O_2 partial pressure (Regime 3). The limiting case of O_2 surface depletion is than represented by pyrolysis and reforming reactions (identified with Regime 4 as above mentioned). Regimes, turn over rates and kinetic dependences for methane and ethane oxidation on Pt are reported in Fig. 1.

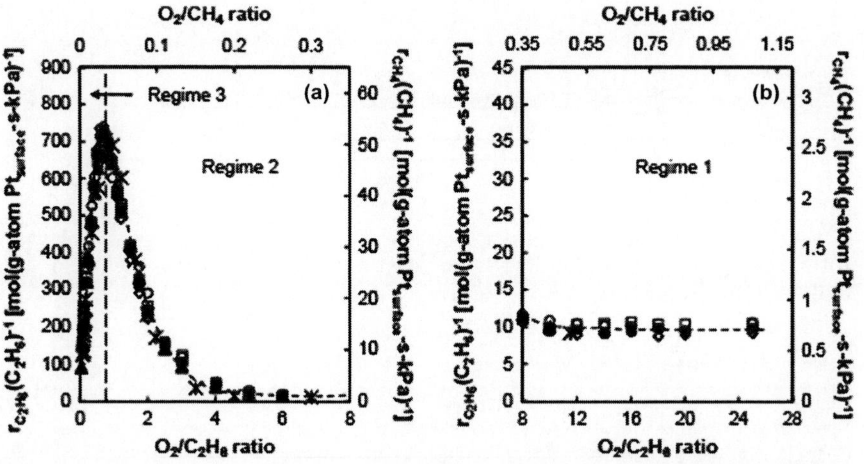

Fig. 1 Measured effect of the O_2/C ratio on the turnover rate of methane and ethane oxidation on Pt by using a infinitely diluted differential microreactor. Reprinted from Journal of Catalysis, vol. 285, M. García-Diéguez, Y. H. C. Chin, and E. Iglesia Catalytic reactions of dioxygen with ethane and methane on platinum clusters: Mechanistic connections, site requirements, and consequences of chemisorbed oxygen, 260–272, Copyright (2012), with permission from Elsevier.

3.2 The annular micro-reactor

The concept of annular reactor was originally introduced by McCarthy[31] to study the catalytic combustion of methane; it was later applied and further improved in our laboratory to study the short contact time oxidative dehydrogenation of light alkenes over Pt[32] as well as the kinetics of CH_4 combustion over Pd.[33,34]

Figure 2 reports a schematic representation. The reactor consists of a catalyst coated ceramic tube, coaxially inserted into a quartz tube giving rise to an annular duct through which the gas stream flows in laminar regime.

The main geometrical parameters of the reactor are listed in Table 1.

The reactor design has been optimized, with the main aim of performing experiments at very high space velocities without appreciable pressure drop, far from thermodynamic constraints, under a limited or even negligible effect of external and internal mass transfer limitations; these, given the simple geometry of the catalyst layer and the known flow field of the reactor, can be easily quantified by the use of well assessed literature correlations. The thermal behavior of the annular reactor has been quantitatively analyzed within a EUROKIN project,[35] which has shown that the mechanism of heat dissipation consists of back-conduction of the heat of reaction from the catalyst layer through the alumina tube and radiation from the entire surface of the tube (the coated portion and the surface upstream) towards the oven wall. Such a mechanism is so efficient that quasi isothermal conditions are easily obtained when the reactor is operated with diluted feed streams. Instead, under concentrated feed streams axial temperature gradients develop, but their extent is largely moderated by the

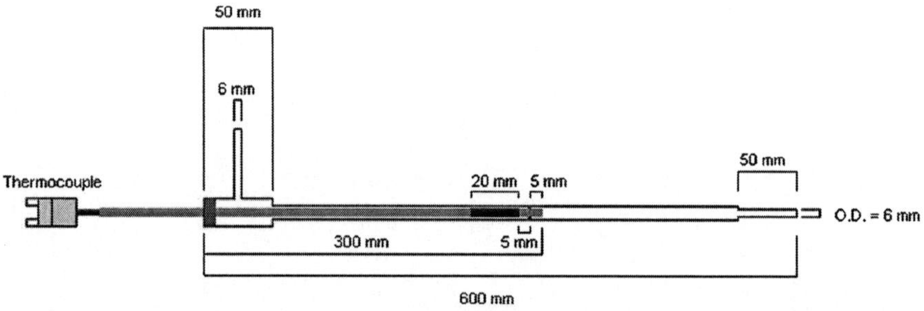

Fig. 2 The annular reactor. Reprinted from Journal of Catalysis, vol. 255, A. Donazzi, A. Beretta, G. Groppi and P. Forzatti, Catalytic partial oxidation of methane over a 4% Rh/ α-Al_2O_3 catalyst: Part I: Kinetic study in annular reactor, 241–258, Copyright (2008), with permission from Elsevier.

Table 1 Geometry of the annular reactor.

Quartz tube: I.D. = 5 mm, O.D. = 8 mm
Dense α-Al_2O_3 Tube: I.D. = 3 mm, O.D. = 4 mm
Channel Height = 0.5 mm
Catalyst weight = 7.5 mg
Catalyst layer length = 20 mm
Catalyst layer thickness = 20 μm
Washcoat density ρ = 1.5 g/cm^{-3}

radiation towards the inner furnace surface. Also, it has been shown that the sliding thermocouple inside the ceramic tube provides a reliable estimate of the actual catalyst temperature.

The design specifically represented in Fig. 1 has been extensively applied to investigate the process of CH_4 activation on Rh with O_2 and/or CO_2 and/or H_2O.[36,37] The wide investigated experimental field is summarized in Table 2. Tests with CH_4/O_2 mixtures (the CPO tests) were performed from 300 to 850 °C; the operation at high space velocity (varied from 8×10^5 to 4.5×10^6) results in a decreased sensitivity of the conversion curves on temperature, so that it enlarges significantly the T-window for the kinetic investigation.

Experiments were purposely performed both in differential and in integral regime, up to complete conversion of the reactants. As more clearly shown in the following, the production of synthesis gas on Rh occurs only consecutively to complete O_2 conversion. Thus, pieces of information on the kinetics of H_2 and CO production can be obtained only at high conversion of reactants.

A model of the annular reactor, accounting for gas-solid concentration gradients and intra-porous mass transfer limitations, was necessary to derive the kinetic dependences from the data.[36]

Table 2 shows that the kinetics of reacting subsystems were also studied independently: CH_4/steam, CH_4/CO_2, CO/H_2, CO_2/H_2, CO/O_2 and H_2/O_2. These data were extremely useful to gain independent kinetic information on the several steps involved in the process.

The investigation was recently extended to study the kinetics of propane partial oxidation on Rh. The explored operating field is summarized in Table 3. Notably, the reactivity of CO/H_2 and CO_2/H_2 systems was further investigated to better elucidate the kinetics of methanation.

Figure 3 shows an example of standard CH_4 CPO and C_3H_8 CPO runs over a 2% Rh/α-Al_2O_3 catalyst at equal inlet C-concentration (3% CH_4 *vs.* 1% C_3H_8 at $O_2/C = 0.56$). Dotted lines represent the thermodynamic equilibria. In both cases, the conversion of reactants started at about 300 °C; O_2 conversion increased steadily with temperature and was complete at about 400 °C in the case of propane and about 500 °C in the case of methane. Up to complete O_2 conversion, CO_2 and H_2O were uniquely observed in the

Table 2 Syngas from CH_4: investigated experimental field.

CH_4 CPO tests:
$CH_4 = 1–27\%$
$O_2/CH_4 = 0.56–1$
H_2O, CO_2, CO co-feed = 1–2%
Steam Reforming tests:
H_2O content = 1.5–2.5%
Dry Reforming tests:
$CO_2 = 1–4\%$, $CH_4 = 1–2\%$
O_2 co-feed = 0.1%, H_2 co-feed = 0.1–1%
RWGS tests:
$CO_2 = 1–2\%$, $H_2 = 1–4\%$
H_2O, CO co-feed = 0.5%
WGS tests
$CO = 1–2\%$, $H_2O = 1–4\%$
CO & H_2-rich combustion tests

Table 3 Syngas from C_3H_8: investigated experimental field.

C_3H_8 CPO tests:
$C_3H_8=0.5–5\%$
$O_2/CH_4=0.56–1$
H_2O co-feed = 1–2%
Steam Reforming tests:
$C_3H_8=0.5–3.0\%$
H_2O content = 1.5–8.0 %
Methanation tests:
$CO_2=1–4\%$, $H_2=1–10\%$
$CO=1–4\%$, $H_2=1–10\%$

product mixture. At higher temperatures, the conversion of the fuel proceeded and H_2 and CO were progressively formed at the expense of the deep oxidation products. At this high space velocity, the fuel conversion was complete at high temperature, close to 850 °C. It is interesting to note that, in the C_3H_8 CPO experiment, methane formation accompanied the production of synthesis gas; its concentration passed through a maximum located at about 500 °C, coherently with the thermodynamic constraints.

Figure 4 illustrates the measured effect of varying the O_2/CH_4 ratio (above the stoichiometric value of 0.5).[36] At lower temperatures, where deep oxidation was uniquely active, CH_4 conversion was insensitive to the increase of O_2 partial pressure, thus confirming the picture of a fully oxygen saturated surface, wherein CH_4 activation is the only rate determining step. Looking at the initial formation of synthesis gas, the data showed that this was shifted at higher temperatures at increasing O_2 feed content.

However, experiments with $CH_4/H_2O/N_2$ fed mixtures, reported in Fig. 5, clearly showed that steam reforming was also active at 300 °C with comparable turnover rates as those measured for the fuel oxidation.[36] It was thus concluded that in the CPO process, steam reforming does not occur until the surface oxygen coverage becomes critically low, and that H_2 and CO do not desorb from the surface in the presence of residual surface O_2.

The steam reforming experiments also showed that the rate of the CH_4 consumption was independent from the amount of H_2O fed (always over-stoichiometric, in these experiments).

CH_4 CPO experiments at increasing concentration of both reactants showed a decrease of conversion in the whole temperature range. This was interpreted as an evidence that surface saturation effects slow down the reactivity of the catalyst and produce overall kinetic orders below unity for both the oxidation and the steam reforming processes.

Another important result from the investigation in annular reactor was the identification of a probable reaction scheme behind the stoichiometry of CO_2 reforming of CH_4.[37] Based on independent data on the reactivity of H_2O/CH_4, $CO_2/CH_4(/O_2)$, CO_2/H_2 reacting systems, it was shown that the apparent kinetics of CO_2 reforming were in reality well explained by the combination of the rates of steam reforming and reverse-water gas shift. The rate determining step (either CH_4 activation, or CO_2 activation) would uniquely depend on the partial pressure of CO_2. In fact, at the stoichiometric CO_2/CH_4 ratio of 1, we found that the CO_2 reforming of CH_4 had the same

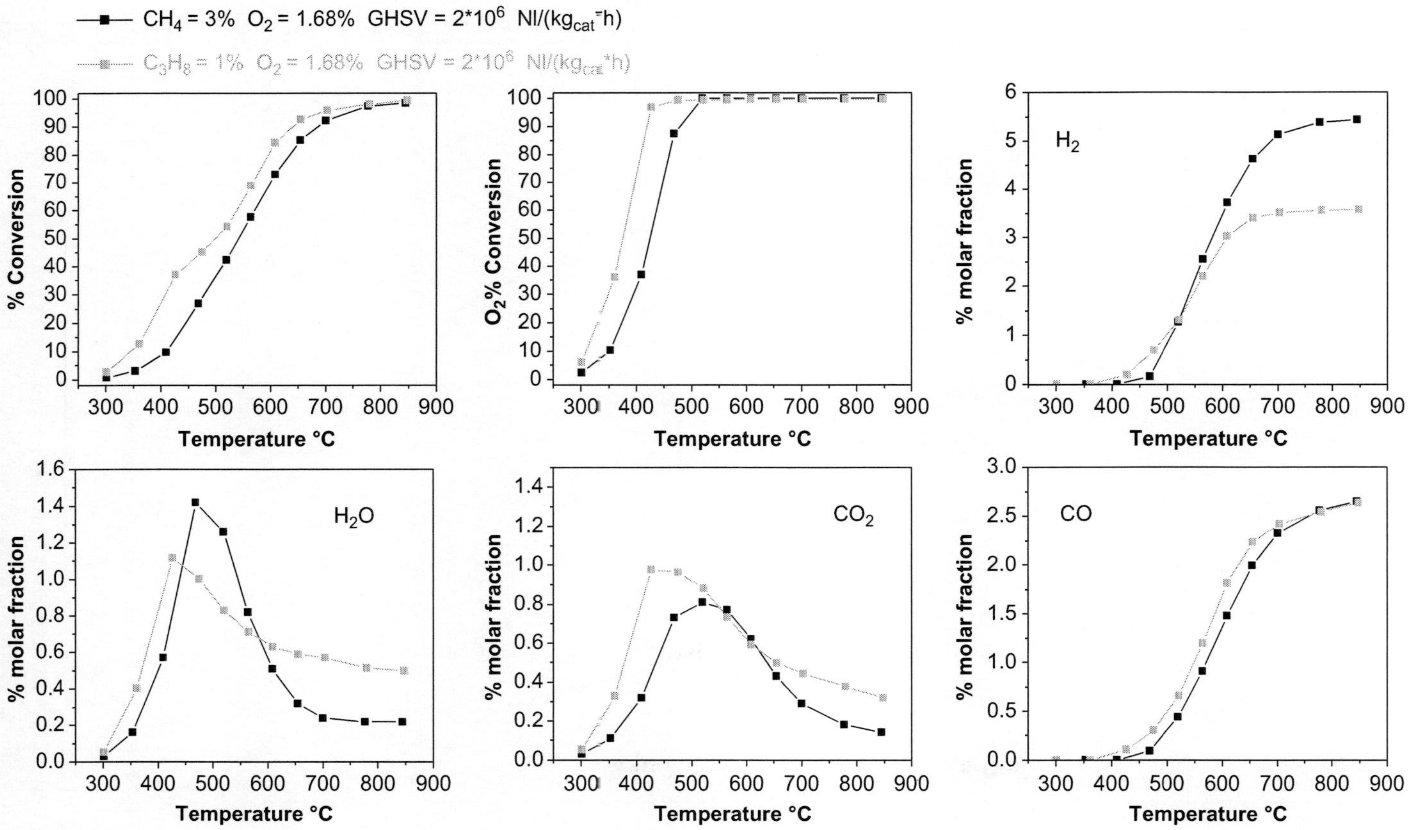

Fig. 3 CPO of CH_4 over a 4% Rh/Al_2O_3 catalyst and CPO of C_3H_8 over a 2% Rh/Al_2O_3 catalyst in annular reactor. $GHSV = 2 \times 10^6$ l(NTP)/kgcat/h. $O_2/C = 0.56$, $CH_4 = 3\%$, $C_3H_8 = 1\%$, balance N_2.

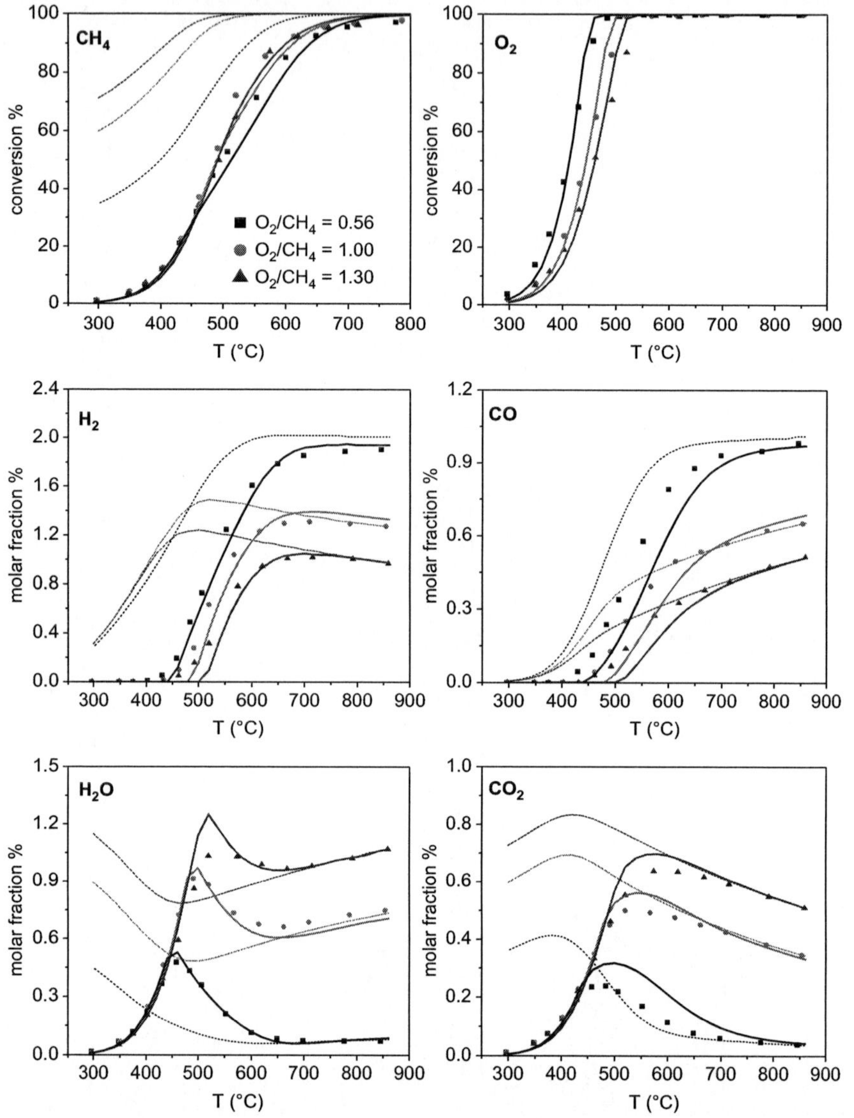

Fig. 4 CH_4 CPO experiments in annular reactor. GHSV $= 2 \times 10^6$ l(NTP)/kgcat/h. $O_2/C = 0.56$, 1.00, 1.30 at constant CH_4 mole fraction of 1%, balance N_2. Reprinted from Journal of Catalysis, vol. 255, A. Donazzi, A. Beretta, G. Groppi and P. Forzatti, Catalytic partial oxidation of methane over a 4% Rh/α-Al_2O_3 catalyst: Part I: Kinetic study in annular reactor, 241–258, Copyright (2008), with permission from Elsevier.

rate as the reverse-water gas shift reaction and was proportional to CO_2 partial pressure, while at CO_2/CH_4 ratios higher than 2, the CO_2 reforming of CH_4 had the same rate as the steam reforming of CH_4 and was thus independent from the amount of the co-reactant. All the data of CO_2 reforming could be very well explained on the assumption that steam reforming and reverse water gas shift were uniquely active, with independent rates. It is worth stressing that the understanding of the kinetic scheme behind CO_2-reforming strongly relied on the independent characterization of the kinetics of the

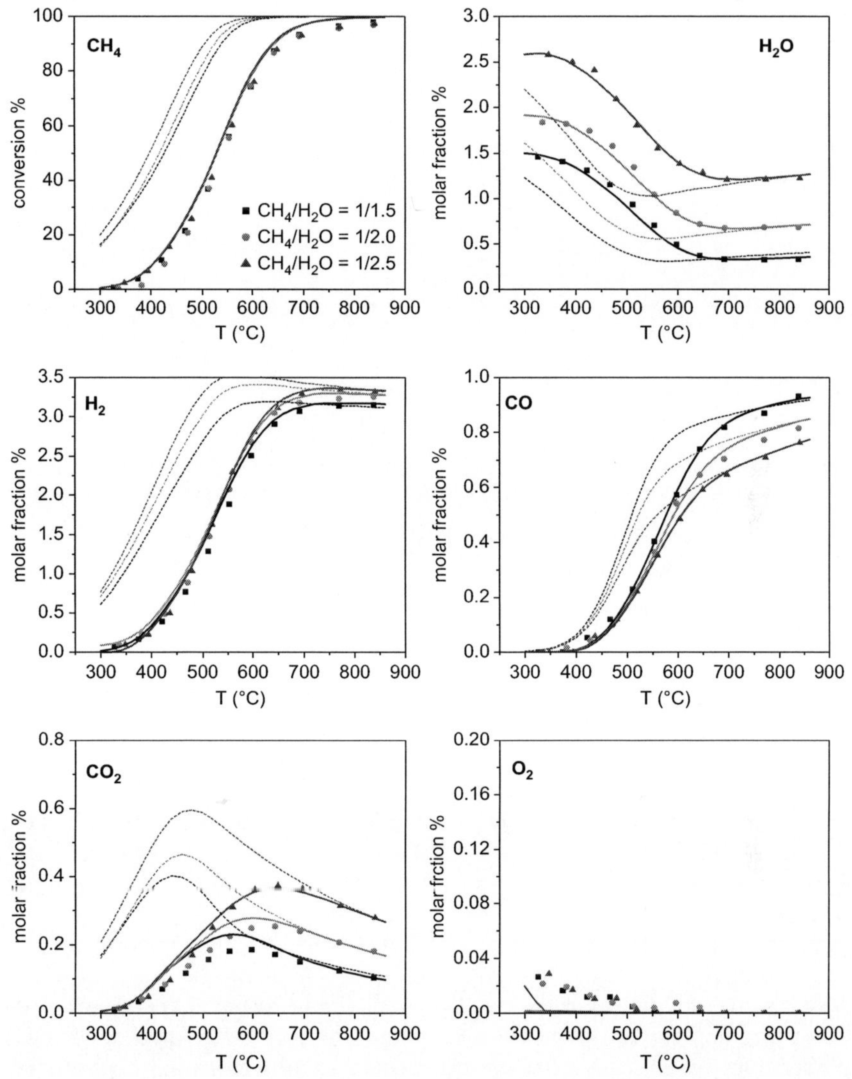

Fig. 5 CH_4 steam reforming experiments in annular reactor, effects of H_2O concentration. GHSV $= 2 \times 10^6$ l(NTP)/kgcat/h. CH_4 mole fraction $= 1\%$, N_2 to balance. Reprinted from Journal of Catalysis, vol. 255, A. Donazzi, A. Beretta, G. Groppi and P. Forzatti, Catalytic partial oxidation of methane over a 4% Rh/α-Al_2O_3 catalyst: Part I: Kinetic study in annular reactor, 241–258, Copyright (2008), with permission from Elsevier.

reverse-WGS reaction. This was possible because of two key experimental conditions: high space velocities (such that the reaction was not equilibrated at low to medium temperatures) and widely varying CO_2 concentrations.

These results were emblematic of the crucial roles that the selection of operating conditions and the width of the kinetic investigation have; these cannot be defined a-priori, since they depend of the intrinsic activity of the catalyst under study (which in turn will depend for instance on the metal dispersion).

Concerning the CPO of C_3H_8, Fig. 6 shows data of propane activation at varying O_2 partial pressure. Similarly to methane partial oxidation, at low temperature we observed a substantial independence of propane

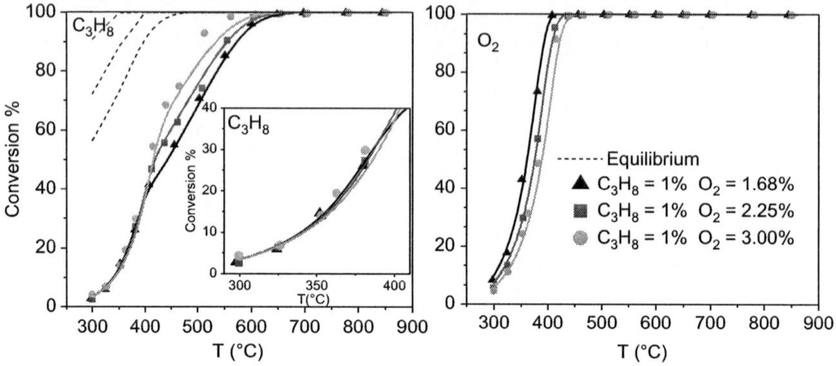

Fig. 6 C_3H_8 CPO experiments over a 2% Rh/Al_2O_3 catalyst in annular reactor, effects of O_2/C ratio. GHSV = 2×10^6 l(NTP)/kgcat/h. C_3H_8 mole fraction = 1%, N_2 to balance. Reprinted from Catalysis Today, vol. 197, D. Pagani, D. Livio, A. Donazzi, A. Beretta, G. Groppi, M. Maestri, E. Tronconi, A kinetic analysis of the partial oxidation of C_3H_8 over a 2% Rh/Al_2O_3 catalyst in annular microreactor, 265–280, Copyright (2012), with permission from Elsevier.

Table 4 Observed kinetic dependences on Rh.

CH_4 or C_3H_8 activation with O_2:	rate first order in p_{fuel} zero order in p_{O2} overall order <1
CH_4 or C_3H_8 activation with H_2O:	rate first order in p_{fuel} zero order in p_{H2O} overall order <1
CH_4 (C_3H_8) activation with CO_2:	rate = steam reforming (first order in p_{fuel}) in excess of CO_2 rate = reverse WGS (first order in p_{CO2}) in defect of CO_2

In oxidation and steam reforming TOF_{C3H8} ~2.5 TOF_{CH4} with lower Eatt (86 kJ/mol *vs.* 92 kJ/mol)

conversion on the amount of O_2 and (from other experiments) the prevailing dependence from propane partial pressure. The data collected at higher temperature, together with dedicated steam reforming experiments, indicated that also the conversion rate to CO and H_2 depends negligibly on the concentration of H_2O, but mostly on the concentration of the fuel.

The main kinetic dependences, based on the experimental data, are summarized in Table 4.

The observed independence of CH_4 and C_3H_8 oxidation from O_2 partial pressure (on a fully oxygen saturated surface) reflects the so-called Regime I, according to Iglesia's findings; in analogy, the independency of steam reforming of CH_4 from H_2O partial pressure is in line with the kinetics of Regime IV in Iglesia's schematization. However, differently from Iglesia's results, we found that both in the case of CH_4 and in the case of C_3H_8, the overall kinetic order of oxidation and steam reforming were below unity so that the conversion of the reactants were observed to decrease at descreasing dilution of the feed mixture. This was interpreted as the evidence of some coverage depences of the surface reactions that only experimental data under integral conditions (high values of conversions) could make visible. Also, notably, under the experimental conditions realized in the annular reactor, the water gas shift reaction was not always equilibrated.

3.3 Other reactor designs

Traditionally, catalyst screening has been performed in diluted packed-bed microreactors;[8,9,38–43] the preparation of this reactor is very straight forward, which is likely the reason of its popularity in the literature. Verykios *et al.*[43] have optimized the reactor configuration in terms of catalyst load and catalyst to diluent ratio in order to guarantee easily reproducible results, but also the best compromise between the need of integral data at high space velocity and the goal of minimum axial temperature gradients. This consisted in a 1 to 5 catalyst to inert volume ratio, with a 50 mg of catalytic powders. We have tested in the past the same reactor configuration; it was noted that, although inter-phase heat and mass transfer limitations have a minimal impact, intraparticle resistances may have a relevant role. Also, axial temperature gradients are important and the gap between maximum temperature (usually located at the reactor inlet) and outlet temperature was as high as 100 °C at complete oxygen conversion. Differently from both the previously described reactor designs, then, the traditional packed-bed micro-reactor does not allow to analyze the kinetics of O_2 conversion; the reactor has typically a sort of light-off behavior, such that after a minimal conversion of O_2, a slight increase of the heating temperature gives rise to a rapid increase of the temperature of the catalyst bed with reach of complete O_2 conversion. This reactor design, though successfully used to compare the reforming activity of various catalyst formulations, in practice are affected by the same irreducible heat transfer limitations as the adiabatic short contact time reactors.

Stagnation flow reactors have also been proposed in the literature to measure the intrinsic activity of Pt and Rh catalysts in CH_4 partial oxidation. Rice *et al.*[44] compared the performances of Pt and Rh foils at low pressure (30 Torr) and confirmed the superiority of Rh for syngas production. Deutschmann and co-workers[45] have recently developed a stagnation flow reactor equipped with a sampling probe which allows the measurement of concentration profiles within the boundary layer; this new equipment is applied for studying the kinetics of partial oxidation and reforming of light alkanes over Rh catalyst. Experimental findings are collected with the aim of verifying and further improving the reliability of microkinetic schemes for light alkanes activation on Rh that were previously developed by the group.

4 A theoretical tool for the kinetic investigation: microkinetic scheme of CH_4 CPO on Rh

In recent years, more attention has been paid to the physical meaning of the rate constant and to the fundamentals of catalysis, by exploiting the great advances in surface science and physical chemistry.[46–48] In essence, rather than fitting a hypothesized rate equation to experimental data, attention has focused on kinetic models that attempt to incorporate the basic surface chemistry involved in the catalytic reaction. Essential to this approach is the application of fundamental theoretical concepts for explaining the catalytic phenomena in terms of chemical physics, avoiding the qualitative ideas typical of classical catalytic kinetics, that cannot be directly translated into theoretically accessible phenomena.[47] Different from the classical approach used in catalysis, microkinetic modeling does not require a priori assumptions about the rate determining step (RDS) or the most abundant reactive intermediate (MARI), and, in principle, can capture the reacting system behavior under significantly different operating conditions. A crucial challenge for corresponding microkinetic models concerns the estimation of the reaction parameters, especially when a large number of reactions are involved. It has been demonstrated in the literature that there is not sufficient information to extract the values of kinetic parameters from experiments and, on the other hand, purely theoretical techniques (such as density functional theory, DFT) cannot be used alone to assess the relative rates of the reaction pathways.

In the following we discuss the development of a microkinetic scheme for CH_4 CPO on Rh and its application to the analysis of CPO data in annular reactor.

Among a variety of approaches reported in the literature, the most common one is to take reaction parameters from wherever available, *e.g.* first-principles calculations, surface science experiments, or fits of selected experimental data (*e.g.*, Refs. 49, 50). In the context of CH_4 partial oxidation, two models fit this category. The first detailed model was proposed by Hickman and Schimdt.[51] In this scheme (19 reactions), H_2 and CO were considered to be primary products and the CH_4 dissociation at the metallic surface was lumped into one single step. Reactions for CO_2 reforming were not considered and steam reforming turned out to be negligible (even though included). The mechanism were then further improved by Deutschmann and co-workers and extended to 38 reactions, by including also coverage-dependent heats of chemisorption for CO and O_2 in order to better capture light-off behavior.[49]

While this type of model can lead to a good prediction of experiment results under specific conditions, there are a number of clear drawbacks, the most prominent being that the model could be thermodynamically inconsistent and that the underlying chemistry may not be correctly accounted for.

To overcome these limitations and to make possible the development of physically sound microkinetic models even for reacting systems characterized by a large number of reactions, Vlachos and co-workers[52–56] have proposed a hierarchical and multi-scale methodology. This approach is a combination of different theoretical techniques (such as UBI-QEP, Unit-Bond-Index-Quadratic-Exponential-Potential theory,[57,58] with heats of chemisorption coming from DFT and/or surface science experiments, and order of magnitude estimates for prexponentials from Transition-State Theory, TST[59]) and proper fitting of selected experimental data. The use of coverage-dependent heats of chemisorption allows generation of coverage-dependent activation energies within the UBI-QEP framework, while the thermodynamic consistency of the model parameters is ensured through an optimization procedure with respect to the gas-phase thermodynamics.[54] Still, an experimentally-driven fine-tuning of the parameters, accounting both for the uncertainty of the calculation methods (*e.g.*, UBI-QEP has an estimated uncertainty of ~3–5 kcal/mol, that can be even larger considering the uncertainty in the heats of chemisorptions)[53] and of the model assumptions (*e.g.*, single site model and mean-field approximation), is always needed in order to make the microkinetic model quantitatively predictive. In this respect, a hierarchical data-driven refinement methodology has been proposed to fine tune the model parameters within the uncertainty range and it has been shown that, if a comprehensive set of kinetic experiments is available for the refinement, this methodology can lead to the development of reliable microkinetic models.[55] Following this new methodology, Maestri *et al.*[55] proposed a new version of the model of Mhadeshwar and Vlachos [53] for C1 chemistry on Rh. This model has been also recently improved by means of first-principles calculations[60] and it was verified that the model is able to predict correctly the behaviour of the several reacting mixtures studied in the annular reactor,[60–62] including CH_4/H_2O (steam reforming, SR) and CH_4/CO_2 (dry reforming, DR) reforming, CH_4/O_2 (CPO), H_2/O_2 and CO/O_2, CO/H_2O (water-gas-shift, WGS), under a wide range of operating conditions in full agreement with independent experimental studies (*e.g.*, Ref. 26).

4.1 Microkinetic analysis of CH_4 CPO data in annular reactor

The possibility of modelling the surface chemistry in detail allows proposing an explanation of the main molecular pathways involved in the process, leading to a deeper understanding of the underlying catalytic mechanisms. Maestri *et al.* have used their microkinetic model in combination with detailed reactor modelling for the microkinetic analysis of CH_4 CPO in an annular reactor at quasi-isothermal conditions.[62]

For the description of the macro-scale regime a 2D model of the annular reactor has been considered.[63] Figure 7 compares model predictions and data and gives an example of how well the model is capable of describing CPO (integral) data.

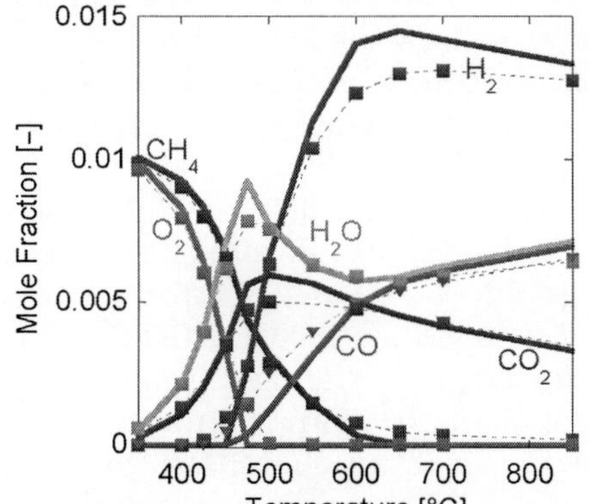

Fig. 7 CH_4 CPO experiments over a 2% Rh/Al_2O_3 catalyst in annular reactor, effects of temperature. GHSV = 2×10^6 l(NTP)/kgcat/h. CH_4 mole fraction = 1%, O_2/C = 0.56, N_2 to balance. Symbols = experiments, solid lines = predictions of the microkinetic model. With kind permission from Springer Science + Business Media: Topics in Catalysis, Dominant reaction pathways in the catalytic partial oxidation of CH_4 on Rh, 53 (13–20), 1983–1988, M. Maestri, D. Vlachos, A. Beretta, G. Groppi and E. Tronconi, Fig. 1.

The model predicted correctly that, at low temperatures, O_2 is only partially converted and deep oxidation occurs (CO_2 and H_2O are the main products); at higher temperatures, O_2 is completely converted, syngas is produced, and steam and CO_2 are being consumed. However, the simulations reveal that at intermediate temperatures, a stratification of the reactor (Fig. 8) may happen, with up to three reaction zones.

Upstream (zone 1), the most abundant reactive intermediate (MARI) on the catalyst surface is O^*. Under these conditions, methane is completely oxidized to combustion products. Reaction path analysis (RPA) (Fig. 9) indicates that, following CH_4 dissociative adsorption, CH_x^* is consumed *via* oxidative dehydrogenation ($CH_x^* + O^* \rightarrow CH_{x-1}^* + OH^*$ or $CH_x^* + OH^* \rightarrow CH_{x-1}^* + H_2O^*$), whereas the parallel pyrolytic path ($CH_x^* + {}^* \rightarrow CH_{x-1}^* + H^*$) is negligible.

This behavior arises because the activation energies of the oxidative paths (*via* OH^* and O^*) are energetically favored at high O^* coverages. C^*, formed *via* oxidative dehydrogenation, is oxidized (with O^*) to CO^*. CO^* is oxidized faster to CO_2^* than it desorbs, mainly according to

$$CO^* + OH^* \rightarrow COOH^* + {}^* \rightarrow CO_2^* + H^* \tag{R1}$$

Similarly, H^*, formed *via* oxidative dehydrogenation of CH_4, is consumed *via* oxidation:

$$O^* + H^* \rightarrow OH^* + {}^* \tag{R2}$$

$$H^* + OH^* \rightarrow H_2O^* + {}^* \tag{R3}$$

giving eventually gaseous water. Paths R2 and R3 dominate over H^* desorption. While H^* and CO^* are key surface intermediates, their oxidation (R1 to R3) is faster at high O^* coverages than their desorption, leading to deep CH_4 oxidation. Under these conditions, the RDS (identified *via* sensitivity analysis[64]) is the CH_4 dissociative adsorption. As O^* is being depleted, more Rh sites become available, and the rate of CH_4 consumption increases due both to mass action kinetics and to the O^* coverage dependence of the energetics of the oxidative paths.

Once O^* has been consumed, H^* and CO^* become most abundant, and the activation energies of the oxidative paths become comparable with the pyrolytic ones (zones 2 and 3 in Figs. 7 and 8). In addition, the surface becomes reasonably clean (Rh vacancies $>70\%$, Fig. 8). As a result, the dominant pathways on the catalyst in zones 2 and 3 change completely. In zone 3, there is no gaseous O_2 left and OH^* forms from adsorption of H_2O and its subsequent decomposition to OH^* and H^*. C^*, formed *via* pyrolysis ($CH_4 \rightarrow C^* + 4H^*$), is oxidized to CO^* and then to CO_2^* *via* OH^*.

Overall, syngas forms *via* the indirect route and SR and WGS (close to equilibrium) are the dominant reactions,[61] in good agreement with independent kinetic studies.[26] In zone 2, the gaseous oxygen concentration is nearly zero at the catalytic surface and, as in zone 3, H^* and CO^* dominate on the surface. However, due to interphase mass transfer limitations, gaseous oxygen is still present in the bulk of the channel. O_2 from the bulk of the gas-phase adsorbs on the catalytic surface and activates different reaction paths: OH^* forms from the reaction between H^* (from CH_4 pyrolysis)

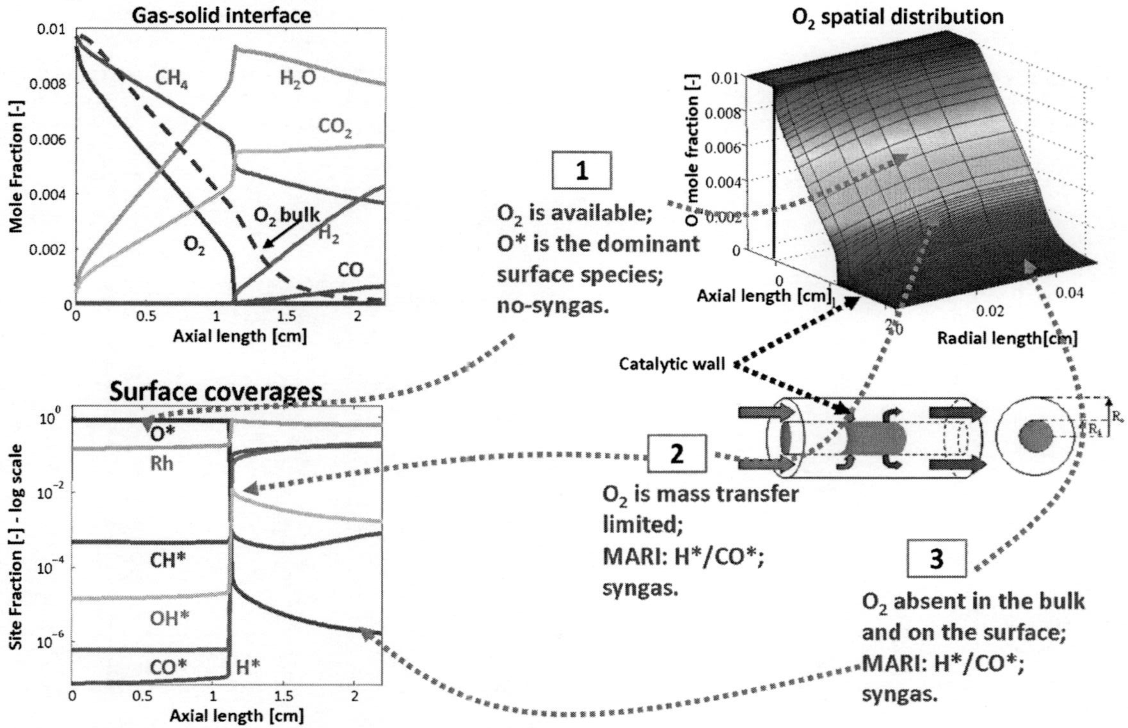

Fig. 8 Modelling analysis of CH_4 CPO in annular reactor, evolution of species concentration at the wall and corresponding surface coverages, oxygen concentration surface and identification of distinct kinetic regimes. GHSV $= 2 \times 10^6$ l(NTP)/kgcat/h. CH_4 mole fraction $= 1\%$, $O_2/C = 0.56$, balance N_2, $T = 500\,°C$. Reprinted from Microkinetic analysis of complex chemical processes at surfaces by M. Maestri, chapter in New Strategies in Chemical Synthesis and Catalysis, B. Pignataro Editor, Figure 10.6, page 236, 2012. Copyright Wiley-VCH Verlag GmbH & Co. KGaA. Reproduced with permission.

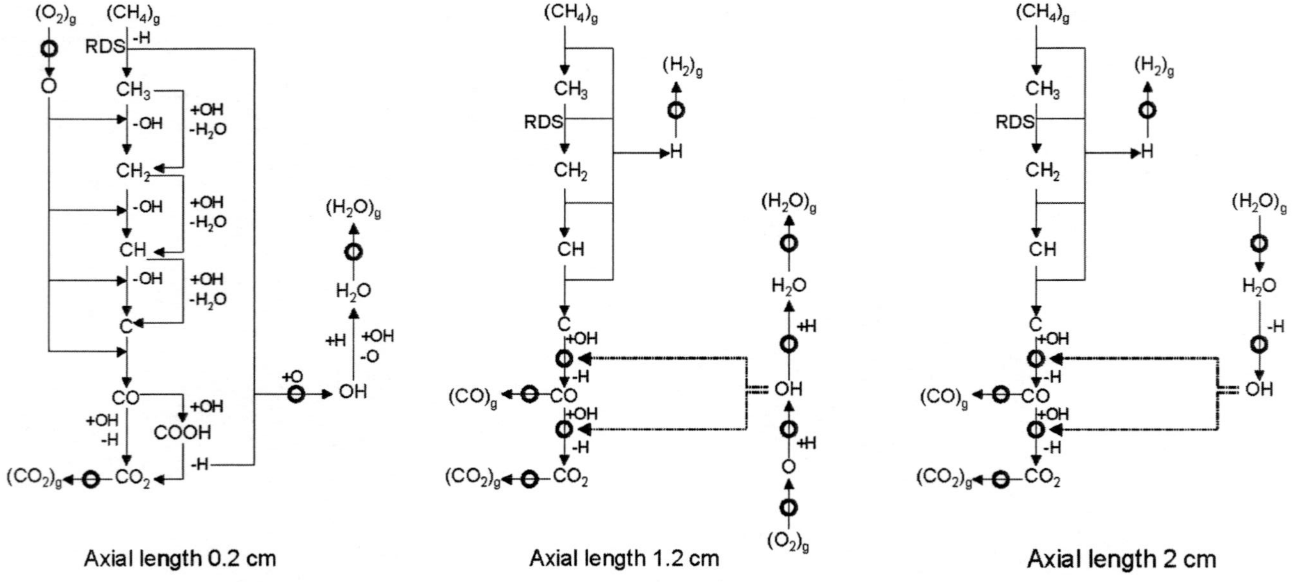

Fig. 9 CH_4 CPO experiments over a 2% Rh/Al_2O_3 catalyst in annular reactor, reaction path analysis at different axial positions. T = 500 °C. GHSV = 2×10^6 l(NTP)/kgcat/h. CH_4 mole fraction = 1%, O_2/C = 0.56, balance N_2. With kind permission from Springer Science + Business Media: Topics in Catalysis, Dominant reaction pathways in the catalytic partial oxidation of CH $_4$ on Rh, 53 (13–20), 1983–1988, M. Maestri, D. Vlachos, A. Beretta, G. Groppi and E. Tronconi, Fig. 3.

and O^* (from dissociation of molecular oxygen that slowly diffuses from the gas-phase). A fraction of the OH^* reacts with H^* giving H_2O^*, leading to gaseous H_2O, and with CO^*, giving eventually rise to CO_2. The slow-down of the CO^* and H^* oxidation reactions makes desorption of syngas competitive with its combustion. As a result, syngas is produced *via* a "direct route", since H_2O adsorption is not needed. In this second zone, a mixed mode with competition between the direct path of methane to syngas and syngas combustion is found. The lack of excess adsorbed oxygen activates the pyrolytic chemistry over the oxidative chemistry. Consequently, zone 2 has many features (in terms of elementary steps) of the steam reforming and water-gas shift zone 3 but with noticeable differences, *e.g.*, OH^* forms from $H^* + O^*$ rather than from H_2O.[61]

An additional key observation of the model analysis is that the main oxidizer is OH^* rather than O^*. This is different from most of the mechanisms reported in the literature, where it is usually *assumed* that the main oxidizer is O^* (coming from $OH^* + {}^* \rightarrow H^* + O^*$) and oxidation *via* OH^* is not considered. According to these calculations, the OH^* dissociation to O^* and H^* is slow, and the parallel consumption paths (*e.g.*, oxidation of C^* *via* OH^*) turn out to be favored at the investigated conditions.

As a whole, the microkinetic analysis revealed that the partial oxidation of methane does not occur at the molecular level at the surface. Rather, the combination of different reacting systems gives rise to up to three reaction zones: a deep combustion zone for methane, followed by a zone where direct formation of syngas in parallel with catalytic combustion occurs, and finally a SR and WGS zone, when oxygen is no longer available. Simulations in the absence of mass transfer have indicated that zone 2 exists only because of mass transfer limitations and the extent of the various zones depends mainly on temperature. Therefore, the analysis revealed that the interplay between mass-transfer effects and surface chemistry is crucial in dictating the dominant reaction pathways leading to syngas.

These findings underscore the importance of the long postulated interplay of transport phenomena and surface chemistry, pointing out that an accurate description of transport effects is required for the identification of the dominant reaction mechanism and the description of the related experimental data. In particular, these results have reconciled, for the first time in the literature, the apparent long-debated mechanistic differences regarding direct versus indirect pathways.

The extent of the three zones in the reactor strictly depends on temperature. At low temperatures, O^* dominates and complete combustion prevails (zone 1). At higher temperatures, zones 2 and 3 prevail and O_2 consumption can be mass transfer controlled. This is usually the case for the steady-state operation of adiabatic CPO reformers, as better illustrated in a following section.

5 Experimental tools for the kinetic investigation of CH_4 CPO in short contact time adiabatic reformers

As already noted in the Introduction, most experimental studies in the literature have investigated the CPO process with the main aims of verifying its potential fuel conversion and H_2 yield. Toward these aims, insulated structured reactors have been used at the laboratory scale, thus exploiting and demonstrating the flexibility of the process to miniaturization. However, as also mentioned above, the characterization of reactor behavior may be extremely complex due the presence of dramatic temperature and concentration gradients. In the attempt to quantify the irreducible transport phenomena, earlier works have mostly focused on the measurements of temperature gradients, but great improvements have been made in the general understanding of the CPO reactors by the introduction of *in situ* and spatially resolved techniques of temperature and concentration.

Some topical examples from the literature are herein illustrated.

5.1 Temperature gradients around gauzes

Several authors[65–68] have extensively studied the partial oxidation of methane over single Pt and Pt/Rh gauzes; since contact times are extremely low, this configuration results in relatively low methane conversions. Still, high CO selectivity were reported over different catalysts; concerning H_2 formation, poor or null values of H_2 selectivity were measured over Pt gauzes, while 30–40% H_2 selectivity were reported over single or layered Pt/10%Rh gauzes. Authors seem to unanimously agree that, over metal gauzes, CO and H_2 are primary products of methane oxidation. Mechanistic TAP studies over the pure Pt and Rh sponges confirmed this hypothesis, indicating for instance that CO and CO_2 were produced in parallel over bulk metals.[69]

Studying CH_4 CPO over gauzes, De Smet *et al.*[67,68] first addressed the quantitative study of the kinetics of CH_4 CPO in the presence of important heat transport limitations between the catalyst and the gas phase. Using a specifically designed reactor, represented in Fig. 10, these were detected by comparing the measurements of thermocouples placed upstream and downstream from the gauze, with the measurement of a Pt-Rh wire ($d_w = 0.10$ mm) spot-welded at the center of the Pt gauze. An optical pyrometer was also used to verify the accuracy of the surface thermocouple.

Examples of the measured catalyst and bulk gas-phase temperatures (an average of the thermocouple measurements upstream and downstream of the gauze catalyst) were reported. It was clearly shown that the catalyst surface temperatures were at least 200 K higher than the average gas-phase bulk temperature, indicating that strong heat transfer limitations occurred around the gauze. This is the reason why methane and oxygen conversion hardly changed with temperature. Still CO selectivity varied considerably with temperature, being the product distribution still controlled by the surface kinetics. Experiments were performed at varying temperature, space velocity and partial pressure of the reactants. They were simulated by using the model of a parallel row of flat plates (after validation against the conversion measurements and the comparison with the prediction of a 3D FLUENT reactor model). By model fit, the kinetics of a 6 steps kinetic scheme were estimated; such scheme assumes that adsorbed CO^* is the primary product of methane activation, leading to CO_2 formation *via* oxidation or CO formation *via* desorption, in line with the TAP measurements.[69]

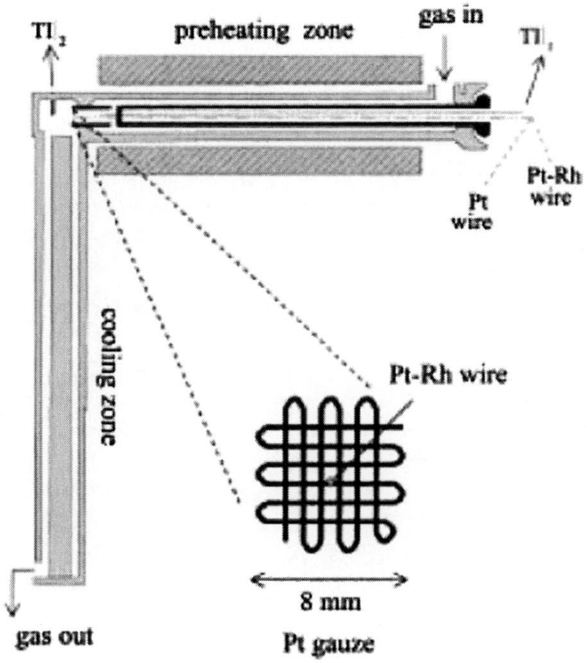

Fig. 10 Schematic drawing of the laboratory reactor used for kinetic measurements over a single Pt gauze. TI1 and TI2 are Nisil/Nicrosil thermocouples, located in quartz thermocouple wells. The surface thermocouple is Pt-10%Rh wire, spot-welded to the centre of the gauze. Reprinted from Applied Catalysis A: General, vol. 187, C. R. H. de Smet, M. H. J. M. de Croon, R. J. Berger, G. B. Marin and J. C. Schouten, An experimental reactor to study the intrinsic kinetics of catalytic partial oxidation of methane in the presence of heat-transport limitations, 33–48, Copyright (1999), with permission from Elsevier.

5.2 Axial temperature profiles in monolith channels and in packed beds

From the early 90' to 2005, the pioneering work of Schmidt and coworkers covered a remarkable variety of fuels (ranging from methane to hexadecane), operating conditions, catalyst compositions (Rh, Pt, Pd, Rh-Ce at different metal loading) and supports (gauze packs, spheres and foams). Such an extensive body of experiments was obtained by means of a millisecond adiabatic reactor with a very simple and practical configuration. The typical set-up consisted of a quartz tube (18 mm inner diameter) wherein the catalytic monolith (or gauze pack) was immediately preceded and followed by inert alumina extruded monoliths, which acted as radiation shields. This stack was closely wrapped into a high temperature alumina-silica cloth, in order to minimize the bypass of the reactant gases around the annular gap between the catalyst sample and the quartz tube. In the reaction zone, the outside of the quartz tube was insulated with thick silica wool, while in the inlet zone, it was wrapped by heating tapes in order to preheat the feed mixture. The temperature was measured by placing one thermocouple upstream of the front heat shield and a second thermocouple close to the catalyst outlet section, right in front of the back heat shield. Although extremely versatile, such a configuration exclusively provided integral data, which led to several mechanistic interpretations and left several unresolved questions concerning the evolution of the process.

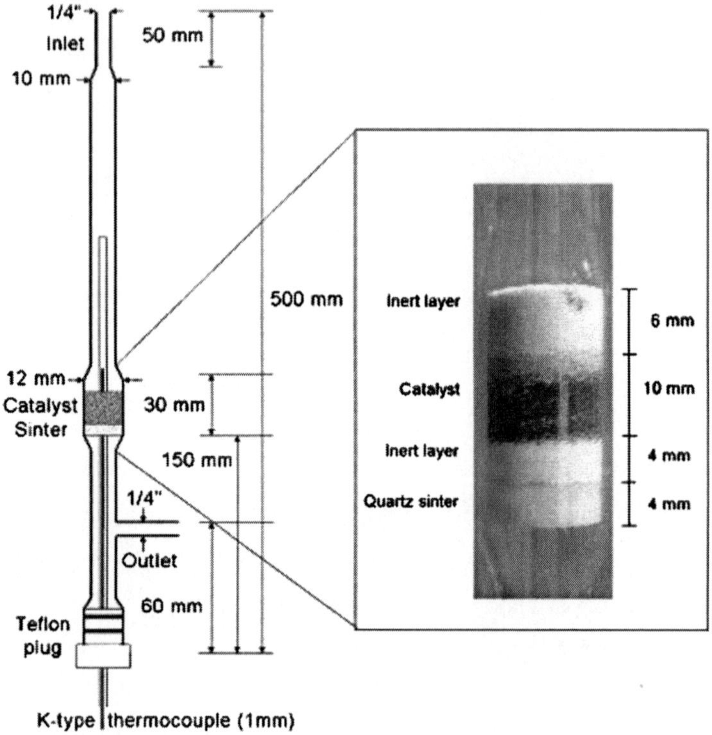

Fig. 11 Reactor configuration adopted at NTNU by Holmen and co-workers; it has been applied to test both monolith catalysts and catalyst coated spheres. Reprinted from Applied Catalysis A: General, vol. 364, B. C. Enger, R. Lødeng and A. Holmen, Evaluation of reactor and catalyst performance in methane partial oxidation over modified nickel catalysts, 15–26, Copyright (2009), with permission from Elsevier.

Holmen and coworkers[70] have studied the partial oxidation of methane over Ni, Pd, Pt and Rh – coated monoliths and first introduced the use of sliding thermocouples to characterize the axial evolution of the process. Cordierite honeycombs monoliths with 400 cpsi were used as supports (cut in pieces with length of 23 mm and diameter of 15 mm, in the middle of which a channel with diameter 3 mm was cut out to host a quartz capillary that served as thermocouple well); metals were deposited according to a washcoating procedure. For testing, the catalyst was inserted in a tubular quartz reactor in between two inert monoliths. A furnace externally heated the reactor, which allowed running blank tests and verifying the negligible reactivity of empty volumes and the inertness of reactor components other than the catalytic monoliths. The reactor set-up is shown in Fig. 11.

It was found that axial temperature profiles presented an important hot-spot at the reactor inlet; a minimum was also visible at high reaction temperatures. An indirect reaction scheme was thus clearly shown, with an initial highly exothermic reaction (the combustion of methane) followed by secondary exothermic reactions (such as methane steam reforming). The comparison of temperature profiles between honeycombs and packed bed configurations indicated that in the packed bed reactor the exothermic region was narrower than in the extruded monolith.

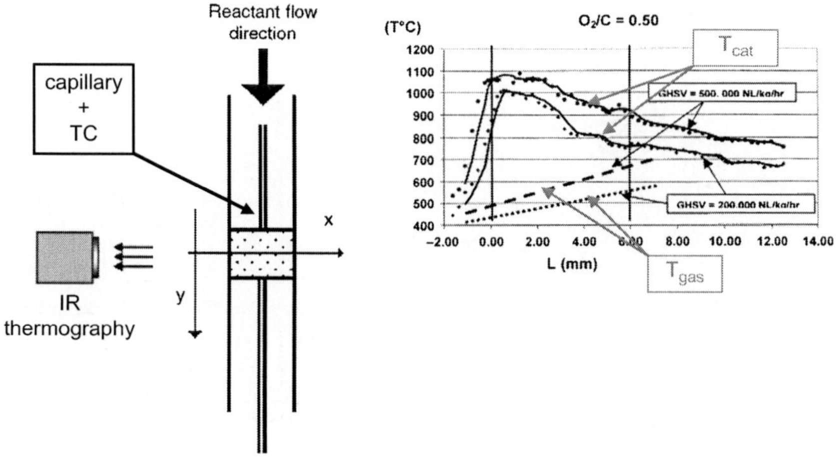

Fig. 12 A schematic drawing of the experimental apparatus for IR temperature measurements and an example of the temperature profiles measured by Basini *et al.* during methane partial oxidation on Rh. Reprinted from Catalysis Today, vol. 64, L. Basini, K. Aasberg-Petersen, A. Guarinoni and M. Østberg, Catalytic partial oxidation of natural gas at elevated pressure and low residence time, 9–20, Copyright (2001), with permission from Elsevier.

Concerning the issue of measuring the catalyst temperature under working conditions, an early solution has been the use an infrared camera directly pointed at the wall of a quartz reactor. In this arrangement, the external radiation is collected and associated with the temperature of the solid by means of a calibration.

One of the first reports comes from Basini[71,72] at Eni; the same apparatus has been recently adopted also by Simeone *et al.*[73,74] A schematic drawing and an example of the results reported by Basini group is given in Fig. 12.

The reactor consisted of quartz tube, into which a quartz thermowell was inserted, and a thermocouple was moved axially. An IR camera with a resolution of 250 microns was used to collect the temperature profile of the solid. The measurement was conducted by quickly removing the insulating layer of the reactor once light off had occurred and by focusing the lens on the catalyst volume, therefore relaxing the constraint of adiabaticity. As shown in Fig. 12, a hot spot was found, followed by a decrease towards the outlet section. Correspondingly, the temperature of the gas phase increased almost linearly, but always maintained lower than the temperature of the solid.

The results were rather unique. However, it must be noted that the CPO process is greatly limited by thermodynamics and an incorrect evaluation of the heat losses from the reactor can lead to important variations in the distribution of the products at the outlet. Rigorous determination of the temperatures should be performed by maintaining the adiabatic character of the reactor, *i.e.* without removing the external insulation. This has been recently achieved by inserting suitable probes inside the insulated reactor, as illustrated in the following paragraph.

5.3 Spatially resolved measurements of temperature and composition in adiabatic reactors

Along with the need for a more detailed picture of the CPO process, a major advance was introduced by Schmidt and his group in 2006[75] by the

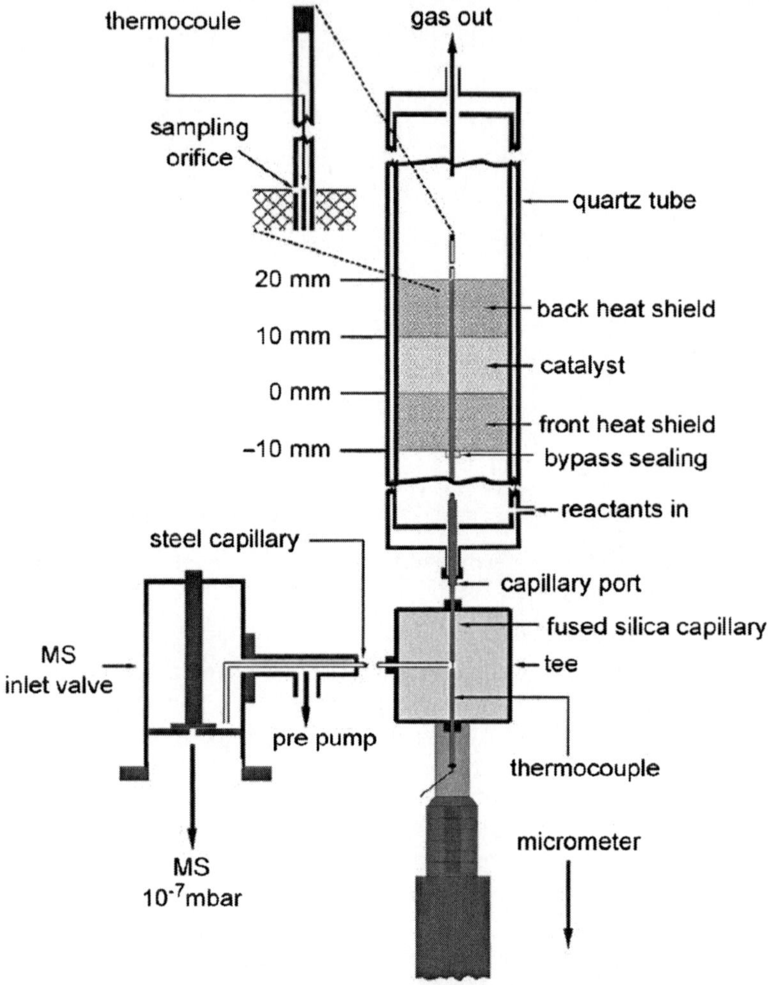

Fig. 13 Reactor set-up for spatially resolved sampling technique. Adapted from: Catalysis Letters, 110, 2006, 169–178, Spatial and temporal profiles in millisecond partial oxidation processes, R. Horn, N. J. Degenstein, K. A. Williams and L. D. Schmidt, Fig. 1, original copyright notice, with kind permission from Springer Science and Business Media.

realization and application of the so-called spatially resolved sampling technique. This technique allows measuring the temperature profile of the gas phase and of the catalyst surface, as well as the composition profile of the gas phase, along the axis of a catalyst working under adiabatic conditions. A picture of the apparatus, which was rapidly optimized in later works,[19,21,76–79] is reported in Fig. 13.

Using a reactor tube with foam supported catalysts (45 or 80 ppi α-Al_2O_3 foams, either Rh- or Pt-coated), a minimally intrusive probe consisting of an inert fused silica capillary (0.55 mm outside diameter) was inserted along the axis of the foams stack by drilling a hole barely larger than the capillary itself (0.75 mm diameter). The resulting gap (0.1 mm) was smaller than the pore diameter of the foams and guaranteed the absence of gas bypass in the annular channel. The capillary was externally connected to a micro-volume tee mounted on a linear translation stage and moved along the axis of the reactor with micrometrical resolution. To measure the composition of the gas phase, a sealed capillary with a ~0.2 mm side orifice was used and the tee was connected to a sampling loop consisting of a mass spectrometer and a gas chromatograph. A vacuum pump was used to draw the gas through the orifice into the sampling loop.

A quartz optical fiber (0.33 mm outer diameter) connected to an infrared pyrometer and a K-type thermocouple (0.254 mm diameter) were used to measure the temperature. The temperature profiles were thus acquired by inserting either the thermocouple or the optical fiber into a transparent sealed capillary. Analysis of the data with numerical models[21,80] demonstrated that the temperature measured by the pyrometer is representative of the temperature of the catalyst surface, whereas the temperature measured by the thermocouple is close to the temperature of the gas phase. Further optimization of this reactor configuration allowed measurement of methane CPO profiles under pressures as high as 11 bar by application of a connection port suitable to introduce the capillary and permit movements while maintaining the reactor gas tight.[19]

Schmidt and coworkers acquired spatially resolved profiles exclusively on foam supported catalysts, whose reticulated structure intrinsically guarantees optimal radial mixing and minimal influence of the measuring probe on the flow pattern. Instead, early works at Oak Ridge National Lab[81,82] and at PSI Institute[83] focused on the application of the spatially resolved measurements with different catalyst supports and reactor configurations, as well as reaction processes. Recently,[84–86] our group examined the application of the spatially resolved technique within honeycomb supported catalysts for the CPO of CH_4 and C_3H_8. The state-of-the-art reactor set-up and internal layout were adopted, while the capillary was introduced into a central channel of the catalytic honeycomb. In this case, non trivial problems became the estimation of the probe influence on the flow pattern and the evaluation of the adequacy of the optical fiber measurement inside an open channel configuration due to its higher axial transparency compared to foams and packed beds of spheres. The case of the CH_4/air CPO reaction ($CH_4 = 27.3\%$ v/v, $O_2/CH_4 = 0.56$) performed over a 400 cpsi cordierite honeycomb coated with a 2 wt% Rh/α-Al_2O_3 catalyst was examined.[84] Special attention was paid to the correct assessment of the surface temperature as related to the geometrical shape of the optical fiber; indeed, depending on the angle of the tip (*i.e.* on its goniometrical characteristics), the acceptance cone for the incoming radiation has different orientations and secondary components that can in principle influence the measurement. The analysis of the temperature profiles and of the goniometric characteristics of the optical fiber clearly indicated that 45° angled fibers (side-looking probes) are preferred for spatially resolved measurements, owing to the capability of looking at a precise spot on the surface of the catalyst support. In contrast, 90° angled fibers (front-looking probes) led to average measurements with a strong underestimation of the hot spot. An example of these results is reported in Fig. 14 (panel a): it can be clearly seen that the hot spot measured with a 90° angled fiber was ~70 °C lower than that obtained with a 45° angled fiber. By comparing the experimental results and

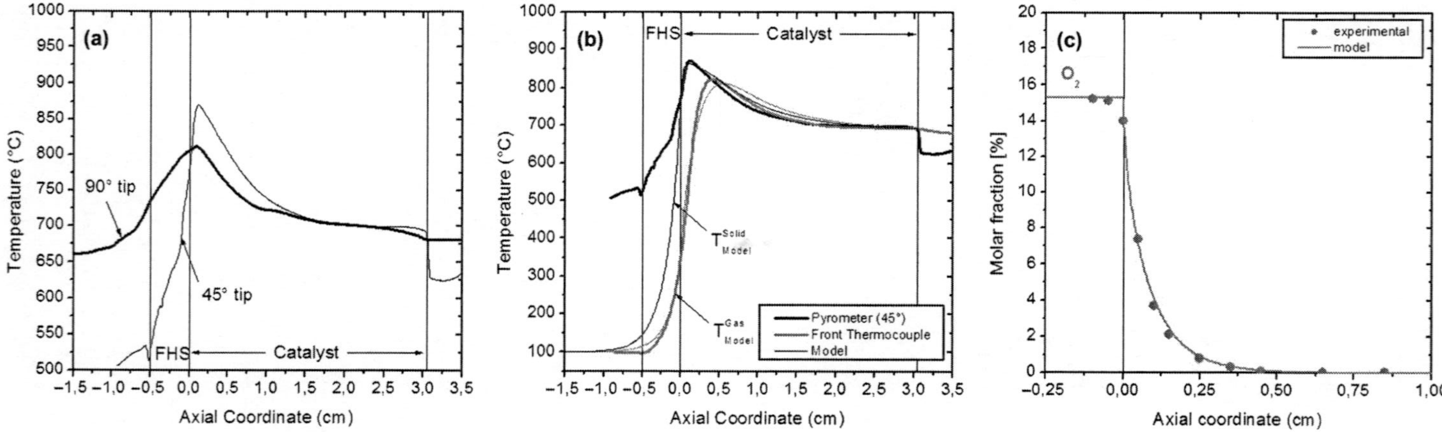

Fig. 14 Effect of the angle of the optical fiber tip. Feed composition: $CH_4 = 27.3\%$, $O_2/CH_4 = 0.56$. Flow rate = 20 Nl min^{-1}, $T_{IN} = 100\,°C$. Panel a: comparison between temperature profiles collected with a 90° fiber (thick lines) and a 45° fiber (thin lines). Panel b: comparison between experimental (thick lines) and model (thin lines) results. Panel c: measured and simulated O_2 concentration profile. Adapted from Applied Catalysis A: General, Volume 402, A. Donazzi, D. Livio, A. Beretta, G. Groppi and P. Forzatti, Surface temperature profiles in CH_4 CPO over honeycomb supported Rh catalyst probed with *in situ* optical pyrometer, 41–49, Copyright (2011), with permission from Elsevier.

the simulations of our predictive numerical model (panel b), a very close match was found and the adequacy of the pyrometer measurements with 45° angled fibers was verified. Additionally, measurement artifacts were identified outside the catalytic monolith (between −0.5 and 0 cm, Fig. 14 panel b), where deviations from the expected temperatures were systematically observed (T^{Solid}_{Model} *vs.* $T_{Pyrometer}$). Overall, the results confirmed that the introduction of the capillary probe in the honeycomb channel had a negligible influence on the temperature profiles. This was explained by considering that the channels are not thermally segregated, due to the conductivity of the walls, so that the final temperature results form a balance between the local profiles and the surrounding unperturbed channels.

Concerning the influence of the sampling probe on the gas composition profiles within honeycomb channels (which are radially segregated from the point of view of mass balance), further work is required, possibly with the application of specific fluid-dynamic computational codes. Nonetheless, it is crucial to note that close agreement was found between the experimental and simulated O_2 consumption profiles (Fig. 14, panel c), by application of rigorous and well-established mass transfer correlations.[87] Indeed, since the consumption of O_2 is strictly limited by mass-transfer (Section VII), the measured O_2 concentration profile depends exclusively on the local flow pattern: hence, the agreement found suggested that the probe had a limited influence.

The spatially resolved measurements provide a most comprehensive and neat picture of a reaction process as a whole, wherein the evolution of the reacting species can be directly observed and associated with that of the products and of the temperature of both the catalyst surface and of the gas phase. The typical evolution of the gas phase composition and of the temperature along the axis of a 400 cpsi cordierite honeycomb coated with a 2 wt% Rh/α-Al_2O_3 catalyst is reported in Fig. 15.[85]

In the inlet portion of the catalyst, (from 0 to 5 mm in the figure), the consumption of O_2 and CH_4 occurs with production of CO, H_2, H_2O and CO_2. Subsequently, once gas phase O_2 is totally depleted, CH_4 is further consumed at a lesser extent with additional production of H_2 and CO. The temperature profile of the solid surface shows a sharp maximum, followed by a decrease up to the outlet section. Notably, in the first millimeters of the catalyst, a significant gap is observed between the gas temperature, closely represented by the thermocouple, and the surface temperature, which confirms the key role of optical measurements for a correct mapping of the heat distribution. The gas temperature reaches a maximum, lower than that of the surface, and then decreases, matching the surface temperature.

Qualitatively, the species and temperature profiles in a monolith channel are very similar to those collected in foams, as those measured by Schmidt and co-workers and reported in Fig. 16.

Recently, Horn and coworkers[88] optimized the spatially resolved set-up to collect profiles in packed beds of spheres, by introduction of a rotational component in the capillary movement and by minimization of the side orifice diameter by laser drilling. Additionally, these authors adopt a conical shape for the tip of the optical fiber (42° angled), which results in a circular and axis-symmetrical radiation acceptance cone. The rotating sampling capillary and the conical tip of the optical fiber guarantee that the obtained

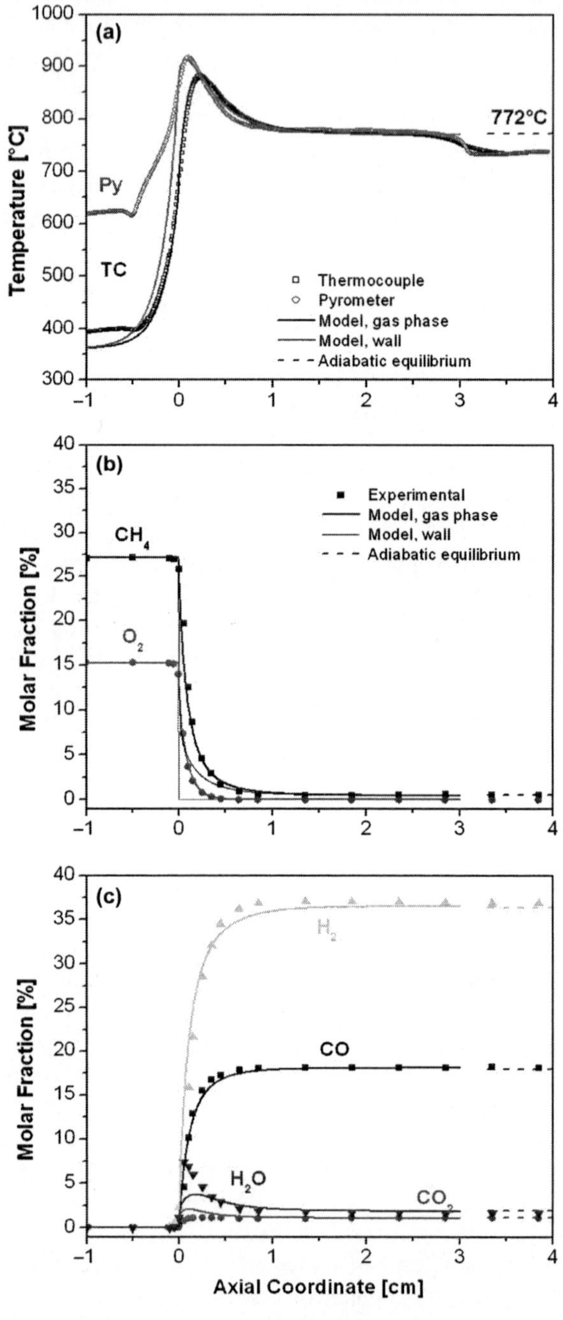

Fig. 15 CH_4 CPO experiment in a Rh coated 400 cpsi cordierite honeycomb support. Total flow rate = 10 l(NTP)/h. CH_4 and air with $O_2/C = 0.56$. Reprinted from Catalysis Today, Vol. 171, A. Beretta, A. Donazzi, D. Livio, M. Maestri, G. Groppi, E. Tronconi and P. Forzatti, Optimal design of a CH_4 CPO-reformer with honeycomb catalyst: Combined effect of catalyst load and channel size on the surface temperature profile, 79–83, Copyright (2011), with permission from Elsevier.

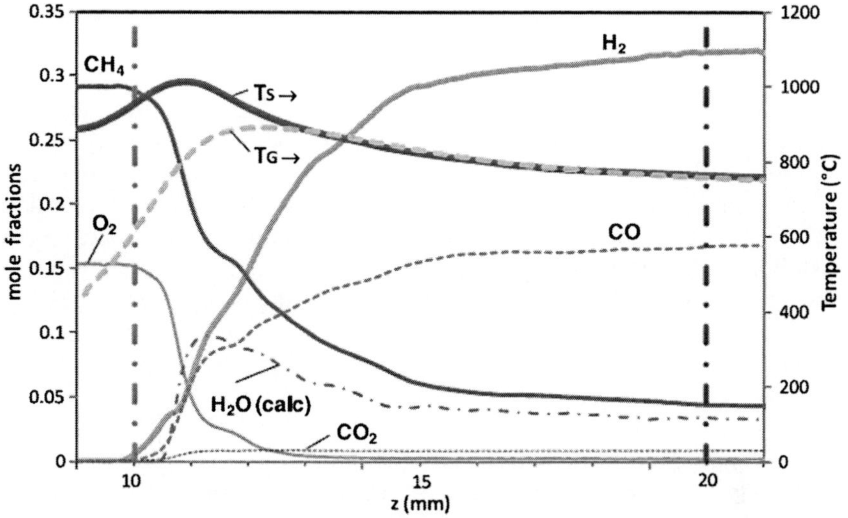

Fig. 16 CH_4 CPO spatially resolved temperature and concentration profiles over 80 ppi α-Al_2O_3 foam. Reprinted from journal of Catalysis, Vol. 277, D. Dalle Nogare, N. J. Degenstein, R. Horn, P. Canu, L. D. Schmidt, Modeling spatially resolved data of methane catalytic partial oxidation on Rh foam catalyst at different inlet compositions and flowrates, 134–148, Copyright (2011), with permission from Elsevier.

spatial measurements better represent the average conditions met in the section of the reactor.

6 Modeling of CPO short contact time adiabatic reformers

As mentioned above, and better illustrated in following analysis of case studies, the quantitative interpretation of CPO tests in adiabatic short contact time reactors needs the means of detailed reactor modeling. Heat and mass transfer limitations between the gas phase and the catalytic washcoat layer, heat conduction along the structured support, radiation from the glowing monoliths faces, heat dispersions from the reactor walls, gas-phase reactions are kinetic phenomena of relevant impact on the final performance of these highly intensified reactors. However, a source of great complexity for the quantitative description of these phenomena together with that of the surface process, is the flow pattern of the reactor.

The detailed reactor modeling based on the coupling of reliable surface and gas-phase schemes with rigorous description of fluodynamics and transport phenomena is still open in the literature. The complexity (and thus the huge computational cost) of detailed modeling is presently addressed by various authors at the expense of some simplifying assumptions, mostly on the reactor model.

Both heterogeneous plug flow reactor (PFR) models and two-dimensional (2D) pseudo-homogenous models have been developed to simulate methane (and light hydrocarbons) oxidation to syngas.[17,49,80,89–91] Nevertheless, given the very high external mass transfer limitations, an accurate description of the interface resistances is mandatory for a detailed analysis of experimental data. In this respect, a dynamic heterogeneous 1D model of a fixed bed adiabatic catalytic reactor was developed by Maestri *et al.*,[92] where specific bed transport properties and experimentally validated heat and mass transfer correlations for each support were included (ref. and references therein). Three different types of ceramic catalyst supports were analyzed (foam monolith, honeycomb monolith and spheres), all successfully applied in the literature for the autothermal partial oxidation of CH_4.

The analysis led to the conclusion that transport phenomena play a decisive role in the process of methane partial oxidation. In particular, a great sensitivity of the reactor behavior upon the support characteristics was found. Thus, the choice of catalyst support can lead to greatly different reactor performances as experimentally observed.[93] This offers a large number of degrees of freedom for reactor design and optimization, by proper tailoring of the type, the geometry and the material of the catalyst support.[62,94]

Mass and, particularly, heat transfer properties markedly affect the reactor dynamic behavior, both at start-up and at steady-state. Concerning the reactor start-up, simulations revealed that better interphase heat transport properties and lower heat capacity are useful to minimize the total start-up time (ignition and propagation). Conversely, more favorable transport properties reduce the maximum flow rate which allows achieving (ignition) and maintaining (blow-out) an ignited steady-state. Consequently, the foam – characterized by the best transport properties – exhibited the shortest start-up time and the lowest value of the blow-out gas flow rate. At steady state, it was found that oxygen conversion is strictly governed by interphase mass transfer. In this respect, foams were found to perform slightly better than the honeycombs, both being significantly better than the packed beds.

The relative rate of mass transfer coefficients among different configurations has however an important impact on the final temperature profile of the catalyst surface; this is discussed in the following section.

7 Case studies: analysis of spatially resolved measurements in adiabatic short contact time CPO reactors

Case-studies were selected from the published experimental and modeling studies, among those mostly informative on the kinetic factors (either chemical or physical) that govern the final performance and the thermal behavior of CPO reformers.

7.1 CH_4 CPO in foams: effect of Rh dispersion

Thanks to their suitable structure, foams were the first supports being investigated with the spatially resolved technique in CPO. In particular, CH_4 CPO experiments over Rh- and Pt-coated foams have been traditionally reported as case studies.

In a recent study, our group (in cooperation with the research groups of Schmidt and Vlachos) has investigated the effect that the increase of the active Rh metal surface area has on the evolution of the reaction, based on experiments performed over 5 wt% Rh-coated 80 ppi α-Al_2O_3 foams, at $C/O = 1$ and 5 slpm flow rate.[21] A ten-fold increase of the metal area was obtained by addition of 3 wt% γ-Al_2O_3 prior to the impregnation of the foam with the Rh precursor. The experiments were predictively simulated by applying the 1D model by Maestri *et al.*,[92] wherein the C1 microkinetic scheme described in Section 4 was incorporated. The correlation of Incera-Garrido *et al.*[23] were adopted to describe heat- and mass-transport in the foam support: though still debated, these correlations were obtained on 80 ppi foams identical to those used in the experiments. In this way, a completely predictive approach was followed, the only input parameter provided to the model being the metal surface area, which was derived with SEM micrographs and by H_2 pulse chemisorption experiments. Notably, in the absence of the washcoat, the extent of the metal surface area (72 cm^2/g) was found comparable to the geometric surface area of the foam, suggesting that a film of Rh formed during the catalyst preparation, which quickly sintered and cracked upon exposure to the reaction atmosphere. In the presence of the washcoat, the metal area was equal to 643 cm^2/g.

The experiments showed that the O_2 consumption profile did not change with Rh dispersion (Fig. 17, panel a). The numerical analysis confirmed that the consumption of O_2 is strictly limited by mass-transfer: at the very inlet of the catalyst, the O_2 wall concentration immediately drops to zero and no relevant O^* coverage is predicted on the surface. It was thus explained that the co-existence of synthesis gas and O_2 in the gas phase is exclusively due to control by external diffusive limitations. In contrast, as experimentally demonstrated in Ref. 78 and confirmed by the model, in the absence of the washcoat, the rate of CH_4 consumption is only partially governed by mass transfer and the chemical kinetics still have an influence; therefore, a strong acceleration in observed rate was experienced when ten times more Rh active sites were available on the surface (panel b). Such an acceleration led the reaction to the thermodynamic equilibrium and produced a significant decrease both of the hot spot and of the catalyst temperature level (panel c), which arise from the endothermic chemistry of CH_4 consumption. Notably, upon a one order of magnitude change of the active area, the model agrees well with the data for both the composition and temperature profiles; the deviations found in the O_2 consumption curve for 1 to 3 mm reactor length (panel a, Fig. 17) were shown to be associated with local variations of the foam structure.

With reference to the mechanism discussed in Section 7.1, the model analysis showed that the reactor entrance (the so-called "oxidation zone") behaved as a zone of type II. Here CH_4 is consumed *via* surface pyrolysis with progressive removal of H. The carbon adatom C^* reacts with OH^* and desorbs as CO or CO_2 depending on the local WGS equilibrium. H^* adatoms either recombine with OH^* to give H_2O or with H^* adatoms producing H_2. The overall molecular stoichiometry can be well approximated as the coupling of H_2 oxidation, steam reforming of CH_4 and WGS. Downstream from this O_2 consumption zone (in the so-called "reforming zone"), a zone III was active. Here the reaction path is almost identical to that of Regime II, but the OH^* species originate from adsorption and dissociation of H_2O, and not by reaction between H^* adatoms and O_2 leaking to the surface. In this case, the overall molecular stoichiometry represents an endothermic process, which is a combination of steam reforming and WGS only. The analysis of the surface coverage revealed that the Rh surface was

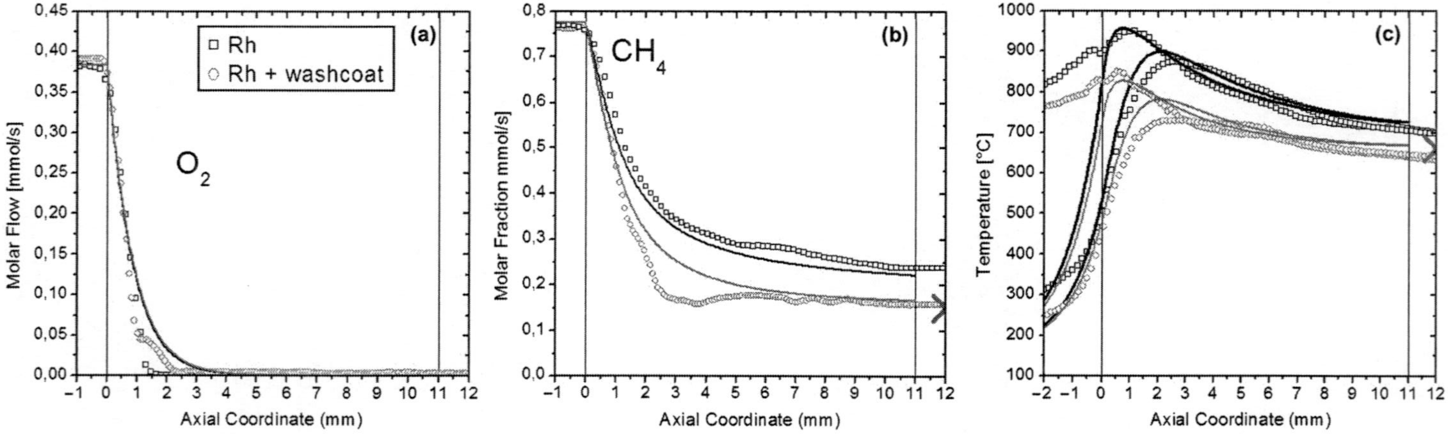

Fig. 17 CH_4 CPO over 5 wt% Rh-coated 80 ppi α-Al_2O_3 foams, comparison between model (solid lines) and data (symbols). The color black refers to the sample with no washcoat, red to the sample with 3 wt% γ-Al_2O_3 washcoat. X is thermodynamic equilibrium. CH_4 = 20% v/v, C/O = 1, Ar to balance, 5 slpm flow rate. Adapted from Journal of Catalysis, Vol. 275, A. Donazzi, M. Maestri, B. C. Michael, A. Beretta, P. Forzatti, G. Groppi, E. Tronconi, L. D. Schmidt and D. G. Vlachos, Microkinetic modeling of spatially resolved autothermal CH_4 catalytic partial oxidation experiments over Rh-coated foams, 270–279, Copyright (2010), with permission from Elsevier.

almost totally bare under all conditions examined, H^* and CO^* being the only adsorbed species found in non-negligible concentrations.

Thus, despite the fact that in the literature an oxidation zone and a reforming zone are typically identified as consecutive zones of the reactor, the present analysis show that while indeed O_2 is consumed with large heat release within a very short entrance length, CH_4 consumption occurs through an endothermic pyrolysis path along the whole length of the reactor. The superposition of exothermic and endothermic steps produces the sharp surface hot spot in the inlet portion of the reactor. Its intensity depends on the local balance between the mass transfer controlled rate of heat release and the mixed diffusion-kinetically controlled rate of heat consumption.

7.2 CH_4 CPO in foams: effect of operating conditions

Dalle Nogare *et al.*[80] numerically analyzed spatially resolved CH_4 CPO experiments carried out over 5 wt% Rh-coated 80 ppi α-Al_2O_3 foams. The effects of ranging the C/O ratio from 0.8 to 1.3 at 5 slpm, and of increasing the flow rate from 2.5 to 10 slpm at C/O = 1 were examined. In this case, the authors used a pseudo-1D, heterogeneous, fixed bed model of the reactor. Homemade mass- and heat-transport correlations were adopted (previously derived with dedicated experiments), and the value of Rh metal surface area was fitted on the basis of a sensitivity analysis. As well, the microkinetic scheme by Deutschmann and coworkers for CH_4 oxidation on Rh[49] was included in the model. Despite these differences, agreement is found upon the main features of the process. The authors confirmed that the consumption of O_2 is strictly mass transfer limited at all the C/O ratios and flow rates investigated. Also, due to the slow kinetics and to the very low surface area, the consumption of CH_4 was always found to occur under a chemically controlled regime, which became mixed chemical-diffusive only at the highest flow rates and temperatures. The analysis of the surface coverage suggested that the majority of Rh sites are free. Though, notably, a significant increase of the CO^* and C^* coverage was predicted upon increasing the C/O ratio, in line with the decreasing temperature and the increasing carbon-rich stoichiometry. When the flow rate was increased, the CO^* coverage is found to prevail on the surface in the second part of the foam, outside of the oxidation zone.

As stated before, spatially resolved CH_4 CPO experiments carried out on Pt-coated foams have not yet been modeled as extensively as in the case of Rh. Recently Horn *et al.*,[88] following an earlier work at the University of Minnesota,[19] reported extremely interesting effects by analysis of the reaction at increasing total pressure. In these works, two sets of experiments were examined: experiments at C/O = 1, 5 slpm flow rate and P = 1–11 bar, performed over 5 wt% Pt-coated 80 ppi α-Al_2O_3 foams;[19] experiments at C/O = 2, 4 slpm flow rate, P = 1–15 bar, performed over 1 wt% Pt-coated 80 ppi α-Al_2O_3 foams. In Ref. 19, the direct comparison with the profiles obtained over 5 wt% Rh-coated foams was also presented. Compared to Rh, over Pt the CPO of CH_4 resulted significantly slower: in particular, upon increasing the pressure, the length of the oxidation zone rapidly decreased. In contrast, under identical operating conditions, the oxidation length maintained constant over Rh, in line with the invariance of the mass transfer rate with respect to pressure. These observations, accompanied by the application of a simple 1D mass balance, suggested that, in the case of Pt, only the progressive increase of the oxidation rate with pressure caused the consumption of O_2 to enter a fully mass transport limited regime. As a consequence, the authors concluded that, under atmospheric pressure, Pt foam catalysts are governed by a chemical kinetic regime. Along with this line, the co-existence of syngas and O_2 observed in the spatially resolved profiles would suggest, on one hand, that a different reactivity characterizes Pt and, on the other hand, that a direct route of CH_4 oxidation to syngas could be present. The slower reactivity of Pt was also associated to the formation of carbon deposits that can block the surface, which would be absent (or formed to a lesser extent) in the case of Rh.

7.3 CH_4 CPO in honeycombs: analysis of spatially resolved measurements and effect of catalyst ageing

Several groups (Canu and coworkers,[80] Russo and coworkers,[95–97] Sadykov and coworkers,[98,99] Veser and coworkers,[100] Mantzaras and coworkers[101–105]) focused their interests on CH_4 CPO performed over honeycomb structured catalysts. From an application viewpoint, honeycombs are the reference supports for the automotive industry. From a scientific viewpoint, the regular channel geometry offers advantages when modeling, since well established mass- and heat-transport correlations are available,[87] so that kinetic and mechanistic studies can be performed under controlled conditions. At steady state, the effects of several parameters have been systematically investigated with respect to syngas selectivity and outlet temperature, namely the feed composition (C/O ratio), the inlet temperature and the gas flow rate. While data have been collected under a variety of contact times and conditions (more or less dissipative), it is commonly found that the reactions reach a thermodynamic equilibrium dependent upon their stoichiometric C/O ratio. Upon increasing the gas flow rate, the reactor approaches adiabatic conditions, due to the lower importance of heat dissipations compared to the enthalpy load supplied with the feed.

The behavior of honeycombs in the case of CH_4 CPO have been extensively analyzed also in our laboratory. The spatially resolved evolution of CH_4 CPO in honeycomb monoliths was explored (technical details are discussed in Section 5.3) over 400 cpsi cordierite honeycombs (30 mm long, 23 mm diameter) washcoated with 2 wt% Rh/α-Al_2O_3 catalysts, at 10 Nl min^{-1} flow rate, $CH_4/O_2 = 0.56$ and autothermal conditions ($T_{IN} = 25\,°C$); data are shown in Fig. 10. The same qualitative trends as those reported for foams were observed. The length of the oxidation zone ($\sim$5 mm) became longer than experiments performed over Rh-coated 80 ppi foams under identical operating conditions, ($\sim$3 mm)[17] consistent with both the lower geometric surface area (a_v is 2800 m^{-1} with 0.7 void fraction for 400 cpsi honeycombs, *vs.* 5500 m^{-1} with 0.8 void fraction for 80 ppi [80,106]) and the lower values of the heat and mass transfer coefficients (estimated at equal gas flow rate). Both the temperature and the concentration profiles became flat downstream from the hot spot within the first 15 mm of the channel, suggesting that thermodynamic equilibrium was reached. Upon increasing the gas flow to 21 Nl min^{-1}, the surface hot spot increased, while the maximum in the gas temperature profile shifted downstream without significant increase.

Deactivation occurred under the most severe conditions (21 Nl min^{-1} flow rate and 420 °C preheat);[94] as shown in Fig. 18, the deactivation process was observed to involve preferentially the first portion of the catalyst, with a marked increase (run after run) of the measured hot spot temperature, leaving the outlet conditions nearly unaltered. The phenomelogy of deactivation was fully explained by a modeling analysis on the impact of a progressive loss of Rh surface. It was found that, because of the full diffusion control of the O_2 conversion, this conversion is practically insensitive to a reduction of the Rh on the surface of as much as one order of magnitude. Instead, because of its more kinetically controlled regime, the rate of CH_4 surface pyrolysis is greatly sensitive to the amount of Rh on the surface. Thus, catalyst ageing produces an imbalance between the rates of heat production and heat removal with an increase of the surface and gas-phase temperatures. Another important conclusion of this analysis was that, because of the temperature increase, the overall performance of the reactor (in terms of outlet CH_4 conversion and syngas yield) is not sensitive to deactivation, at least in its initial stages. However, the progressive increase of temperature has a dramatic autocatalytic effect on deactivation.

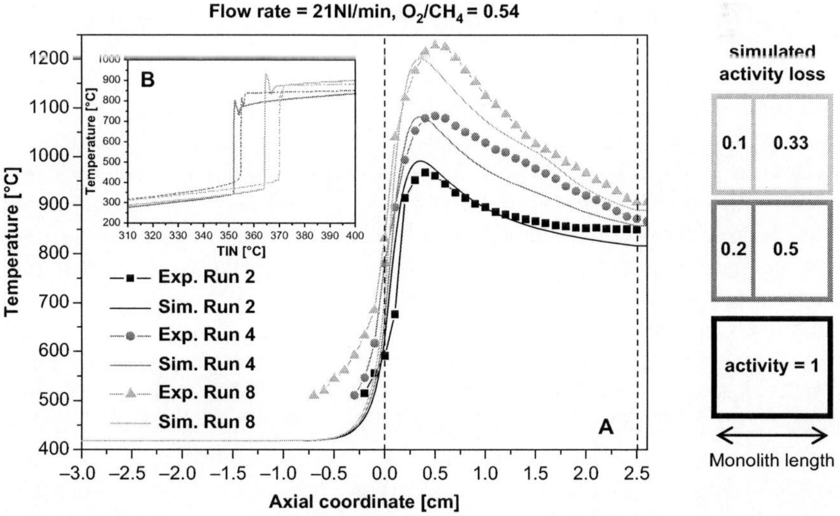

Fig. 18 Ageing of Rh-coated monolith, as measured by the evolution of temperature profiles. Simulations (solid lines) assume a reduction of the Rh surface through an "activity factor", which is 1 at initial conditions and decreases to 0.2 and 0.1 in initial third of the reactor, while it reduces to 0.5 and 0.33 in the remaining part of the reactor. Adapted with permission from A. Beretta, G. Groppi, M. Lualdi, I. Tavazzi and P. Forzatti, Experimental and Modeling Analysis of Methane Partial Oxidation: Transient and Steady-State Behavior of Rh-Coated Honeycomb Monoliths, *Ind. Eng. Chem. Res.*, **48**(8), 3825–3836. Copyright (2009) American Chemical Society.

While it starts in the inlet hot portion of the reactor, it eventually extends to the whole reactor volume.

7.4 C_3H_8 CPO in honeycombs: effect of homogeneous reactions

Recently, our group applied the spatially resolved sampling technique to the CPO of C_3H_8 within 400 cpsi cordierite honeycombs coated with a 2 wt% Rh/α-Al_2O_3 catalyst.[86,107] The experiments were performed at 5 and 10 Nl min^{-1} flow rate, with C_3H_8 concentration ranging from 4 to 11% v/v, corresponding to the stoichiometric C_3H_8/air concentration at $O_2/C = 0.56$. At the highest flow rate, the results indicated that catalyst deactivation occurred after prolonged exposure to the stoichiometric mixture, due to the high temperature reached at the hot spot (>960 °C). Stable operations were achieved under more heat-dissipative conditions, that is, either by lowering the gas flow rate or by adopting an optimized configuration for the internal reactor layout (see Section VIII). In this way, the severity of the hot spot was moderated and thermal degradation was avoided. Results are reported in Fig. 19. The spatially resolved measurement of the gas composition provided novel pieces of evidence concerning the evolution of C_3H_8 CPO. Aside from the typical profiles of syngas and total oxidation products, other intermediate species were observed, namely C_2H_4, C_3H_6, and C_2H_6, which peaked at the catalyst inlet and were completely consumed before the outlet (Fig. 19c). A complex evolution was revealed also in the case of CH_4; the initial peak was followed by consumption within the first half of the catalyst and again by formation in the remaining part.

The numerical analysis of the experiments followed the predictive approach previously discussed for CH_4 CPO. Molecular kinetic schemes of C_3H_8 CPO, independently derived on the basis of experiments performed on the annular microreactor, were introduced in the model. The detailed homogeneous reaction schemes by Ranzi *et al.*[108] were included as well. An independent, experimentally estimated value of the Rh active metal area was introduced in the simulations. The numerical analysis indicated that gas-phase activation of C_3H_8 by cracking and dehydrogenation was responsible for the formation of CH_4, C_2H_4, and C_3H_6 at the catalyst inlet. In contrast, the predicted contribution of homogenous oxidative pyrolysis was negligible; the consumption of O_2 was controlled by external mass transfer to the catalyst wall. After peaking, the intermediate species were consumed on the catalyst surface by heterogeneous steam reforming with formation of H_2 and CO. With respect to CH_4, the gas-phase production by C_3H_8 cracking was followed by steam reforming and then by further production by methanation, favored by decreasing temperatures and a high partial pressure of the syngas. Overall, the occurrence as a synergistic hetero–homogenous process emerged from the experiments, wherein the conversion of C_3H_8 to syngas takes place on the surface by a heterogeneous route, while a homogenous by-pass route is also active, producing additional reactive species that adsorb on the Rh surface and are also reformed to syngas. The homogeneous route makes the catalytic conversion even faster. However, the intermediate olefins produced in the gas-phase are coke precursors, which suggests the possibility of catalyst fouling.

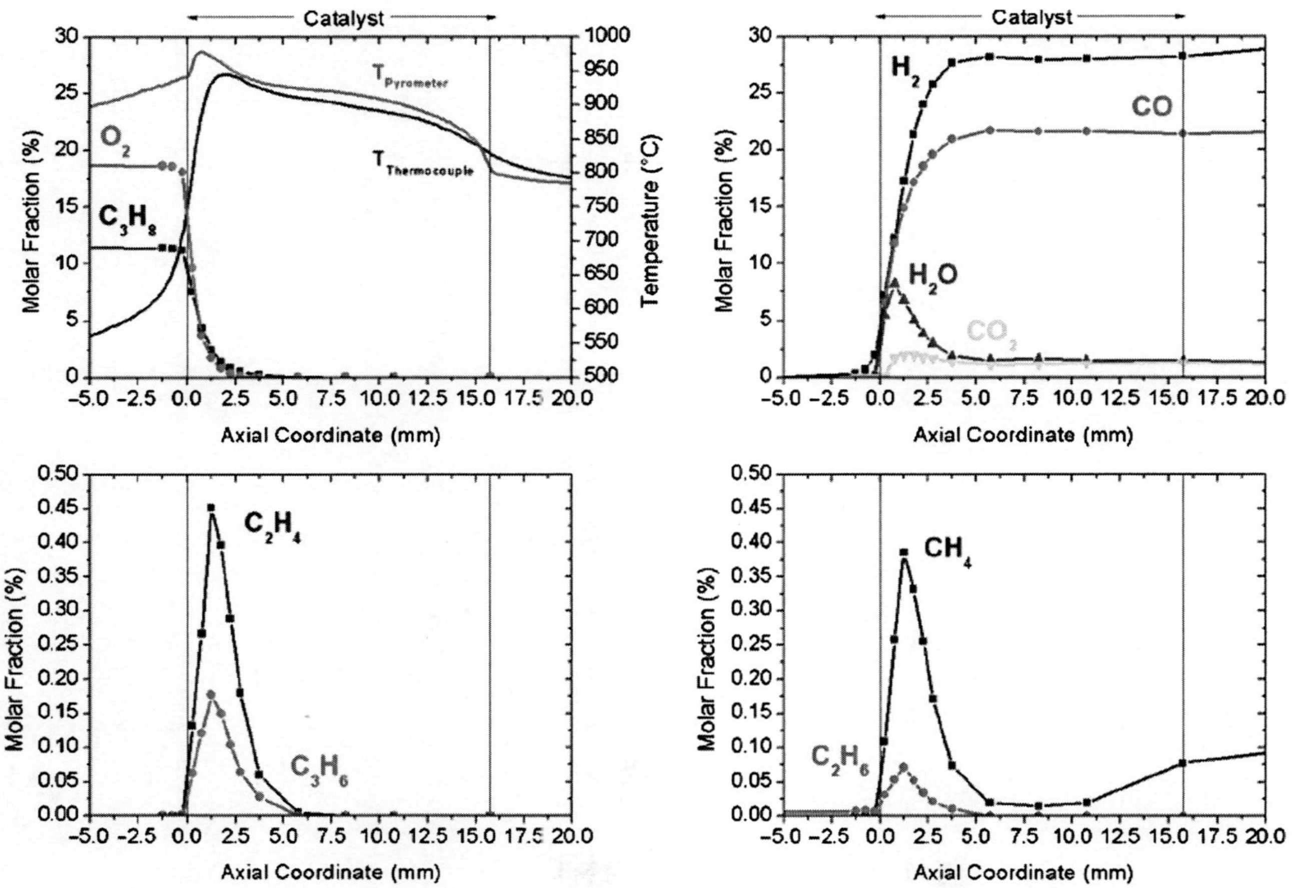

Fig. 19 An experiment of C_3H_8 CPO in a Rh-coated 400 cpsi honeycomb monolith, spatially resolved temperature and concentration profiles. Flow rate = 5 l(NTP)/kgcat/h, feed is propane + air with $O_2/C = 0.56$.

8 Optimization of the thermal behavior of short contact time CPO reformers

The comprehension of the kinetic factors controlling the CPO reactor, gained through experimental and modelling studies as those described above, and in particular the understanding of how transport phenomena and surface reaction affect the reactor behavior, made us recognize that the feasibility of small scale production of syngas *via* CH_4 CPO relies on the thermal management of the short contact time reactor, and not on obtaining high syngas yields (which is not a challenging task).

In dedicated modelling and experimental studies we have addressed the optimization of the reactor design, with the specific objective of minimizing the surface hot spot temperature. Beretta *et al.*[94] presented a sensitivity analysis which showed the positive effects of increasing the monolith channel size and increasing the Rh load of the washcoat. Because of the full mass transfer control of the O_2 consumption, the rate of heat release can effectively be slowed down by increasing the diffusion length of O_2 from the bulk to the wall. Conversely, because CH_4 conversion occurs in a mixed diffusion-kinetic regime, the increase of the catalyst Rh load is expected to promote selectively the endothermic reactions of CH_4 pyrolysis. Concepts and model predictions were demonstrated in the laboratory.[85] Figure 20 shows the results of a CH_4 CPO experiment performed at the same flow rate and feed composition as that of Fig. 15, but using a monolith with larger channel opening (1.8 mm *vs.* 1.0 mm) and increased catalyst load (washcoat thickness of about 40 μm *vs.* 15 μm). The measurements of spatially resolved temperature and concentration profiles and the model predictions showed that this configuration realized both a reduction of the rate of O_2 conversion and a selective enhancement of the rate of CH_4 conversion through endothermic reactions. The resulting temperature profile was significantly flatter than in the standard reactor configuration, almost eliminating the problem of the catalyst inlet hot spot.

Overheating is an even more critical issue when performing the CPO of C_{2+} fuels. Livio *et al.*[107] performed spatially resolved CPO experiments with stoichiometric C_3H_8/air mixtures and found that temperatures as high as 950 °C were reached on the catalyst surface, which caused rapid deactivation. The comparison with reference CPO experiments carried out with stoichiometric CH_4/air mixtures revealed that much higher temperatures were reached and sharper gradients were present in the case of C_3H_8. The reasons for these differences were rationalized and catalyst design criteria for moderating the hot spot were analyzed. In the case of C_3H_8 CPO the rate of heat release is larger than in the case of CH_4 CPO due to the higher O_2 concentration of the inlet stoichiometric mixture, whereas the rate of heat removal is slowed down because of the higher external diffusive resistances that affect the steam reforming reaction of propane. In light of these results, strategies different from CH_4 CPO are required to minimize the hot spot. Indeed, C_3H_8 CPO simulations showed that enlarging the channel opening of the honeycomb does not produce beneficial effects, while only a moderate temperature decrease (55 °C) is obtained by increasing the

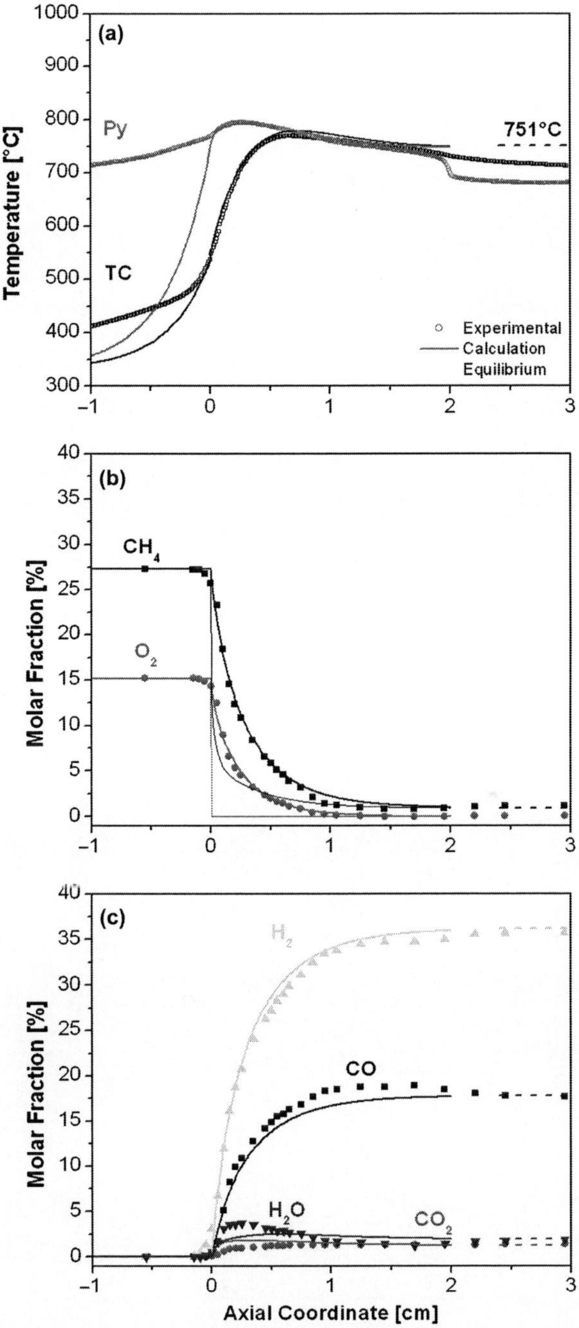

Fig. 20 CH_4 CPO experiment in a Rh-coated 110 cpsi monolith reactor. Operating conditions as in Figure W.11 and channel opening = 1.8 mm. Reprinted from Catalysis Today, Vol. 171, A. Beretta, A. Donazzi, D. Livio, M. Maestri, G. Groppi, E. Tronconi and P. Forzatti Optimal design of a CH_4 CPO-reformer with honeycomb catalyst: Combined effect of catalyst load and channel size on the surface temperature profile, 79–83, Copyright (2011), with permission from Elsevier.

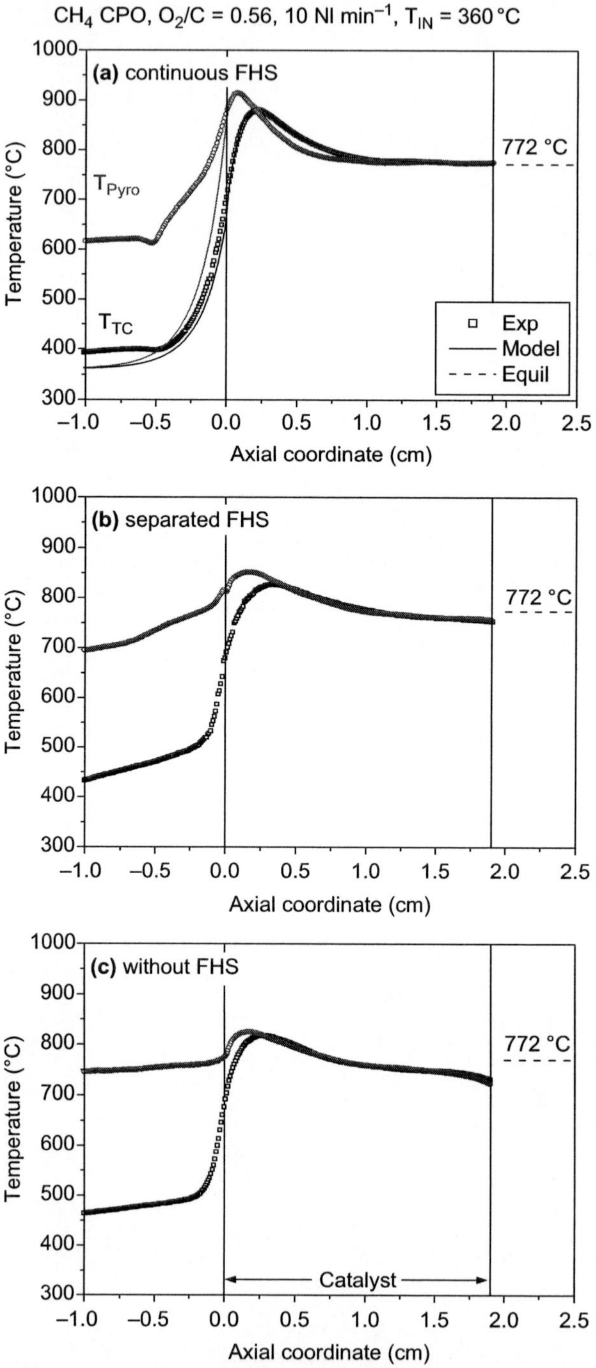

Fig. 21 CH_4 CPO experiments in Rh-coated 400 cpsi honeycomb monolith. Effect of the position of the front heat shield (FSH). (a) FHS is a continuum with the catalytic monolith, (b) FHS is at a certain distance from the monolith, (c) FHS is removed. Adapted from: Topics in Catalysis, 54, 2010, 866, Optimal design of a CPO-reformer of light hydrocarbons with honeycomb catalyst: Effect of frontal heat dispersions on the temperature profiles, D. Livio, A. Donazzi, A. Beretta, G. Groppi and P. Forzatti, Fig. 2 original copyright notice, with kind permission from Springer Science and Business Media.

catalyst load. The moderation of the hot spot can instead be accomplished by reducing either the total flow rate or the aspect ratio of the reactor.

Combined with these solutions, the adoption of a non perfectly adiabatic reactor configuration is also suggested to optimize the axial temperature profile, for instance by separating the front heat shield from the catalytic monolith in order to enhance heat dissipation by radiation.

As shown in Fig. 21 an effective moderation of the inlet temperatures was in fact observed upon passing from the perfectly adiabatic configuration, where the FHS is contiguous with the catalytic monolith, to the slightly dissipative configurations of a separated FHS and the removal of the FHS.[109] We observed a deviation in the approach to the adiabatic temperature, but a preferential moderation of the temperature of the glowing inlet face.

The experiment reported in Fig. 21 was performed with a CH_4/air mixture, still it was extremely instructive about the possible exploitation of a moderately more dissipative reactor configuration to control the inlet temperature rise even with fuels heavier than methane.

9 Summary and open issues

We have here presented an overview of the experimental and modeling studies that have concurred to explain the kinetics of CH_4 and light hydrocarbons partial oxidation on Rh, leading to the development of engineering tools (kinetic schemes, reactor models) that can be presently applied to design CPO-reformers for various applications.

Kinetically relevant data were obtained in isothermal micro- reactors, where temperature gradients were negligible or minimized by a large rate of heat dispersion. We have illustrated with some detail the data obtained in our laboratory with the use of the annular reactor; some of the major findings that have an immediate impact on the modeling of CPO-reformers are the following:

– under chemically controlled conditions, fuel oxidation gives rise to CO_2 and H_2O; the reaction rate is proportional to the concentration of the fuel and independent of the concentration of O_2. This finding refers to O/C ratios higher than 1 and are coherent with Regime 1, defined by Chin, Iglesia *et al.*[30] as the condition of an oxygen saturated surface where the unique kinetically relevant step is the rupture of the first C-H bond.

– the rate of the steam reforming of the fuel is proportional to the concentration of the fuel and independent of the concentration of H_2O (Regime 4 in Iglesia's definition); thus, co-feed of water in CPO reformers does not enhance the rate of steam reforming where it is kinetically controlled, but it plays a role when the reaction approaches the thermodynamic equilibrium.

– both reactions (oxidation and steam reforming) manifest an overall kinetic order below 1; this result does not find a comparison with studies from Iglesia's group, where coverage effects were always found to be negligible. Experiments were however performed under strictly differential regime, while data in the annular reactor were collected under integral conditions; these could be likely responsible of a higher impact of competitive adsorption phenomena. In practice, experiments (or kinetics) under diluted conditions tend to overestimate the reactivity of fuel/air mixtures.

– the kinetics of CO_2 reforming of methane can be well explained by a combination of the kinetics of the reverse water gas shift reaction and the steam reforming of CH_4; this finding is not in contrast with the findings from Iglesia's group since their data were obtained under WGS-equilibrated conditions.

A key step forward towards the engineering of the process has been the development of reliable kinetic schemes. Among the several schemes proposed in the literature, we have proposed a thermodynamically consistent microkinetic model of CH_4 CPO that has been validated against a broad variety of data on various reacting systems; the model allows identification of the prevalent reaction paths behind the observed dependences in CPO experiments in annular reactor. Most importantly, by coupling the surface kinetic scheme with the CPO reactor model, we could recognize that different regimes (or zones) can establish in the reactor and have an impact on the surface chemistry; these depend on the specific conditions of temperature and concentration that locally characterize the gas-solid interphase. These "reactor driven" regimes (or zones) were distinguished depending on the presence of O_2 at the surface (zone I), the onset of a mass transfer limited diffusion of O_2 at the wall (zone II), or the absence of O_2 in the gas phase (zone III). It was found that in the presence of O_2 on the surface, the typical case of CPO reactors at low temperature, CH_4 activation (H-abstraction by a surface O-species) is the rate determining step and CO_2 and H_2O are uniquely formed. Under O_2 mass transfer limitation, which takes place at the reactor inlet at high surface temperatures, CH_4 activation is still the rate determining step but it proceeds according to a pyrolytic path, while O_2 is rapidly consumed by surface H^* species giving rise to OH^*, which acts as the oxidizing species and is responsible for the formation of CO. Once O_2 is completely consumed, similar reaction paths occur , but OH^* is provided by re-adsorpiton of H_2O.

CH_4 CPO experiments at the macro-scale of short contact time insulated reactors were performed by implementing state of the art techniques for the axially resolved measurements of concentration and temperature of both the gas and wall phases. Both foams and honeycomb monoliths were studied. The experiments were interpreted in the literature and in our laboratory based on 1D models of the CPO-reactors, implementing adequate correlations for describing heat and mass transfer from the gas phase to the wall. Because of the high reaction temperatures, both in monoliths and in foams, zones II and III can explain most of the reactor behavior and of the surface kinetics under operating conditions of practical interest. The initial consumption of O_2 is in fact fully controlled by external mass transfer and concentrated within a short inlet zone. Most of the CH_4 conversion occurs instead though an endothermic reforming path along the whole length of the structured reactor. This explains why a sharp hot-spot is established in the entrance region at the catalyst surface. Its intensity results from the balance between mass transfer controlled exothermic steps and kinetically controlled endothermic steps.

Concerning the design of the CPO-reformer, we have focused our efforts towards the minimization of the hot-spot temperature, recognized as *the* critical feature of the process. The interplay between surface kinetics and transport phenomena offered several degrees of freedom; we showed, for instance, that increasing the Rh content and enlarging the channel size are highly beneficial in decreasing the inlet surface temperature by promoting the endothermic reactions and by slowing down the diffusion of O_2 at the wall, respectively.

The CPO of propane in autothermal reactors was also studied. In the case of propane, the formation of hydrocarbon species was detected and their evolution along the reactor could be quantitatively explained by a synergy between homogeneous cracking reactions of propane and heterogeneous consumption of the formed hydrocarbon species *via* steam reforming to synthesis gas. The thermal behavior of a C_3H_8-CPO reformer is characterized by even higher hot-spot temperatures, which cannot be reduced by the same strategies that were used for CH_4 CPO because of the greater diffusional resistances of propane. However, the reactor configuration (*e.g.* the positioning of the front heat shield) and the operating conditions (*e.g.* the total flow rate) can be designed in order to maintain surface temperatures within an acceptable range, thus preventing rapid deactivation of the catalyst.

In the end, it is interesting to highlight those issues that could deserve further consideration in the future:

1) intrinsic kinetics. We have seen that the kinetic investigation needs specifically designed tools, but the complete elimination of physical artifacts is a great challenge when studying a very fast catalytic process as the CPO of light hydrocarbons on Rh. Iglesia and co-workers have proposed an infinitely diluted packed bed-reactor but the applicability to differential conditions is an intrinsic limitation, especially for a process characterized by a complex kinetic scheme. Other reactor designs can overcome this disadvantage, but they require the medium of reactor modeling for data analysis.

2) micro-kinetic modeling. Presently, the microkinetic schemes proposed in the literature assume that the active site for CH_4 activation is a metallic Rh site. However, Grunwaldt *et al.*[110–112] have observed the presence of oxidized Rh species at the entrance of a CPO short contact time reactor, while the fraction of reduced sites was found to increase along the longitudinal coordinate with the progressive consumption of oxygen. Apparently, thus, a gap exists between kinetic studies and mechanicistic studies that has not been addressed in the literature yet.

3) kinetics of CO_2 reforming – CO_2 reforming has been widely recognized as a negligible route in CH_4 CPO[105,113] and minor attention has been paid on the chemical role of CO_2 in CH_4/O_2 reacting systems. On the opposite a very wide literature is devoted to develop catalysts for the CO_2 reforming of methane and light hydrocarbons (see for example ref. 45 and papers cited therein). In a dedicated analysis of the relative rate of CO_2 activation and methane activation, we have explained the reactivity of CO_2/CH_4 mixtures

as a result of the combined kinetics of the reverse-WGS reaction and the steam reforming reactions. It is interesting to note that, in the literature of CO_2-reforming, little care is usually taken (and more could be spent) to the characterization of the rate of reverse-water gas shift that, in our view, is the kinetic bottleneck of the apparent reaction between CH_4 and CO_2 under conditions of non-equilibration.

4) the sampling technique – the axially resolved sampling techniques are extremely powerful, however CFD analyses are needed to better comprehend the fluodynamics around and within the sampling capillary, thus supporting on a theoretical basis that the intrusion of the capillary and of the sampling procedure on the reactor is negligible.

5) a look inside the reactor – Several groups have developed experimental systems that combined the catalyst testing with in-situ characterization techniques, with the aim of gaining insight into the nature the reaction intermediates,[71,114,115] the detailed composition of the gas-phase,[104,113] the state of Rh.[110–112] These techniques and their contribution to unravel mechanicistic issues have been illustrated in other papers.[8] A very recent and new proposal is the exploitation of optical fibers to record the Raman spectra from the catalyst under working conditions. Horn and co-workers[26,47] have applied this technique, in combination with the measurements of surface and gas-phase temperature profiles as well as axial concentration profiles) to analyze the formation of C-deposits on the surface of Pt-coated monoliths under CH_4-CPO conditions, providing rather unique pieces of evidence of the specific reactivity of Pt. We may expect that this technique will have wide applications in the future.

6) CPO reformers – It is a strong belief of the authors that producing high yields of syngas *via* CPO of light hydrocarbons is not a challenge by itself, since Rh catalysts are extremely active and selective; notably, the reaction approaches equilibrium under adiabatic conditions. The formation of a hot-spot is instead a highly undesired aspect of this process, since Rh catalysts cannot stand temperatures as high as 1000 °C for prolonged time on stream. Thermal management needs to exploit the degree of freedom that the reactor design offers. In this respect, the minimization of the active elements in order to reduce the cost of the catalyst is a false objective. We have clearly shown that hot spot temperatures can be reduced either decreasing the rate of O_2 diffusion in the reactor, or increasing the catalyst activity. The tailoring of the catalyst structured support (nature and conductivity, as well as geometrical configuration) is believed a promising field which offers open room for the reactor optimization.

As a concluding comment, we observe that strengths and weaknesses of the CPO-process have been better understood in recent years and this, as documented in this review, greatly relied on the combined and synergistic use of experiments and modeling. Several independent pieces of evidence contributed to cover the gap from surface site to reactor design, to form an harmonic picture of the process, to identify and quantify the several kinetic drivers that control the CPO of light hydrocarbons to synthesis gas under different operating conditions and reactor configurations and finally to develop reliable engineering tools.

References

1 D. K. Lee, K. Y. Koo, D. J. Seo and W. L. Yoon, *Renewable Energy*, **42**, 234.

2 D. C. Walther and J. Ahn, *Progress in Energy and Combustion Science*, **37**, 583.

3 J. Ki, D. Kim and S. Honavara-Prasad, *Journal of Fuel Cell Science and Technology*, **9**.

4 D. Scognamiglio, L. Russo, P. L. Maffettone, L. Salemme, M. Simeone and S. Crescitelli, *International Journal of Hydrogen Energy*, **37**, 263.

5 N. U. Andersen, H. Olsson and H. Topsøe, *Hydrocarbon Engineering*, **16**, 48.

6 N. S. Kaisare and D. G. Vlachos, *Progress in Energy and Combustion Science*, **38**, 321.

7 T. V. Choudhary and V. R. Choudhary, *Angewandte Chemie Int. Ed.*, 2008, **47**, 1828–1847.

8 B. C. Enger, R. Lødeng and A. Holmen, *Applied Catalysis A: General*, 2009, **364**, 15.

9 A. P. E. York, T. C. Xiao, M. L. H. Green and J. B. Claridge, *Catalysis Reviews - Science and Engineering*, 2007, **49**, 511.

10 D. A. Hickman and L. D. Schmidt, *Journal of Catalysis*, 1992, **138**, 267.

11 A. S. Bodke, S. S. Bharadwaj and L. D. Schmidt, *Journal of Catalysis*, 1998, **179**, 138.

12 R. P. O'Connor, E. J. Klein and L. D. Schmidt, *Catalysis Letters*, 2000, **70**, 99.

13 J. J. Krummenacher, K. N. West and L. D. Schmidt, *Journal of Catalysis*, 2003, **215**, 332.

14 L. D. Schmidt, E. J. Klein, C. A. Leclerc, J. J. Krummenacher and K. N. West, *Chemical Engineering Science*, 2003, **58**, 1037.

15 K. A. Williams and L. D. Schmidt, *Applied Catalysis a-General*, 2006, **299**, 30.

16 B. J. Dreyer, I. C. Lee, J. J. Krummenacher and L. D. Schmidt, *Applied Catalysis a-General*, 2006, **307**, 184.

17 D. D. Nogare, N. J. Degenstein, R. Horn, P. Canu and L. D. Schmidt, *Journal of Catalysis*, 2011, **277**, 134.

18 P. J. Dauenhauer, B. J. Dreyer, N. J. Degenstein and L. D. Schmidt, *Angewandte Chemie-International Edition*, 2007, **46**, 5864.

19 A. Bitsch-Larsen, R. Horn and L. D. Schmidt, *Applied Catalysis A: General*, 2008, **348**, 165.

20 P. J. Dauenhauer, J. L. Colby, C. M. Balonek, W. J. Suszynski and L. D. Schmidt, *Green Chemistry*, 2009, **11**, 1555.

21 A. Donazzi, M. Maestri, B. C. Michael, A. Beretta, P. Forzatti, G. Groppi, E. Tronconi, L. D. Schmidt and D. G. Vlachos, *Journal of Catalysis*, 2010, **275**, 270.

22 P. Mulheims and B. Kraushaar-Czarnetzki, *Industrial & Engineering Chemistry Research*, 2011, **50**, 9925.

23 G. Incera Garrido, F. C. Patcas, S. Lang and B. Kraushaar-Czarnetzki, *Chemical Engineering Science*, 2008, **63**, 5202.

24 G. Jones, J. G. Jakobsen, S. S. Shim, J. Kleis, M. P. Andersson, J. Rossmeisl, F. Abild-Pedersen, T. Bligaard, S. Helveg, B. Hinnemann, J. R. Rostrup-Nielsen, I. Chorkendorff, J. Sehested and J. K. Norskov, *Journal of Catalysis*, 2008, **259**, 147.

25 J. Perez-Ramirez, R. J. Berger, G. Mul, F. Kapteijn and J. A. Moulijn, *Catalysis Today*, 2000, **60**, 93.

26 O. Korup, C. F. Goldsmith, G. Weinberg, M. Geske, T. Kandemir, R. Schlogl and R. Horn, *Journal of Catalysis*, 2012.

27 Y. H. Chin, C. Buda, M. Neurock and E. Iglesia, *Journal of the American Chemical Society*, 2011, **133**, 15958.
28 Y. H. Chin, C. Buda, M. Neurock and E. Iglesia, *Journal of Catalysis*, 2011, **283**, 10.
29 Y. H. Chin and E. Iglesia, *Journal of Physical Chemistry C*, 2011 **115**, 17845.
30 M. García-Diéguez, Y. H. C. Chin and E. Iglesia, *Journal of Catalysis*, 2012, **285**, 260.
31 J. G. McCarthy, *Catalysis Today*, 1995, 283.
32 A. Beretta, P. Baiardi, D. Prina and P. Forzatti, *Chemical Engineering Science*, 1999, **54**, 765.
33 G. Groppi, W. Ibashi, E. Tronconi and P. Forzatti, *Catalysis Today*, 2001, **69**, 399.
34 W. Ibashi, G. Groppi and P. Forzatti, *Catalysis Today*, 2003, **83**, 115.
35 R. J. Berger and H. Stitt, *EUROKIN project - Reactor designs for testing fast or strongly exothermic heterogeneous reactions*, 2011.
36 A. Donazzi, A. Beretta, G. Groppi and P. Forzatti, *Journal of Catalysis*, 2008, **255**, 241.
37 A. Donazzi, A. Beretta, G. Groppi and P. Forzatti, *Journal of Catalysis*, 2008, **255**, 259.
38 P. D. F. Vernon, M. L. H. Green, A. K. Cheetham and A. T. Ashcroft, *Catalysis Letters*, 1990, **6**, 181.
39 M. García-Diéguez, I. S. Pieta, M. C. Herrera, M. A. Larrubia and L. J. Alemany, *Applied Catalysis A: General*, 2010, **377**, 191.
40 F. Basile, G. Fornasari, M. Gazzano, A. Kiennemann and A. Vaccari, *Journal of Catalysis*, 2003, **217**, 245.
41 R. López-Fonseca, C. Jiménez-González, B. de Rivas and J. I. Gutiérrez-Ortiz, *Applied Catalysis A: General*, 2012, **437–438**, 53.
42 L. Majocchi, G. Groppi, C. Cristiani, P. Forzatti, L. Basini and A. Guarinoni, *Catalysis Letters*, 2000, **65**, 49.
43 V. A. Tsipouriari, Z. Zhang and X. E. Verykios, *Journal of Catalysis*, 1998, **179**, 283.
44 S. F. Rice, A. H. McDaniel, E. S. Hecht and A. J. J. Hardy, *Industrial and Engineering Chemistry Research*, 2007, **46**, 1114.
45 N. E. McGuire, N. P. Sullivan, O. Deutschmann, H. Zhu and R. J. Kee, *Applied Catalysis A: General*, 2011, **394**, 257.
46 J. A. Dumesic, G. W. Huber, and M. Boudart, in 'Microkinetics: Rates of Catalytic Reactions', ed. G. Ertl, H. Knozinger, F. Schuth and J. Weitkamp, New York, 2008.
47 O. Korup, R. Schlogl and R. Horn, *Catalysis Today*, 2012, **181**, 177.
48 P. Stoltze and J. K. Norskov, in 'Theoretical modeling of catalytic reactions', ed. G. Ertl, H. Knozinger, F. Schuth, and J. Weitkamp, New York, 2008.
49 R. Schwiedernoch, S. Tischer, C. Correa and O. Deutschmann, *Chemical Engineering Science*, 2003, **58**, 633.
50 J. Xu, R. Clayton, V. Balakotaiah and M. P. Harold, *Applied Catalysis B-Environmental*, 2008, **77**, 395.
51 D. A. Hickman and L. D. Schmidt, *AIChE J.*, 1993, **39**, 1164.
52 P. Aghalayam, Y. K. Park and D. G. Vlachos, *AIChE J.*, 2000, **46**, 2017.
53 A. B. Mhadeshwar and D. G. Vlachos, *Journal of Physical Chemistry B*, 2005, **109**, 16819.
54 A. B. Mhadeshwar, H. Wang and D. G. Vlachos, *Journal of Physical Chemistry B*, 2003, **107**, 12721.

55 M. Maestri, D. G. Vlachos, A. Beretta, G. Groppi and E. Tronconi, *AIChE J.*, 2009, **55**, 993.
56 M. Maestri, in 'Microkinetic Analysis of Complex Chemical Processes at Surfaces', ed. B. Pignataro, Weinheim, 2012.
57 E. Shustorovich and H. Sellers, *Surface Science Reports*, 1998, **31**, 5.
58 M. Maestri and K. Reuter, *Angewandte Chemie-International Edition*, 2011, **50**, 1194.
59 J. A. Dumesic, D. F. Rudd, L. M. Aparicio, J. E. Rekoske and A. A. Trevino, 'The microkinetics of heterogeneous catalysis', American Chemical Society, 1993.
60 M. Maestri and K. Reuter, *Chemical Engineering Science*, 2012, **74**, 296.
61 M. Maestri, D. G. Vlachos, A. Beretta, G. Groppi and E. Tronconi, *Journal of Catalysis*, 2008, **259**, 211.
62 M. Maestri, D. G. Vlachos, A. Beretta, P. Forzatti, G. Groppi and E. Tronconi, *Topics in Catalysis*, 2009, **52**, 1983.
63 M. Maestri, A. Beretta, T. Faravelli, G. Groppi, E. Tronconi and D. G. Vlachos, *Chemical Engineering Science*, 2008, **63**, 2657.
64 C. T. Campbell, *Journal of Catalysis*, 2001, **204**, 520.
65 K. H. Hofstad, O. A. Rokstad and A. Holmen, *Catalysis Letters*, 1996, **36**, 25.
66 K. H. Hofstad, T. Sperle, O. A. Rokstad and A. Holmen, *Catalysis Letters*, 1997, **45**, 97.
67 C. R. H. De Smet, M. H. J. M. De Croon, R. J. Berger, G. B. Marin and J. C. Schouten, *Applied Catalysis A: General*, 1999, **187**, 33.
68 C. R. H. De Smet, M. H. J. M. De Croon, R. J. Berger, G. B. Marin and J. C. Schouten, *AIChE Journal*, 2000, **46**, 1837.
69 K. Heitnes Hofstad, J. H. B. J. Hoebink, A. Holmen and G. B. Marin, *Catalysis Today*, 1998, **40**, 157.
70 M. Fathi, K. Heitnes Hofstad, T. Sperle, O. A. Rokstad and A. Holmen, *Catalysis Today*, 1998, **42**, 205.
71 L. Basini, A. Guarinoni and A. Aragno, *Journal of Catalysis*, 2000 **190**, 284.
72 L. Basini, K. Aasberg-Petersen, A. Guarinoni and M. Østberg, *Catalysis Today*, 2001, **64**, 9.
73 M. Simeone, L. Menna, L. Salemme and C. Allouis, *Experimental Thermal and Fluid Science*, 2010, **34**, 381.
74 M. Simeone, L. Salemme and L. Menna, *International Journal of Hydrogen Energy*, 2012, **37**, 9049.
75 R. Horn, N. J. Degenstein, K. A. Williams and L. D. Schmidt, *Catalysis Letters*, 2006, **110**, 169.
76 R. Horn, K. A. Williams, N. J. Degenstein, A. Bitsch-Larsen, D. D. Nogare, S. A. Tupy and L. D. Schmidt, *Journal of Catalysis*, 2007, **249**, 380.
77 A. Bitsch-Larsen, N. J. Degenstein and L. D. Schmidt, *Applied Catalysis B: Environmental*, 2008, **78**, 364.
78 A. Donazzi, B. C. Michael and L. D. Schmidt, *Journal of Catalysis*, 2008, **260**, 270.
79 R. Horn, K. A. Williams, N. J. Degenstein and L. D. Schmidt, *Chemical Engineering Science*, 2007, **62**, 1298.
80 D. Dalle Nogare, S. Salemi, P. Biasi and P. Canu, *Chemical Engineering Science*, 2011, **66**, 6341.
81 J. S. Choi, W. P. Partridge and C. S. Daw, *Applied Catalysis a-General*, 2005, **293**, 24.
82 J. S. Choi, W. P. Partridge, J. A. Pihl and C. S. Daw, *Catalysis Today*, 2008, **136**, 173.
83 M. Bosco and F. Vogel, *Catalysis Today*, 2006, **116**, 348.

84 A. Donazzi, D. Livio, A. Beretta, G. Groppi and P. Forzatti, *Applied Catalysis A: General*, 2011, **402**, 41.
85 A. Beretta, A. Donazzi, D. Livio, M. Maestri, G. Groppi, E. Tronconi and P. Forzatti, *Catalysis Today*, 2011, **171**, 79.
86 A. Donazzi, D. Livio, M. Maestri, A. Beretta, G. Groppi, E. Tronconi and P. Forzatti, *Angewandte Chemie - International Edition*, 2011, **50**, 3943.
87 R. K. Shah and A. L. London, 'Laminar flow forced convection in ducts', Academic Press, 1978.
88 O. Korup, S. Mavlyankariev, M. Geske, C. F. Goldsmith and R. Horn, *Chemical Engineering and Processing: Process Intensification*, 2011, **50**, 998.
89 O. Deutschmann and L. D. Schmidt, *A.I.Ch.E. Journal*, 1998, **44**, 2465.
90 O. Deutschmann, R. Schwiedernoch, L. I. Maier and D. Chatterjee, *Natural Gas Conversion VII - Studies in Surface Science and Catalysis*, 2001, **136**, 251.
91 D. K. Zerkle, M. D. Allendorf, M. Wolf and O. Deutschmann, *Journal of Catalysis*, 2000, **196**, 18.
92 M. Maestri, A. Beretta, G. Groppi, E. Tronconi and P. Forzatti, *Catalysis Today*, 2005, **105**, 709.
93 K. L. Hohn and L. D. Schmidt, *Applied Catalysis A: General*, 2001, **211**, 53.
94 A. Beretta, G. Groppi, M. Lualdi, I. Tavazzi and P. Forzatti, *Industrial & Engineering Chemistry Research*, 2009, **48**, 3825.
95 S. Cimino, G. Landi, L. Lisi and G. Russo, *Catalysis Today*, 2005, **105**, 718.
96 S. Cimino, G. Landi, L. Lisi and G. Russo, *Catalysis Today*, 2006, **117**, 454.
97 S. Cimino, G. Landi, P. S. Barbato, L. Lisi and G. Russo, *Catalysis Today*, 2010, **155**, 27.
98 E. L. Gubanova, Y. Schuurman, V. A. Sadykov, C. Mirodatos and A. C. van Veen, *Chemical Engineering Journal*, 2009, **154**, 174.
99 L. Bobrova, N. Vernikovskaya and V. Sadykov, *Catalysis Today*, 2009, **144**, 185.
100 T. Liu, C. Snyder and G. Veser, *Industrial and Engineering Chemistry Research*, 2007, **46**, 9045.
101 C. Appel, J. Mantzaras, R. Schaeren, R. Bombach, A. Inauen, N. Tylli, M. Wolf, T. Griffin, D. Winkler and R. Carroni, *Proceedings of the Combustion Institute*, 2005, **30**, 2509.
102 A. Schneider, J. Mantzaras and P. Jansohn, *Chemical Engineering Science*, 2006, **61**, 4634.
103 S. Eriksson, M. Wolf, A. Schneider, J. Mantzaras, F. Raimondi, M. Boutonnet and S. Järås, *Catalysis Today*, 2006, **117**, 447.
104 A. Schneider, J. Mantzaras, R. Bombach, S. Schenker, N. Tylli and P. Jansohn, *Proceedings of the Combustion Institute*, 2007, **31**, 1973.
105 S. Eriksson, A. Schneider, J. Mantzaras, M. Wolf and S. Järås, *Chemical Engineering Science*, 2007, **62**, 3991.
106 L. Giani, G. Groppi and E. Tronconi, *Industrial & Engineering Chemistry Research*, 2005, **44**, 4993.
107 D. Livio, A. Donazzi, A. Beretta, G. Groppi and P. Forzatti, *Industrial and Engineering Chemistry Research*, in press.
108 E. Ranzi, A. Sogaro, P. Gaffuri, G. Pennati, C. K. Westbrook and W. J. Pitz, *Combustion and Flame*, 1994, **99**, 201.
109 D. Livio, A. Donazzi, A. Beretta, G. Groppi and P. Forzatti, *Topics in Catalysis*, 2011, **54**, 866.
110 J.-D. Grunwaldt, B. Kimmerle, A. Baiker, P. Boye, C. G. Schroer, P. Glatzel, C. N. Borca and F. Beckmann, *Catalysis Today*, 2009, **145**, 267.
111 S. Hannemann, J.-D. Grunwaldt, B. Kimmerle, A. Baiker, P. Boye and C. Schroer, *Topics in Catalysis*, 2009, **52**, 1360.

112 J. D. Grunwaldt, M. Beier, B. Kimmerle, A. Baiker, M. Nachtegaal, B. Griesebock, D. Lützenkirchen-Hecht, J. Stötzel and R. Frahm, *Physical Chemistry Chemical Physics*, 2009, **11**, 8779.
113 A. Schneider, J. Mantzaras and P. Jansohn, *Chemical Engineering Science*, 2006, **61**, 4634.
114 J. D. Grunwaldt, L. Basini and B. S. Clausen, *Journal of Catalysis*, 2001, **200**, 321.
115 J. D. Grunwaldt and B. S. Clausen, *Topics in Catalysis*, 2002, **18**, 37.

Theoretical tools for studying gold nanoparticles as catalysts for oxidation and hydrogenation reactions

Mercedes Boronat
DOI: 10.1039/9781849737203-00050

In this contribution, the ability of small isolated gold NP to dissociate O_2 and generate a reactive surface oxide layer, the nature of the new gold active sites generated, and their implication in the mechanism of alcohol oxidation to aldehydes has been analyzed from a theoretical point of view. The nature of the active sites involved in H_2 dissociation and the possible ways in which Au/TiO_2 catalysts can be modified in order to increase their activity toward hydrogenation of nitroaromatics without modifying their high chemoselectivity is also explored.

1 Introduction

Nowadays, the trend towards a sustainable development implies the design of new and more efficient chemical processes, that optimize the use of raw materials and energy, and minimize the generation of by-products and waste. To achieve this, stoichiometric reactions usually carried out in homogeneous phase with the associated problems of product separation and catalyst recycling, should be replaced by heterogeneously catalyzed processes, able to selectively produce the desired molecules while working under mild reaction conditions. Catalyst selectivity is associated to a uniform distribution of well defined single site active centres, such as those existing in homogeneous catalysts, *i.e.*, cations and anions in solution, transition metal complexes and organocatalysts. One of the major drawbacks of solid catalysts is that they usually contain a distribution of active sites not exactly equivalent, with the corresponding negative impact on selectivity.

To design an efficient heterogeneous catalyst for a given reaction, a detailed knowledge of the reaction mechanism and the nature of the active sites involved in each transformation forming part of the global process is required. In this sense, theoretical modelling is becoming a powerful tool that provides information about the geometrical and electronic properties of solid materials, and their interactions with reactant and/or intermediate species. Theoretical studies combined with surface science and kinetic studies allow to identify the active sites involved in each elementary step of the mechanism, to determine the rate-limiting step and to predict the properties that should be introduced in a catalyst to achieve the highest activity and selectivity towards the desired reaction.[1–6]

In the search for green processes, oxidation reactions with molecular O_2 or air, and selective hydrogenations using molecular H_2, are among the most pursued objectives. The finding that small gold nanoparticles dispersed on metal oxides are highly active catalysts for CO oxidation with O_2 at low

Instituto de Tecnología Química (UPV-CSIC), Av. de los Naranjos s/n, 46022, Valencia, Spain. E-mail: boronat@itq.upv.es

temperature[7,8] opened a new field in heterogeneous catalysis, and the number of papers published since then related with catalysis by gold is continuously increasing. In these twenty five years, it has been established that the catalytic activity of gold depends on several factors such as particle size, particle morphology, surface concentration of coordinatively unsaturated Au atoms, or oxidation state of gold.[9–20] There is also agreement about the influence of the metal oxide or hydroxide support and of special Au sites at the metal-support interface.[21–27] On the other hand, there is recent evidence that Au nanoparticles (NP) supported on inert materials such as C, SiO_2, BN, cellulose, or stabilized by polymers, are active in oxidations if their diameter is smaller than 2 nm.[28–34]

It is really difficult from an experimental point of view to analyze separately the influence of all these factors. The concentration of accessible low coordinated Au atoms increases with decreasing particle size, and also depends strongly on particle morphology. The metal oxide support might on one hand stabilize preferentially gold nanoparticles of a given size and/or shape, but on the other hand might also be involved in electron density transfer from or to the supported gold nanoparticle, thus stabilizing cationic or anionic Au sites. Finally, it should be taken into account that some reactants might modify the gold catalyst during the reaction, either by changing the oxidation state of Au species, or by weakening the metal-support interaction leading to sintering of the gold nanoparticles and catalyst deactivation. For all these reasons, in the last years theoretical modeling has become a powerful tool to help in the task of identifying the active and selective sites for a target reaction and to finally design highly efficient catalysts. To achieve this objective, however, an adequate choice of the catalyst model and of the theoretical level used to describe the electronic properties of the catalyst-adsorbate system is fundamental, as well as a continuous validation of models and methods by comparison of theoretical results with experimental available data.

2 Oxidation reactions

From an industrial point of view, and besides CO oxidation, the value of gold is mainly based on its ability to catalyze organic reactions involved in the synthesis of fine chemicals, such as selective oxidation of hydrocarbons and alcohols, at atmospheric pressure and mild reaction temperature.[35–41] One of the key steps in the mechanism of gold catalyzed oxidation reactions, and yet one of the least understood, is the activation and dissociation of molecular O_2.

2.1 O_2 activation and dissociation

It is known that molecular O_2 does not interact with perfect flat gold surfaces, but it has been experimentally demonstrated that it adsorbs on small Au_n^- ($n = 2, \ldots, 22$) clusters having an odd number of electrons.[42,43] This ability of small gold clusters to adsorb, activate and dissociate O_2 has been interpreted on the basis of theoretical calculations. The mechanism of O_2 binding to gold involves electron density transfer from the HOMO of the metal to the empty π^* molecular orbital of O_2, resulting in a weakening of the O–O bond and in formation of an activated superoxo species. In agreement with this bonding mechanism, interaction energies between O_2

and small Au_n ($n = 2$–8) anionic clusters containing an excess of electrons were calculated to be larger than between O_2 and neutral or cationic clusters.[44,45] DFT calculations on Au NPs larger than 1 nm diameter[46] and on stepped and defective surfaces[47,48] indicate that, while O_2 does not bind to flat gold surfaces, it interacts with step edges and gold clusters deposited on extended surfaces. The reason is that the degree of HOMO(Au)$\rightarrow\pi^*(O_2)$ electron density transfer, and therefore the strength of the gold-oxygen interaction, depends on the relative energies of the two orbitals involved and on the degree of spatial overlap between them. Thus, while the HOMO of the Au(111) surface is delocalized over all atoms of the surface, which is not favorable for orbital overlap, the HOMO of small planar Au_n clusters has a large lobe sticking out in the vacuum at the edge of the cluster, that allows a good overlap with the π^* orbital of O_2. In the case of stepped or rough gold surfaces, the HOMOs are localized around the steps or defects and have some orbital component pointing toward the vacuum, which facilitates orbital overlap and O_2 binding to gold.

Another interesting information provided by DFT calculations is the finding that activation of molecular O_2 and its subsequent dissociation into O atoms is directly related to the mode of adsorption of O_2 on the gold nanoparticles.[49] Three main ways of adsorption were reported and investigated: an end-on mode with only one oxygen atom directly bonded to a gold atom and with O_2 molecule approximately perpendicular to the gold surface (**eo**), a top-bridge-top conformation in which each oxygen atom in O_2 is directly coordinated to a gold atom (**tbt**), and a bridge-bridge adsorption mode in which each oxygen atom in O_2 is bonded to two different gold atoms (**bb**), being the four gold atoms involved in the adsorption arranged forming a square (see Fig. 1).

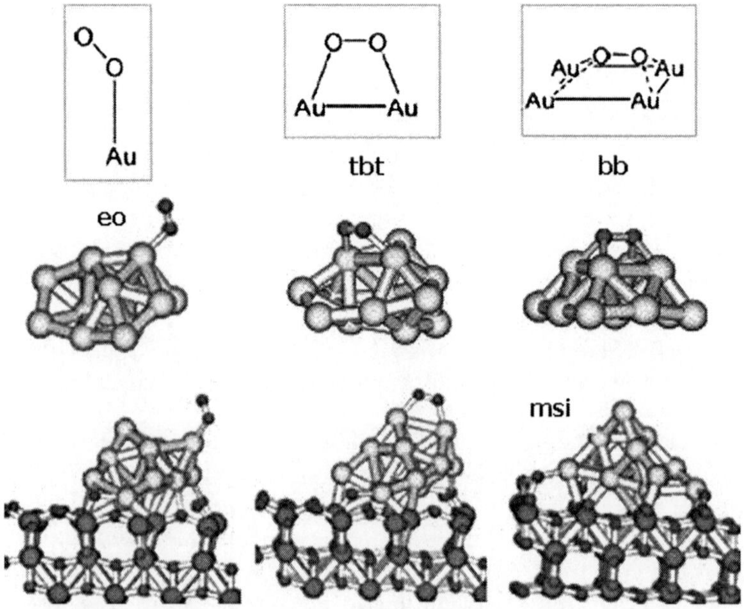

Fig. 1 Adsorption modes of molecular O_2 over gold catalyst models.

O_2 adsorption and dissociation was theoretically investigated over different models of gold catalysts, including extended Au(111) and Au(100) surfaces, small isolated Au NPs of different size, and Au NPs supported on TiO_2. O_2 adsorption on extended surfaces is only slightly exothermic in one case, a **bb** conformation on the Au(100) surface. The **eo** adsorption mode was located on small Au_{13} clusters, both isolated and supported on TiO_2, having low coordinated sites. But it was found to be considerably less stable than the other ways of adsorption, and it could not be obtained on a cuboctahedral Au_{38} nanoparticle. On the other hand, the **tbt** mode with O_2 adsorbed at an edge was located on all systems investigated, with calculated interaction energies between 20 and 25 kcal/mol. A **bb** conformation of similar stability was also found on the (100) facets of the cuboctahedral Au_{38} nanoparticle, but not on smaller Au_{13} clusters not containing this particular arrangement of gold atoms. Finally, in the case of supported gold clusters and nanoparticles, new sites appear at the metal-support interface that have been labelled as **msi** sites (see Fig. 1). According to different theoretical studies, O_2 is highly activated when it is placed at the interface between the gold nanoparticle and the metal oxide support.[50–53] One of the oxygen atoms of molecular O_2 interacts with a gold atom of the nanoparticle and the other one is directly bonded to a titanium atom of the support, forming a Au–O–O–Ti bridge. This position is the most stable of all considered on the Au/TiO_2 models, which suggests that O_2 will preferentially adsorb on the support or at the metal-support interface, and not on the Au NPs.

However, adsorption energy is not necessarily a good predictor of molecular O_2 activation. As described above, the mechanism of O_2 binding to gold involves electron density transfer from the HOMO of gold to the empty π^* molecular orbital of O_2, leading to a weakening of the O–O bond, to an increase in the O–O bond length, and to a shift in the νOO stretching vibration frequency. These three properties, *i.e.*, charge transfer to O_2, optimized O–O bond length, and νOO vibration frequency, were obtained from DFT calculations and the relationship between them is shown in Fig. 2: the νOO stretching frequency decreases almost linearly with

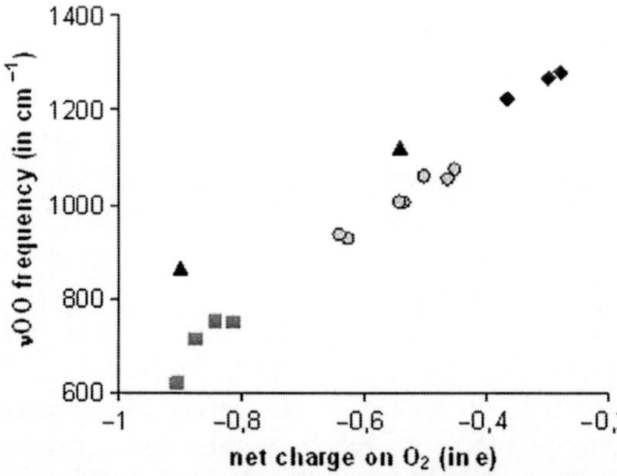

Fig. 2 Correlation between calculated νOO stretching vibration frequency and total charge on O_2 in gold-dioxygen complexes differing in the mode of adsorption of O_2: squares, circles, diamonds and triangles represent bb, tbt, eo and msi adsorption modes, respectively.

increasing O–O distance and with increasing net charge on O_2. And, what is more important, this plot puts forward the large influence of the mode of adsorption of O_2 on the way it is activated. Thus, adsorption in an **eo** mode allows a charge transfer of only ~0.3 e and causes an increase in the O–O bond length of less than 0.1 Å, while the larger degree of O_2 activation is found in the **bb** conformation, with a charge transfer of almost 1 e and optimized O–O distances close to 1.5 Å. It is also interesting to remark that the points corresponding to O_2 adsorption at the metal-support interface **(msi)** in Au/TiO_2 models are close to the most activating **bb** mode when the TiO_2 surface where the Au NP is supported contains oxygen vacancy defects.

Once the most activating mode of adsorption was established, the complete reaction path for O_2 dissociation was calculated, and a direct correlation between activation energy barriers and adsorption mode was found. When starting from O_2 adsorbed in a **tbt** conformation, the reaction coordinate involves not only the dissociation of the O–O bond but also some reorganization of Au atoms, and the calculated barriers were always larger than 20 kcal/mol. However, when starting from O_2 adsorbed on in a **bb** mode, the reaction coordinate involves only the breaking of the O–O bond, that is already highly activated, and the activation energy barriers were calculated to be lower than 10 kcal/mol on all catalyst models. The most remarkable conclusion provided by this theoretical study was that particle morphology plays a key role in O_2 activation, and that on small Au NPs isolated or supported on inert materials, the active sites for O_2 dissociation are arrangements of Au atoms exposing the (100) facet. These findings are in agreement with the recently reported activity of Au NPs smaller than 2 nm, either supported on inert materials (C, SiO_2, BN, cellulose) or stabilized by polymers, for different oxidation reactions.[28–34]

2.2 Oxidized nanoparticles

Once demonstrated that small Au_n clusters with more than 3 atoms[44,45] and Au NPs containing from 13 to 79 Au atoms[17,46,49] are able to dissociate molecular O_2 with low activation barriers, the possibility of generating a surface oxide layer by successive dissociation of an increasing number of O_2 molecules was theoretically investigated and experimentally confirmed.[54] Au NPs within a narrow size distribution (1.1 ± 0.5 nm) were synthesized and stabilized onto multiwalled carbon nanotubes (MWCNT), and their ability to dissociate molecular O_2 and become reversibly oxidized was confirmed by XPS and $^{16}O_2/^{18}O_2$ isotopic exchange experiments.

The calculated energy profile for the process is depicted in Fig. 3. The first O_2 molecule adsorbs and dissociates with a small activation barrier on the (100) facet present in small Au NPs. After dissociation, the two O atoms are placed in separated 3-fold hollow positions, and the activation energy to spill over to other 3-fold hollow positions is also low, so that it can be assumed that O atoms move easily on the particle surface. Moreover, after dissociation of the first O_2 molecule, the active site consisting of four Au atoms arranged as in the (100) facet are free to adsorb and dissociate a second O_2 molecule with a low activation barrier. Assuming a cuboctahedral shape for the Au NP of 1 nm diameter, and considering that there are six of these active sites on each NP and that each active site can dissociate at

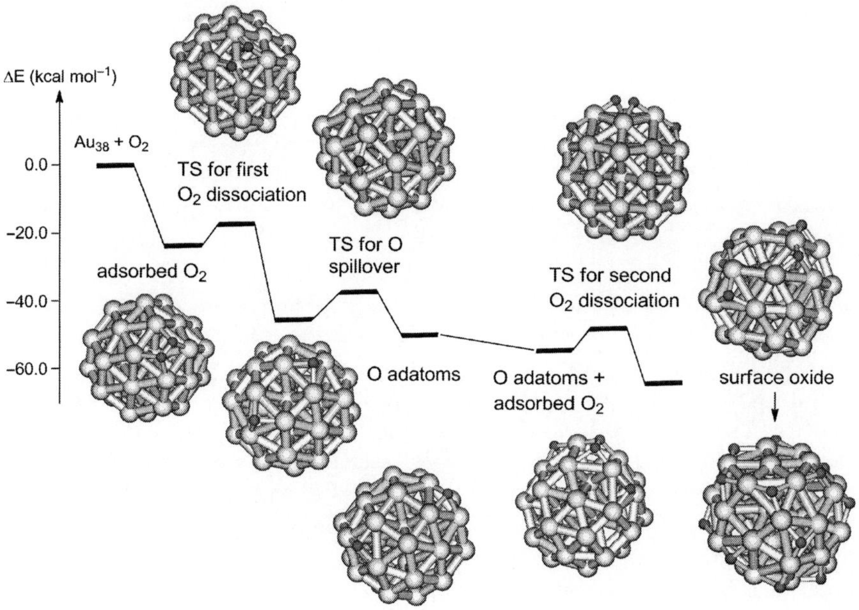

Fig. 3 Calculated energy profile for successive dissociation of molecular O_2 and generation of a surface oxide layer on a Au_{38} NP.

least two O_2 molecules with low activation barriers, it was concluded that is energetically possible to reach a coverage of one monolayer of surface oxide, that is, sixteen O atoms on a Au NP with thirty-two surface atoms. The energy profile also provides information about the reverse process, the recombination of two O atoms to form molecular O_2. Since dissociation is exothermic, the intrinsic reaction barriers for recombination are higher, and oxygen coverage dependent. The activation energy necessary to form molecular O_2 from two oxygen atoms adsorbed on a clean Au NP is larger than 30 kcal/mol, but decreases to 16 kcal/mol when the particle is partially covered by a surface oxide layer. This theoretical result was confirmed by $^{16}O_2/^{18}O_2$ isotopic exchange measurements on Au NP having 1.1 nm diameter.[54] The concentration of $^{16}O_2$ and $^{18}O_2$ decreased with time at 25 °C, indicating that O_2 dissociation already occurs at this temperature, but formation of $^{16}O^{18}O$ was only observed at 80 °C, indicating that more energy is necessary for recombination of atomic oxygen into molecular O_2.

To bring some light on the nature of the oxygen and gold species present in oxidized gold nanoparticles, the adsorption of an increasing amount of O atoms on the Au_{38} NP having 1 nm diameter was theoretically investigated, and the net atomic charges on Au and O atoms analyzed. Isolated O atoms preferentially adsorb on 3-fold hollow positions on the (111) facets on the Au_{38} NP, and after adsorption they bear a negative charge of −0.8 e, while the Au atoms in contact with O become positively charged by 0.2 e. When the number of adsorbed O atoms increases and they are placed close to each other, linear O–Au–O bonds similar to those found in Au^{I} complexes are formed, and the positive charge on the central Au atom increases to 0.5 e. In a $Au_{38}O_{16}$ system, all oxygen atoms are forming the most stable linear

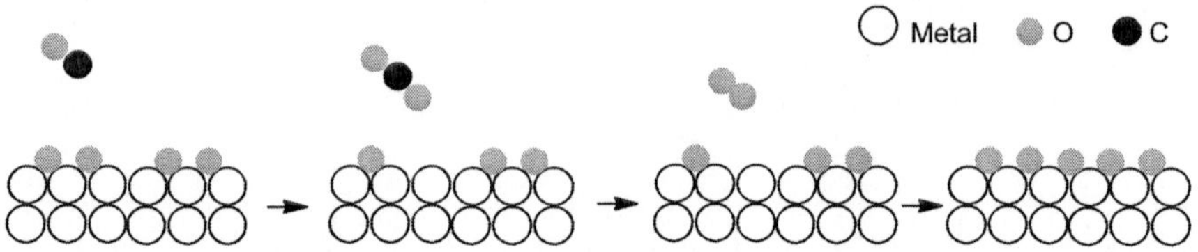

Scheme 1 Schematic representation of a mechanism for CO oxidation involving reaction with O atoms from the gold oxide over layer.

O–Au–O structure and are negatively charged by −0.8 e. Adding more O atoms to the Au_{38} NP results in a decrease in the interaction energy per O atom and in the appearance of highly positively charged gold atoms (+0.8 e) that are directly bonded to three O atoms and that could be assimilated to Au^{III} species. The oxidized Au NPs were characterized by XPS measurements, and the shift in the BE of the $4f_{7/2}$ electrons to 85.8 eV suggested the presence of positive gold species.

If really small gold NPs can be reversibly oxidized by molecular O_2, we could consider an oxidation pathway that could be, in some sense, assimilated to a Mars van Krevelen mechanism, in which the O atoms from the gold oxide over layer would be transferred to the reactant that is oxidized, and then molecular oxygen would dissociate again regenerating the gold oxide over layer, as shown in Scheme 1.

The calculated energy profile for CO oxidation on an oxide-covered Au NP agrees with this proposed mechanism (see Fig. 4). CO does not bind to any positively charged Au site, but directly takes one O from the oxide over layer with an activation energy of 18 kcal/mol to form a CO_2 molecule, being the process highly exothermic. The situation is different at low O coverage, where several gold species differing in oxidation state and in coordination ability due to geometric restrictions, are present on the NP surface. CO does not bind to cationic Au sites directly bonded to two O atoms, but interacts strongly with either neutral Au^0 atoms not in direct contact with any O atom, or with slightly positive gold sites directly bonded to only one O atom. In these two cases the resulting adsorption complex is highly stable and there is no direct pathway for CO_2 formation, suggesting that only gold nanoparticles covered by a surface oxide monolayer are active for CO oxidation. This theoretical finding was confirmed by FTIR spectroscopy of CO_2 evolution at different temperatures on oxidized Au NP having 1.1 nm diameter (see Fig. 5).[54] As temperature increased from 25 °C to 60 °C and 120 °C, the amount of CO_2 formed by reaction of CO with the oxidized Au NPs continuously decreased. Since oxygen coverage decreases with increasing temperature, these results confirm that CO reacts with the oxygen atoms of the oxide monolayer covering the gold nanoparticles. Moreover, when the same experiment was done on Au NPs having 4 nm diameter, no CO_2 evolution was observed, suggesting again that only really small NPs are able to dissociate molecular O_2 and generate the gold oxide monolayer that can participate in oxidation reactions.

It should be mentioned at this point that the mechanism for CO oxidation based on initial O_2 dissociation and subsequent reaction of CO with O atoms described above is only one of the several possible pathways

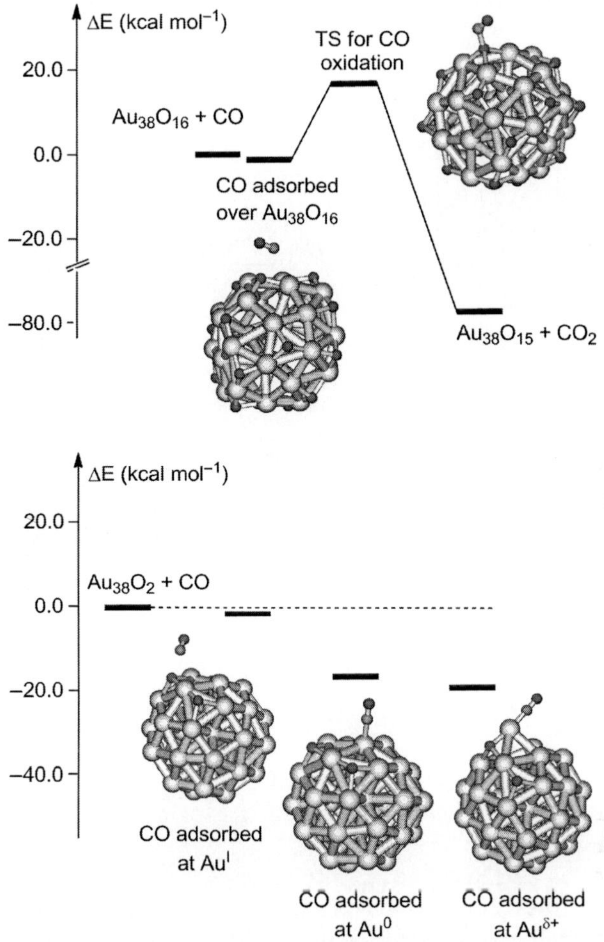

Fig. 4 Calculated energy profiles for CO oxidation over an oxide-covered (top) and a partially oxidized Au NP (bottom).

proposed for this reaction. At low temperatures, where the amount of O_2 that can dissociate is expected to be low, a direct reaction of CO with molecular O_2 has been found to be energetically preferred on different systems: small gold clusters, stepped surfaces, and at gold-support interface sites. According to this mechanism, an O_2 molecule reacts with adsorbed CO to form an adsorbed peroxyformate OCOO intermediate that in a second step produces CO_2 and an O adatom. This adsorbed O atom reacts with a second CO molecule with a very low energy barrier, thus closing the catalytic cycle.[52,55–57] On the other hand, the influence of CO coverage on the reaction pathway has been theoretically and experimentally analyzed, and it has been reported that CO preadsorption enhances the ability of small gold clusters to coadsorb O_2.[58] Moreover, a slightly different mechanism in which an activated O_2 molecule simultaneously reacts with two adsorbed CO forming two CO_2 molecules has been proposed on the basis of theoretical calculations.[59]

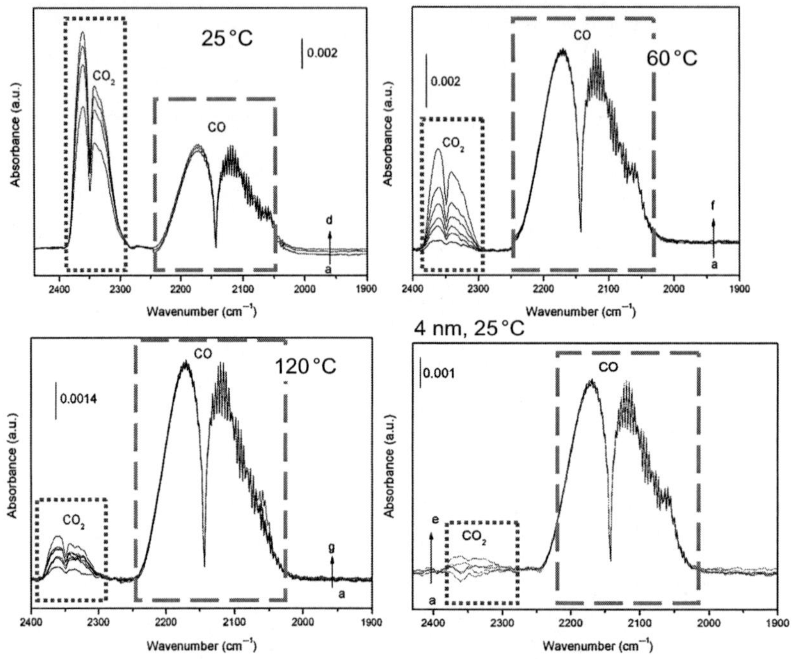

Fig. 5 FTIR spectra of CO_2 evolution with time after dosing 31 mbar CO on a catalyst containing Au NPs of 1.1 nm diameter at different temperatures and on a catalyst containing Au NPs of 4.0 nm diameter at 25 °C.

2.3 Alcohol oxidation

The selective oxidation of alcohols to aldehydes or ketones is an important step in the synthesis of fine chemicals, and in recent years much effort has been done to find gold heterogeneous catalysts able to perform the reaction using molecular O_2 and generating H_2O as the only by-product.[29–39] It is accepted that the mechanism for metal-catalyzed alcohol oxidation involves two steps: an initial deprotonation and formation of a metal-alkoxide intermediate that, in a second step, suffers a β-hydride elimination yielding the carbonylic product and the metal surface covered with hydrogen atoms that, with molecular O_2, form H_2O.[37,41,60] Although in some cases the alkoxide intermediate is not formed on the metal nanoparticle but on the metal oxide support or at the metal-support interface, there are other examples showing that naked Au NP in colloidal solutions, stabilized by polymers, or supported on C or SiO_2, are highly active for alcohol oxidation.[61–65] In these cases, the activity of gold could be initially related to the presence of low coordinated atoms placed at corner or edge positions, an hypothesis that was confirmed by a theoretical study of the mechanism of ethanol oxidation to ethanal on a series of catalyst models having Au atoms with different coordination number, as depicted in Fig. 6.

Similar structures and elementary steps were found over all catalyst models investigated, see Fig. 7. First, ethanol adsorbs on the catalyst surface with the O atom on top of a Au atom, and the OH bond is broken generating a surface ethoxy intermediate and a hydrogen atom adsorbed on

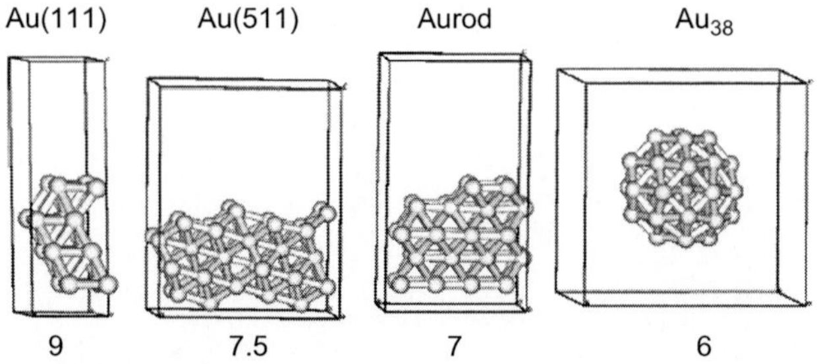

Fig. 6 Unit cells of gold catalyst models differing in surface roughness. The numbers indicate the average coordination number of the Au atoms involved in the mechanism of alcohol oxidation.

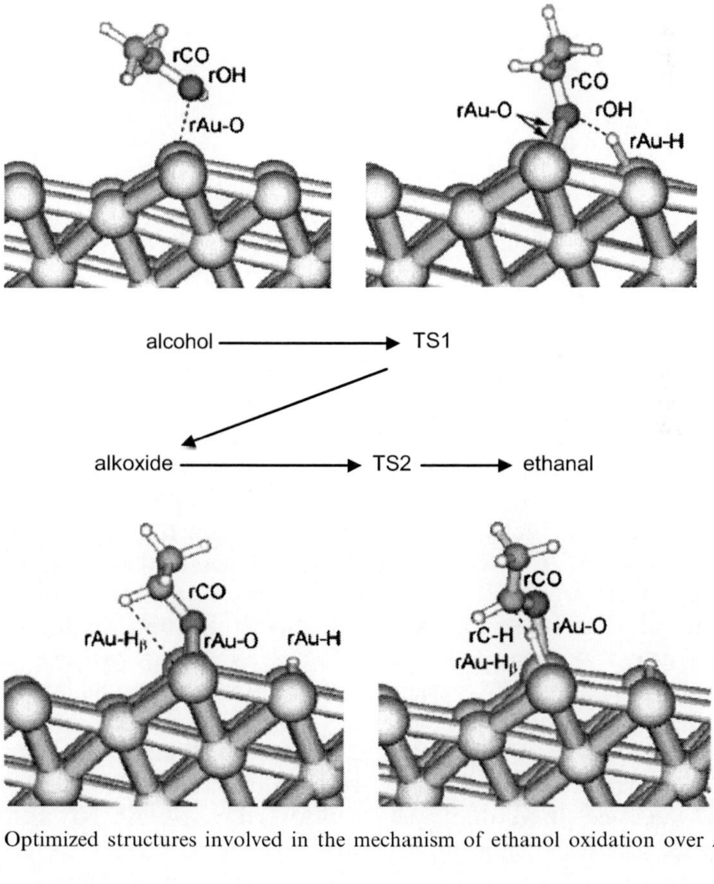

Fig. 7 Optimized structures involved in the mechanism of ethanol oxidation over Au(511) surface.

the gold surface. In a second step, the C–H bond next to the O atom dissociates into an ethanal molecule and a second hydrogen atom that remains adsorbed on the gold surface. The rate determining step of the process is the first deprotonation of ethanol, and the calculated activation energy (E_{ACT1}) decreases from 33 kcal/mol on the Au(111) surface to 19 kcal/mol on the Au NP containing low coordinated Au atoms. Calculated ethanol adsorption

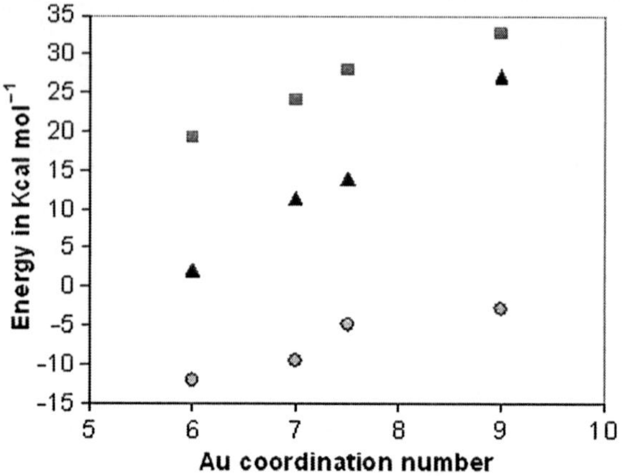

Fig. 8 Calculated adsorption (cercles), activation (squares) and reaction (triangles) energies for ethanol oxidation versus coordination number of Au atoms.

energies (E_{ads}), activation energies for the rate determining step (E_{ACT1}), and reaction energies (ΔE) are plotted in Fig. 8 as a function of coordination number of Au atoms. The trends with increasing surface roughness or decreasing coordination number of Au atoms is clear: ethanol adsorption becomes more favourable, reactants and products are stabilized, and activation barriers considerably decrease, indicating that alcohol oxidation is a structure-sensitive reaction.

Besides surface roughness, adsorbed oxygen atoms and cationic gold species that could exist close to adsorbed oxygen or at the metal-support interface, might play a role on the selectivity of gold catalyzed alcohol oxidation. This issue was theoretically investigated,[66] and the calculated energy profiles for ethanol oxidation over a clean Au NP containing only metallic Au^0 species and a Au NP with one adsorbed O atom, whose presence generates slightly positive $Au^{\delta+}$ sites, are compared in Fig. 9. The full reaction path on the naked Au_{38} cluster is similar to that described above on the Au(511) surface. First, ethanol adsorbs with the O atom on top of a corner Au atom releasing 9 kcal/mol. In the rate determining step, the hydroxyl group is deprotonated with an activation energy barrier of 24 kcal/mol, and the resulting ethoxy fragment and hydrogen atom bind to different sites on the particle surface. Then, the C–H bond in β position dissociates with an activation energy barrier of 12 kcal/mol, yielding ethanal and two H atoms that remain adsorbed on the Au NP surface. To simulate the metal-support interface, an O atom was chemisorbed on the Au NP, generating slightly positive $Au^{\delta+}$ sites. Ethanol adsorbs at these sites as strongly as on neutral corner Au^0 atoms, but in this case a hydrogen bond is formed between the proton of the ethanol OH group and the adsorbed O atom. This new interaction facilitates ethanol deprotonation, that proceeds with a negligible activation energy barrier. The resulting ethoxy fragment does not occupy a bridge position between two gold atoms as it was found on the clean Au_{38} cluster, but it is directly attached to a corner Au^0 atom, with the H_β atom oriented towards the surface hydroxyl group. Dissociation of the C–H_β bond and transfer of the H_β atom to the hydroxyl group

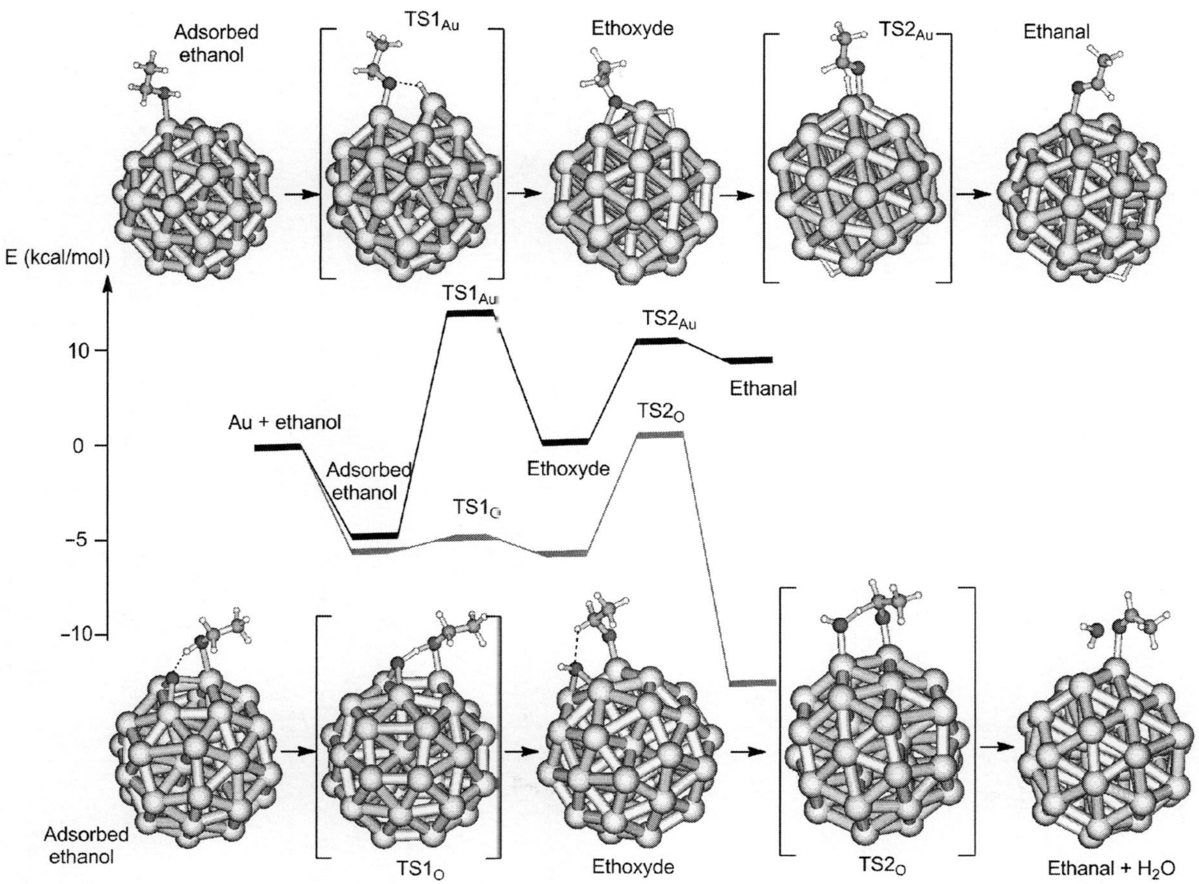

Fig. 9 Optimized structures involved in the mechanism of ethanol oxidation over a metallic Au NP containing only Au^0 atoms (top) and over a Au NP with an adsorbed O atom whose presence generates slightly positive $Au^{\delta+}$ atoms (bottom).

generating ethanal and water requires an activation energy of 13 kcal/mol, similar to that calculated for this step on the clean Au_{38} NP. These results indicate that the presence of O atoms on the Au NP or at the metal-support interface has a positive effect on the first step of the mechanism, decreasing the activation barrier for alcohol deprotonation, but does not facilitate the C–H dissociation that becomes the rate determining step. Anyhow, the calculated activation energy for this step, 13 kcal/mol, is considerably lower than that obtained for the rate determining step on the clean Au NP, 24 kcal/mol, demonstrating the positive role of adsorbed O atoms or O atoms at the metal-support interface on the catalytic activity of Au NP.

It should also be remarked at this point that hydroxyl groups play a key role in the selective oxidation of alcohols in aqueous phase, as has been theoretically and experimentally demonstrated.[4,67] Thus, Au/C catalysts that are inactive for glycerol oxidation in the absence of a base, become active at high pH. The reason of this enhanced activity at high pH is the lowering of the calculated activation barriers for alcohol deprotonation when the process is assisted by hydroxyl groups adsorbed on the catalyst surface, in a similar way as adsorbed O facilitates this elementary step (see Fig. 9).

3 Hydrogenation reactions

Gold nanoparticles supported on metal oxides are also active catalysts for hydrogenation reactions with molecular H_2. One of the most interesting features of hydrogenation gold based catalysts is their high selectivity, that allows to discriminate between different groups or positions within one molecule. Thus, the nitro group is selectively hydrogenated in molecules containing C=C, C=O, Cl- or C≡N groups,[68–70] the carbonyl group is selectively hydrogenated in α,β-unsaturated aldehydes yielding enols,[71–75] and highly unsaturated hydrocarbons like acetylene and 1,3-butadiene are selectively hydrogenated to the corresponding alkenes.[76–79]

3.1 H_2 activation and dissociation

The first and usually rate determining step of gold catalyzed hydrogenation reactions is the dissociation of molecular H_2, and in order to determine which are the active sites involved in this elementary step, the influence of particle size and nature of the support has been investigated both from theory[80–83] and experiment.[74,84–87] Experimental studies showed that molecular H_2 is dissociated by gold atoms in corner or edge positions, and it was suggested that besides particle size, the shape or morphology of the gold nanoparticles plays a key role on the catalyst activity for this process.[88]

On the other hand, the mechanism for H_2 adsorption and dissociation was studied using DFT methods over a series of catalyst models containing Au atoms with different coordination number: perfect Au(111) and Au(100) extended surfaces, which were found to be inactive; a defective gold surface containing a monoatomic row that simulated Au atoms at a particle edge; isolated gold nanoparticles of different size and morphology, Au_{13}, Au_{25} and Au_{38} containing low coordinated corner and edge atoms; and a Au_{13} cluster supported on stoichiometric and reduced TiO_2.[81,82] At difference

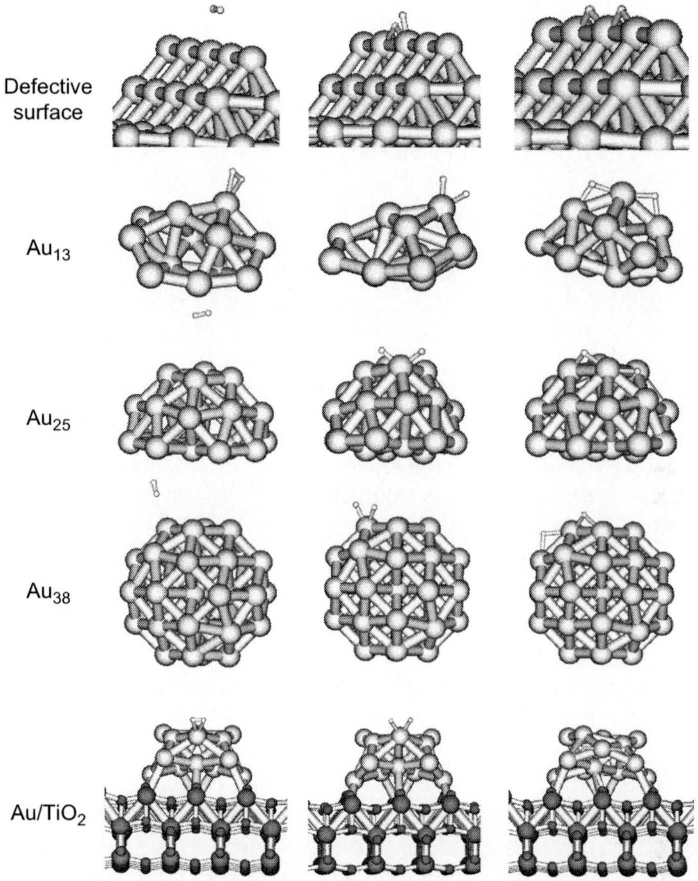

Fig. 10 Optimized structures involved in the mechanism H_2 dissociation over a defective surface and over isolated and TiO_2-supported Au NPs of different size.

with O_2, molecular H_2 is not activated by electron density transfer from/to the metal, but just by weak interaction with low coordinated Au atoms. On all catalyst models investigated, the process involves reactant, transition state and product structures similar to those depicted in Fig. 10. Molecular H_2 adsorbs weakly on a low coordinated Au atom, forming a complex in which the two H atoms are bonded to the same Au atom at a distance of ~ 1.9 Å, while the H–H bond is slightly elongated from 0.75 Å in the gas phase to 0.80 Å. In the transition state for the dissociation process, the two H atoms are directly attached to the same Au atom, and the H–H bond length has increased to ~ 1.5 Å. After dissociation, the two H atoms form nearly symmetrical Au–H–Au bridges, sharing the central Au atom to which they were initially bonded. The energy profiles calculated for H_2 dissociation over the different gold catalyst models described above are plotted together in Fig. 11. Adsorption energies range between −2 kcal/mol on an isolated Au_{13} cluster and −13 kcal/mol over a defective gold surface, while intrinsic activation barriers calculated as the difference in energy between the transition state and the H_2 adsorption complex are in all cases

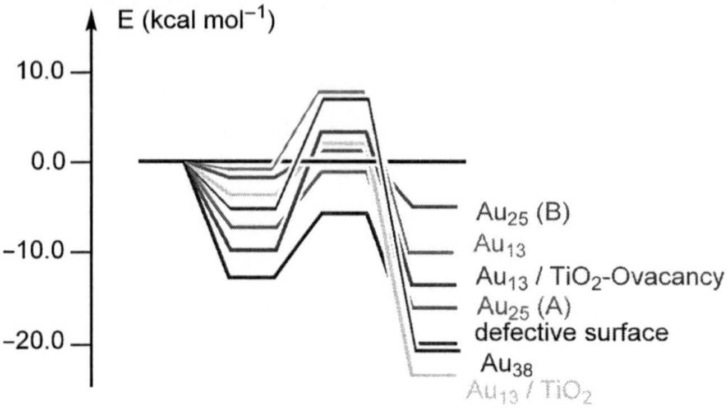

Fig. 11 Calculated energy profiles for H_2 dissociation over isolated and TiO_2-supported Au NPs of different size.

around 8 kcal/mol. This means that molecular H_2 can be dissociated almost without additional energy by low coordinated Au atoms, regardless whether these atoms are located at corner or edge positions of small nanoparticles or belong to an extended defective surface. In the case of Au NPs supported on TiO_2 it was initially concluded, on the basis of DFT calculations and assuming that only Au atoms able to bind molecular H_2 in a rather strong way will also be able to dissociate it, that the active sites for H_2 dissociation are neutral Au atoms located at corner or edge low-coordinated positions and not directly bonded to the support.[82] This conclusion was in apparent contradiction with an experimental study of H_2/D_2 exchange over Au/TiO_2 catalysts,[87] in which analysis of particle size influence on activity indicated that the active sites for H_2 dissociation are placed at the perimeter interface in contact with the oxide support. A very recent theoretical study of H_2 dissociation at the perimeter sites of Au/TiO_2 catalysts and of spill over of H atoms from the Au NPs to the support showed that, although heterolytic dissociation of H_2 at the perimeter sites is energetically favoured, these sites are readily passivated by the formed OH, suggesting that further H_2 dissociation may occur at pure low coordinated Au sites on top of the Au NPs.[83]

3.2 Identification of active sites by IR of CO adsorption

A helpful tool to identify active sites in gold catalysts is IR spectroscopy of CO adsorption, because the CO vibrational stretching mode, νCO, is highly sensitive to the nature and electronic state of the Au atoms where it binds.[89] It is generally accepted that CO adsorbs weakly on top of gold atoms in single crystal gold surfaces, with stretching frequencies around 2110 cm^{-1}. IR studies of CO adsorption on Au NPs supported on TiO_2 show in most cases a single band at 2100–2110 cm^{-1} assigned to CO adsorbed on top of metallic Au^0 sites. When the experiments are performed in the presence of O_2, the band appears at 2120–2130 cm^{-1}, and is assigned to CO co-adsorbed with oxygen on the gold particles or to CO adsorbed on gold atoms at the metal-support interface. Finally, a band appearing at 2140–2160 cm^{-1} has been attributed to CO adsorbed on Au^+ sites, although the exact nature and origin of these sites is not known.[90–94]

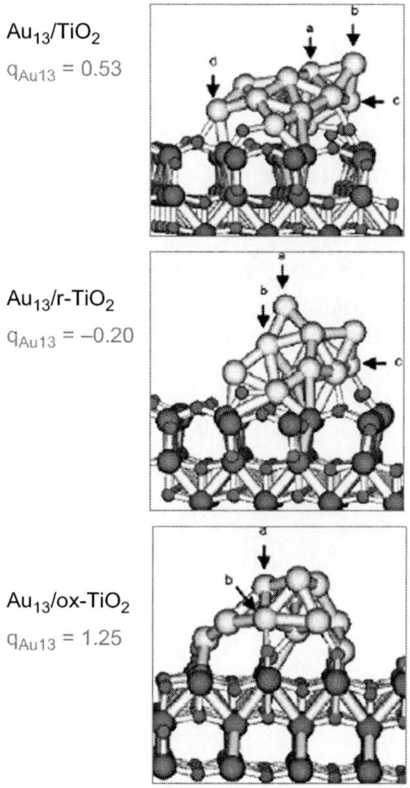

Fig. 12 Optimized structures and total charge on the Au NP obtained for the Au_{13}/TiO_2, $Au_{13}/r\text{-}TiO_2$ and $Au_{13}/ox\text{-}TiO_2$ catalyst models.

To clearly identify the different Au sites present in Au/TiO_2 catalysts and responsible for the several bands observed in the IR spectra, the interaction of CO with three Au/TiO_2 catalyst models was theoretically investigated, and the changes in the νCO stretching frequency associated to under coordination, localization or charging of the Au sites were analyzed in depth.[95] The catalyst models consisted of a Au_{13} cluster supported on: a stoichiometric TiO_2 surface (Au_{13}/TiO_2), a reduced TiO_2 surface containing an oxygen vacancy defect ($Au_{13}/r\text{-}TiO_2$), and an over-oxidized TiO_2 surface generated by adding an extra O adatom ($Au_{13}/ox\text{-}TiO_2$) (see Fig. 12). Due to metal-support interactions, Au NPs become positively charged when supported on stoichiometric and oxidized TiO_2 surfaces, and negatively charged when adsorbed on reduced surfaces. Moreover, the net charge on the Au NP is not equally distributed among all atoms, but clearly localized on some of them. In general, most Au atoms in all the supported NPs are neutral, those Au atoms directly bonded to an O atom of the support forming a Au–O–Ti bridge exhibit a net positive charge, and only medium coordinated Au atoms bonded to a Ti atom of the reduced surface are negatively charged.[82] The Au_{13} irregular clusters resulting from geometry optimization of the Au_{13}/TiO_2, $Au_{13}/r\text{-}TiO_2$ and $Au_{13}/ox\text{-}TiO_2$ systems contain Au atoms with different coordination number, geometrical arrangement, net atomic charge and neighboring atoms. CO was adsorbed close to several of these sites on each Au/TiO_2 model, and after geometry optimization of the resulting system, the νCO stretching frequencies were calculated.

The most relevant results are summarized in Fig. 13, that shows the optimized structures after CO adsorption and the calculated νCO stretching frequencies in cm^{-1}. The first clear conclusion is that the total initial charge on the Au NP due to the different state of the underlying oxide support has no influence on the final νCO values. On all system models, the most stable mode of adsorption corresponds to CO linearly attached on top of an under coordinated Au atom, with optimized Au–C distances of ~2 Å and with calculated νCO vibration frequencies between 2077 and 2103 cm^{-1}. It is important to remark that interaction with CO causes a reorganization of the electron density in the particle and, in some cases, important deformation of the particle shape that remains after CO desorption, thus causing an irreversible change of the particle morphology.[95] CO also adsorbs on Au atoms that bear a net positive charge and that are directly bonded to O atoms of

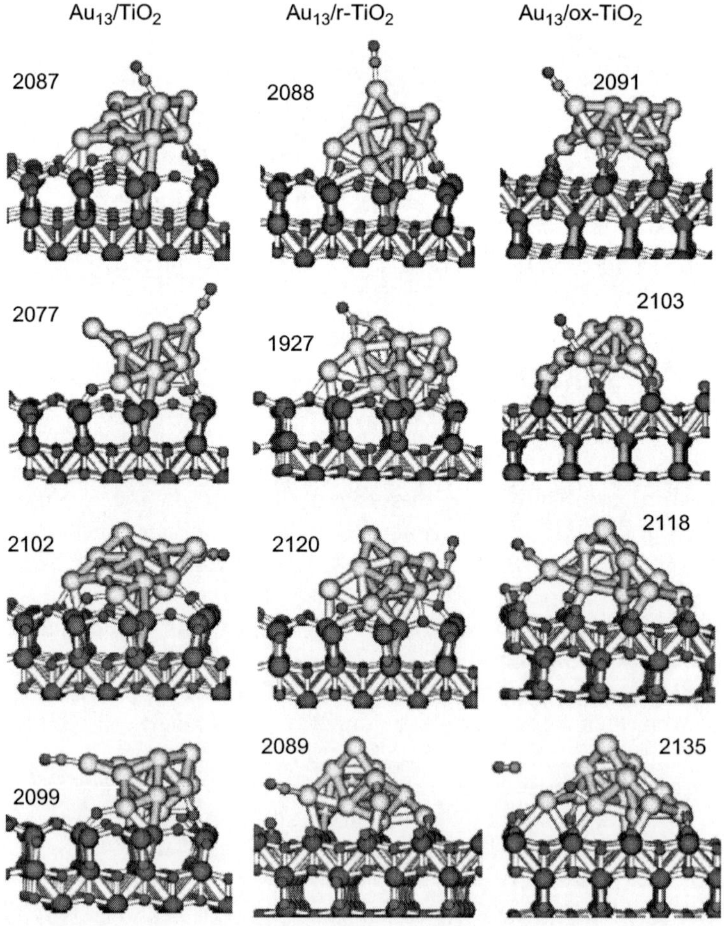

Fig. 13 Optimized structures and calculated νCO stretching frequencies for CO adsorption at different sites of Au_{13}/TiO_2, $Au_{13}/r\text{-}TiO_2$ and $Au_{13}/ox\text{-}TiO_2$ catalyst models.

the support through Au–O–Ti bridges. In these cases, the calculated νCO vibration frequencies are around 2120 cm^{-1}. In some other cases, CO is placed in a bridge position between two gold atoms, resulting in a system with a longer CO bond and therefore with a calculated νCO vibration frequency that appears at 1927 cm^{-1}. However this adsorption mode is considerably less stable than the others and therefore its concentration, if it was formed, should be low. Finally, CO adsorption close to a highly positive Au atom directly bonded to two O atoms was considered, and a weak interaction and a calculated νCO stretching frequency of 2135 cm^{-1} were obtained. With this information, all bands observed in IR spectra of CO adsorption on Au/TiO_2 catalysts could be assigned to particular gold species: IR bands between 2070–2110 cm^{-1} correspond to CO adsorbed on low coordinated neutral Au^0 atoms, bands at ~2120 cm^{-1} were assigned to CO interacting with slightly positively charged $Au^{\delta+}$ atoms involved in Au–O–Ti bonds at the metal-support interface, and the IR band appearing at 2135 cm^{-1} was tentatively related to CO weakly adsorbed on positively charged Au^+ atoms in contact with O adatoms.

To finally elucidate the nature of Au sites for hydrogenation, a number of Au/TiO_2 catalysts with similar gold particle size but different pre-treatment conditions were synthesized and characterized by IR spectroscopy of CO adsorption, and different distributions of sites related to differences in particle morphology were observed. The rate of H/D exchange was measured on these catalysts and plotted against the concentration of the different Au species identified. As shown in Fig. 14, there is indeed a linear correlation between the rate of H/D exchange and the amount of metallic gold sites responsible for the experimental bands at 2077–2110 cm^{-1}, but there is no correlation at all with the amount of positively charged $Au^{\delta+}$ gold sites associated to the bands at 2125 and 2135 cm^{-1}. Thus, it can be concluded that the active sites for H_2 dissociation are low coordinated metallic Au^0 atoms not involved in stable Au–O–Ti linkages.

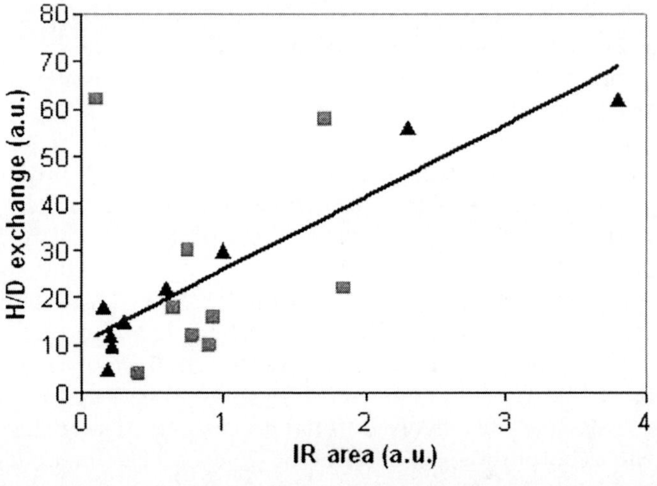

Fig. 14 Relationship between H/D exchange and concentration of metallic (triangles) and cationic (squares) Au sites on different Au/TiO_2 catalysts.

3.3 Selectivity: hydrogenation of nitro group in substituted nitroaromatics

As previously mentioned, the most interesting feature of gold catalysts in hydrogenation reactions is their high selectivity, and the chemoselective hydrogenation of substituted nitroaromatics to the corresponding aromatic amines is a good example of this property.[68–70] Functionalized anilines are valuable intermediates in the industrial production of agrochemicals, pharmaceuticals, pigments and dyes.[96,97] The reduction of nitrobenzene with molecular H_2 is easily carried out with commercial catalysts based on Pd, Ni or Pt, but the selective reduction of a nitro group in a molecule containing other reducible groups such as carbonyl, cyan or C=C double bonds is considerably more difficult and it usually requires stoichiometric amounts of reducing agents or the addition of soluble metals (V, Fe) and other modifiers (P, S, halogen containing compounds).[98–101] Not long ago, it was found that Au NPs supported on TiO_2 are active and very selective catalysts for the hydrogenation of the nitro group in substituted nitroaromatics.[68] It was reported that Pd and Pt supported on C or TiO_2 are considerably more active than any Au catalyst, but they are not selective, and all groups present in the molecule are completely reduced. On the other hand, Au/C or Au/SiO_2 show some selectivity, but a very low activity, and only Au supported on Fe_2O_3 and specially on TiO_2 are both active and highly selective.

The origin of the high chemoselectivity of Au/TiO_2 was explained by a combination of *in situ* IR spectroscopy, quantum chemical modelling and kinetic experiments.[102] It was considered that the preferential activation of the nitro group in nitrostyrene can be related either to a higher intrinsic reaction rate for hydrogenation of the nitro group as compared to the C=C bond, or to preferential adsorption of the nitro group at the active sites. To check the first hypothesis, hydrogenations of styrene and nitrobenzene were carried out separately on Au/TiO_2, Pt/TiO_2 and Au/SiO_2. It was found that, on Pt/TiO_2 and Au/SiO_2, styrene is hydrogenated 4–5 times faster than nitrobenzene, while on Au/TiO_2 the TOF for hydrogenation of nitrobenzene is 2 times larger than for styrene. This result already indicates that Au/TiO_2 should be more selective than Pt/TiO_2 and Au/SiO_2, but it is not enough to explain the 96% selectivity observed.

The second factor influencing selectivity, that is, preferential adsorption of the nitro group at the active sites in Au/TiO_2 catalysts, was studied by comparing the reactivity of styrene and nitrobenzene in the absence and presence of the other reactant. As described above, when only one of the reactants was fed, hydrogenation of nitrobenzene was found to be 2 times faster than that of styrene. But when a mixture of the same quantity of both reactants was fed, the TOF of nitrobenzene hydrogenation did not change, while that of styrene was considerably reduced.[102] This result indicates not only a faster hydrogenation of the nitro group, but also that there is a competitive adsorption on the active sites, which are mostly occupied by nitrobenzene. The question to answer at this point was why this occurs and which are the active sites involved in the process. To answer this question, nitrostyrene adsorption over a series of gold catalyst models including perfect Au(111) and Au(100) surfaces, a defective gold surface containing a monoatomic row, an isolated Au_{38} nanoparticle, and a Au_{13} cluster supported on stoichiometric and reduced TiO_2, was theoretically investigated using DFT methods.

As expected, nitrostyrene does not interact with extended Au(111) and Au(100) surfaces, but it does with low coordinated Au atoms on defective surfaces and Au_{38} isolated nanoparticles. In these two cases, there are two possible ways of nitrostyrene adsorption that are depicted in Fig. 15 for the case of the isolated Au_{38} cluster. Nitrostyrene adsorption through the nitro group is thermodynamically favorable and causes a significant lengthening of one of the NO distances from 1.24 Å in the gas phase to 1.30 Å. But adsorption through the olefin group leading to formation of two new Au–C bonds and causing an important activation of the C=C bond is energetically preferred. Therefore, Au NPs isolated or supported on inert materials such as C or SiO_2 would result in a mixture of partially hydrogenated products, as was indeed experimentally observed.[68,102]

Next, nitrostyrene adsorption on the TiO_2 support was explored and it was found that, regardless the initial relative position of nitrostyrene with respect to the oxide surface, the repulsion between the electrons in the aromatic ring and the O atoms of the oxide is so strong that geometry optimizations converged in all cases to a structure in which nitrostyrene is perpendicular to the surface. The interaction occurs *via* the nitro group, with the two O atoms of the nitro group directly bonded to two Ti atoms of the surface. However, the large adsorption energy value calculated for this structure suggests that strongly chemisorbed nitrostyrene molecules are not involved in the reaction, but might be just spectators. Finally, when

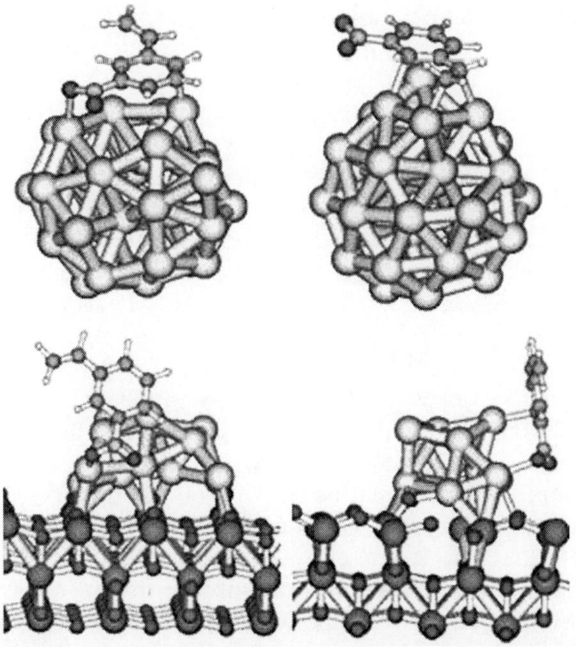

Fig. 15 Optimized geometries of nitrostyrene interacting with a Au_{38} NP through the nitro group (top, left) and through the C=C bond (top, right) and adsorbed at the metal-support interface on a Au/TiO_2 catalyst model (bottom).

nitrostyrene was adsorbed close to the supported Au NP, it remained perpendicular to the oxide surface with the O atoms of the nitro group interacting with the Au atoms at the metal support interface and also weakly with Ti atoms of the support (see Fig. 15, bottom). In this situation, the nitro group is activated to react, while the C=C bond remains far from the Au NP and the support, not accessible and inactive to undergo hydrogenation. Altogether, the high chemoselectivity obtained with Au/TiO_2 catalysts was explained by a cooperative effect between gold and the oxide support, so that H_2 is dissociated on the gold nanoparticles and nitrostyrene adsorbs very selectively through the nitro group at the metal-support interface.

3.4 Activity: role of doping

Despite the high chemoselectivity obtained with the Au/TiO_2 catalyst, its activity towards hydrogenation is too low for practical applications,[103,104] and attempts are being done to design a more efficient catalyst. The complete mechanism of nitrostyrene hydrogenation over Au/TiO_2 was investigated and it was proposed that H_2 dissociation is the rate determining step of the reaction.[105,106] So, efforts should center on improving H_2 dissociation ability without loosing chemoselectivity, and a promising method to achieve this consists of alloying gold with more active transition metals, such as Pd[107,108] or Pt.[109] The catalytic behavior towards hydrogenation of Pt NPs is quite different from that of Au,[110–112] and the understanding of these differences by means of quantum chemical modeling allowed to design highly active and selective bimetallic $Au\text{-}Pt/TiO_2$ catalysts for the hydrogenation of substituted nitroaromatics.[113,114]

In a first step, H_2 dissociation over small Au_{13}, Pt_{13} and $Au_{12}Pt$ clusters, both isolated and supported on TiO_2, was investigated by means of DFT calculations.[114] The energy profiles obtained for the isolated metallic clusters (see Fig. 16) indicate that, as previously described, H_2 dissociation by pure Au clusters involves a non-negligible activation energy. However, the presence of just one Pt atom in the Au nanoparticle is enough to modify the energy profile so that the dissociative chemisorption becomes highly exothermic and barrier less, as it is on pure Pt clusters. These results are in agreement with experimental data[85,109] indicating that the Au–H interaction is weaker than the Pt-H interaction, and that the H_2 dissociation on gold is activated. A similar picture was obtained for larger Au_{38}, Pt_{38} and $Au_{37}Pt$ NPs, as well as for Au_{13}, Pt_{13} and $Au_{12}Pt$ clusters supported on TiO_2. In all cases, doping with Pt leads to an enhancement of the interaction between H_2 and the metal NP and to barrier less dissociation, while the support does not play any active role in the dissociation step. At this point, it seemed that the best way to improve hydrogenation activity was to directly replace gold with platinum, and synthesize Pt/TiO_2 catalysts.

On the other hand, since selectivity is governed by the preferential adsorption and proper activation of the nitro group of nitrostyrene at the metal-support interface, it would be desirable to maximize the number of Pt atoms at the metal-support interface, by decreasing as much as possible the particle size. However, a theoretical study of nitrostyrene adsorption at the metal-support interface between Pt nanoparticles and TiO_2 support showed

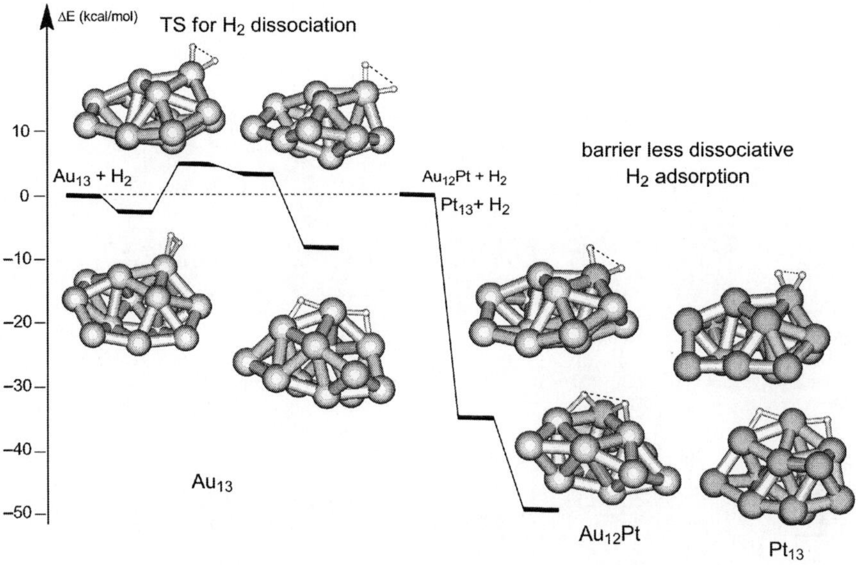

Fig. 16 Energy profile for H_2 dissociation over Au_{13}, $Au_{12}Pt$ and Pt_{13} isolated nanoparticles.

that the cooperative effect between the metal and the support is lost when Au is replaced with Pt. Nitrostyrene was initially adsorbed at several different positions on the P_{13}/TiO_2 model: on the TiO_2 support, at the metal-support interface, and on top of the Pt_{13} NP. In all cases, the strongest interaction was found between the aromatic ring of nitrostyrene and the Pt NP, and structures such as those depicted in Fig. 17 were obtained after geometry optimization. In some conformations the C=C bond was also activated by interaction with Pt, but the nitro group was always far from the metal NP and interacting weakly with Ti atoms of the support. These results showed that, no matter how small the Pt particles are, nitrostyrene always adsorbs on top of the particles, in such a way that the C=C bond is preferentially activated, thus explaining the lack of selectivity of Pt catalysts.

However, when nitrostyrene was adsorbed at the interface between a $Au_{12}Pt$ cluster and the TiO_2 support in a $Au_{12}Pt/TiO_2$ model (see Fig. 17), the structure obtained after geometry optimization was similar to that previously described for nitrostyrene adsorption on Au/TiO_2. The molecule remains perpendicular to the oxide surface, with the two O atoms of the nitro group interacting with both the Ti atoms of the support and the Au atoms of the supported nanoparticle, and with the nitro group being slightly more activated that on Au_{13}/TiO_2 model. This showed that the high chemoselectivity of Au/TiO_2 can be preserved in Au-Pt/TiO_2 catalysts if the Au/Pt ratio is high enough as to keep Pt atoms isolated and not at the active metal-support interface. Since H_2 dissociation is highly enhanced by the presence of just one Pt atom in the Au_{13}/TiO_2 catalyst model, it can be concluded that bimetallic Au-Pt/TiO_2 catalysts should show an activity towards hydrogenation comparable to that of Pt/TiO_2, while preserving the high chemoselectivity of Au/TiO_2. In agreement with this proposal, bimetallic catalysts prepared by impregnating increasing amounts of Pt onto

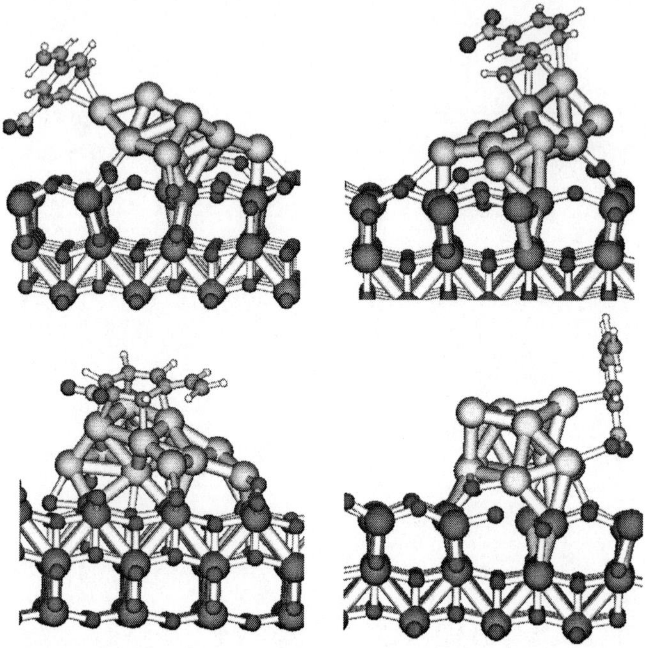

Fig. 17 Optimized geometries of nitrostyrene adsorbed on Pt_{13}/TiO_2 (top and bottom, left) and $Au_{12}Pt/TiO_2$ (bottom, right) catalyst models.

the Au/TiO_2 catalyst provided by the World Gold Council were tested in the hydrogenation of nitrostyrene, and an optimum composition with 1.5%Au and 0.01% Pt was found that enhanced the catalytic activity of Au/TiO_2 by one order of magnitude while maintaining the high chemoselectivity of gold.[105]

4 Conclusions

The exceptional activity of gold catalysts depends on a number of parameters such as particle size and shape, surface concentration of coordinatively unsaturated sites, oxidation state of Au atoms, charge transfer from or to the metal oxide or hydroxide support, and presence of special Au sites at the metal-support interface. It is difficult to unequivocally establish by experimental techniques the direct and specific role played by each of this factors, and theoretical modeling has become in the last years a powerful tool to help in the task of identifying the selective active sites for a target reaction and to finally design highly active and selective catalysts.

In this contribution, the ability of small isolated gold NP to dissociate O_2 and generate a reactive surface oxide layer, the nature of the new gold active sites generated, and their implication in the mechanism of alcohol oxidation to aldehydes has been analyzed from a theoretical point of view. On the other hand, the high chemoselectivity reported for the Au/TiO_2 catalyst in the hydrogenation of substituted nitroaromatics with molecular H_2 has been attributed to a cooperation between the gold NP, where H_2 is dissociated, and the metal/support interface, where nitrostyrene adsorbs selectively through the nitro group. Since H_2 dissociation is the rate determining step of the reaction, we have also analyzed from a theoretical point of view the nature of the active sites involved in H_2 dissociation, and the possible ways in which Au/TiO_2 catalysts can be modified in order to increase their hydrogenation activity without modifying their high chemoselectivity.

Acknowledgements

Financial support was provided by the Spanish MICINN (CONSOLIDER Ingenio 2010-MULTICAT) and Generalitat Valenciana (PROMETEO project 2088/130)) Computational resources and technical assistance provided by *Red Española de Supercomputación* (RES) and *Centre de Càlcul de la Universitat de València* are gratefully acknowledged. Thanks to Avelino Corma, Francesc Illas, Patricia Concepción and Pedro Serna for fruitful discussions.

References

1 K. Honkala, A. Hellman, I. N. Remediakis, A. Logadottir, A. Carlsson, S. Dahl, C. H. Christensen and J. K. Norskov, *Science*, 2005, **307**, 555.
2 F. Studt, F. Abild-Pedersen, T. Bligaard, R. Z. Sorensen, C. H. Christensen and J. K. Norskov, *Science*, 2008, **320**, 1320.
3 I. Lee, F. Delbecq, R. Morales, M. A. Albiter and F. Zaera, *Nature Materials*, 2009, **8**, 132.
4 B. N. Zope, D. D. Hibbitts, M. Neurock and R. J. Davis, *Science*, 2010, **330**, 74.
5 M. B. Boucher, S. Goergen, N. Yi and M. Flytzani-Stephanopoulos, *Phys. Chem. Chem. Phys.*, 2011, **13**, 2517.
6 R. Coquet, K. L. Howard and D. J. Willock, *Chem. Soc. Rev.*, 2008, **37**, 2046.
7 M. Haruta, T. Kobayashi, H. Sano and N. Yamada, *Chem. Lett.*, 1987, 405.
8 M. Haruta, *Chem. Rec.*, 2003, **3**, 75.
9 M. Haruta, *Catal. Today*, 1997, **36**, 153.
10 G. C. Bond and D. T. Thomson, *Catal. Rev. Sci. Eng.*, 1999, **41**, 319.
11 M. S. Chen and D. W. Goodman, *Acc. Chem. Res.*, 2006, **39**, 739.
12 M. D. Hughes, Y. J. Xu, P. Jenkins, P. McMorn, P. Landon, D. I. Enache, A. F. Carley, G. A. Attard, G. J. Hutchings, F. King, E. H. Stitt, P. Johnston, K. Griffin and C. J. Kiely, *Nature*, 2005, **437**, 1132.
13 B. Yoon, H. Häkkinen, U. Landman, A. S. Wörz, J. M. Antonietti, S. Abbet, K. Judai and U. Heiz, *Science*, 2005, **307**, 403.
14 T. M. Bernhardt, U. Heiz and U. Landman, In *Nanocatalysis*, U. Heiz and U. Landman, Eds., Springer-Verlag, Berlin, 2007.
15 S. Carrettin, P. Concepción, A. Corma, J. M. López-Nieto and V. F. Puntes, *Angew. Chem. Int. Ed.*, 2004, **43**, 2538.
16 J. Guzman, S. Carrettin and A. Corma., *J. Am. Chem. Soc.*, 2005, **127**, 3286.
17 N. López, T. V. J. Janssens, B. S. Clausen, Y. Xu, M. Mavrikakis, T. Bligaard and J. K. Nørskov, *J. Catal.*, 2004, **223**, 232.
18 T. V. J. Janssens, B. S. Clausen, B. Hvolbæk, H. Falsig, C. H. Christensen, T. Bligaard and J. K. Nørskov, *Top. Catal.*, 2007, **44**, 15.
19 N. Weiher, A. M. Beesley, N. Tsapatsaris, L. Delannoy, C. Louis, J. A. van Bokhoven and S. L. M. Schroeder, *J. Am. Chem. Soc.*, 2007, **129**, 2240.
20 Th. Risse, Sh. Shaikhutdinov, N. Nilius, M. Sterrer and H. J. Freund, *Acc. Chem. Res.* 2008, **41**, 949.

21 A. A. Herzing, C. J. Kiely, A. F. Carley, P. Landon and G. J. Hutchings, *Science*, 2008, **321**, 1331.
22 D. Widmann and R. J. Behm, *Angew. Chem. Int. Ed.*, 2011, **50**, 10241.
23 J. Guzman, S. Carrettin, J. C. Fierro-Gonzalez, Y. Hao, B. C. Gates and A. Corma, *Angew. Chem. Int. Ed.*, 2005, **44**, 4778.
24 C. Harding, V. Habibpour, S. Kunz, A. N. S. Farnbacher, U. Heiz, B. Yoon and U. Landman, *J. Am. Chem. Soc.*, 2009, **131**, 538.
25 T. Takei, I. Okuda, K. K. Bando, T. Akita and M. Haruta, *Chem. Phys. Lett.*, 2010, **493**, 207.
26 I. X. Green, W. Tang, M. Neurock and J. T. Yates, *Angew. Chem. Int. Ed.*, 2011, **50**, 10186.
27 J. A. Rodriguez, L. Feria, T. Jirsak, Y. Takahashi, K. Nakamura and F. Illas, *J. Am. Chem. Soc.*, 2010, **132**, 3177.
28 M. Turner, V. B. Golovko, O. P. H. Vaughan, P. Abdulkin, A. Berenguer-Murcia, M. S. Tikhov, B. F. G. Johnson and R. M. Lambert, *Nature*, 2008, **454**, 981.
29 H. Tsunoyama, H. Sakurai, Y. Negishi and T. Tsukuda, *J. Am. Chem. Soc.*, 2005, **127**, 9374.
30 H. Tsunoyama, N. Ichikuni, H. Sakurai and T. Tsukuda, *J. Am. Chem. Soc.*, 2009, **131**, 7086.
31 J. Huang, E. Lima, T. Akita, A. Guzman, C. Qi, T. Takei and M. Haruta, *J. Catal.*, 2001, **278**, 8.
32 T. Ishida, H. Watanabe, T. Bebeko, T. Akita and M. Haruta, *Appl. Catal. A: General*, 2010, **337**, 42.
33 H. Liu, Y. Liu, Y. Li, Zh. Tang and H. Jiang, *J. Phys. Chem. C*, 2010, **114**, 13362.
34 Y. Liu, H. Tsunoyama, T. Akita, S. Xie and T. Tsukuda, *ACS Catal.*, 2011, **1**, 2.
35 A. S. K. Hashmi and G. J. Hutchings, *Angew. Chem. Int. Ed.*, 2006, **45**, 7896.
36 A. Corma and H. Garcia, *Chem. Soc. Rev.*, 2008, **37**, 2096.
37 T. Mallat and A. Baiker, *Chem. Rev.*, 2004, **104**, 3027.
38 S. Biella and M. Rossi, *Chem. Commun.*, 2003, 378.
39 C. Della Pina, E. Falletta, L. Prati and M. Rossi, *Chem. Soc. Rev.*, 2008, **37**, 2077.
40 A. Abad, A Corma and H. García, *Chem. Eur. J*, 2008, **14**, 212.
41 A. Abad, P. Concepción, A. Corma and H. García, *Angew. Chem. Int. Ed.*, 2005, **44**, 4066.
42 B. E. Salisbury, W. T. Wallace and R. L. Whetten, *Chem. Phys.*, 2000, **262**, 131.
43 J. Hagen, L. D. Socaciu, M. Elijazyfer, U. Heiz, T. M. Bernhardt and L. Wöste, *Phys. Chem. Chem. Phys.*, 2002, **4**, 1707.
44 G. Mills, M. S. Gordon and H. Metiu, *Chem. Phys. Lett.*, 2002, **359**, 493.
45 B. Yoon, H. Häkkinen and U. Landman, *J. Phys. Chem. A*, 2003, **107**, 4066.
46 A. Roldan, S. González, J. M. Ricart and F. Illas, *ChemPhysChem*, 2009, **10**, 348.
47 G. Mills, M. S. Gordon and H. Metiu, *J. Chem. Phys.*, 2003, **118**, 4198.
48 M. Mavrikakis, P. Stoltze and J. K. Norskov, *Catal. Lett.*, 2000, **64**, 101.
49 M. Boronat and A. Corma, *Dalton Trans.*, 2010, **39**, 8538.
50 M. M. Schubert, S. Hackenbertg, A. C. van Veen, M. Muhler, V. Plzak and J. Behm, *J. Catal.*, 2001, **197**, 113.
51 N. López and J. K. Nörskov, *J. Am. Chem. Soc.*, 2002, **124**, 11262.
52 Z. P. Liu, P. Hu and A. Alavi, *J. Am. Chem. Soc.*, 2002, **124**, 14470.
53 G. J. Wang and B. Hammer, *Top. Catal.*, 2007, **44**, 49.
54 L. Alves, B. Ballesteros, M. Boronat, J. R. Cabrero-Antonino, P. Concepción, A. Corma, M. A. correa-Duarte and E. Mendoza, *J. Am. Chem. Soc.*, 2011, **133**, 10251.

55 J. L. C. Fajin, M. N. D. S. Cordeiro and J. R. B. Gomes, *J. Phys. Chem. C*, 2008, **112**, 17291.
56 M. Molina, M. D. Rasmussen and B. Hammer, *J. Chem. Phys.*, 2004, **120**, 7673.
57 L Molina and B. Hammer, *Phys. Rev. B*, 2004, **69**, 155424.
58 J. Hagen, L. D. Socaciu, M. Elijazyfer, U. Heiz, T. M. Bernhardt and L. Wöste, *Phys. Chem. Chem. Phys.*, 2002, **4**, 1707.
59 L. M. Molina, A. Lesarri and J. A. Alonso, *Chem. Phys. Lett.*, 2009, **468**, 201.
60 M. Boronat, A. Corma, F. Illas, J. Radilla, T. Ródenas and M. J. Sabater, *J. Catal.*, 2011, **278**, 50.
61 K. Shimizu, K. Sugino, K. Sawabe and A. Satsuma, *Chem. Eur. J*, 2009, **15**, 2341.
62 J. D. Grunwaldt, M. Caravati and A. Baiker, *J. Phys. Chem. B*, 2006, **110**, 25586.
63 D. Ferri, C. Mondelli, F. Krumeich and A. Baiker, *J. Phys. Chem. B*, 2006, **110**, 22982.
64 M. Comotti, C. Della Pina, R. Matarrese and M. Rossi, *Angew. Chem. Int. Ed.*, 2004, **43**, 5812.
65 S. Kanaoka, N Yagi, Y Fukuyama, S. Aoshima, H. Tsunoyama, T. Tsukuda and H. Sakurai, *J. Am. Chem. Soc.*, 2007, **129**, 12060.
66 M. Boronat and A. Corma, *J. Catal.*, 2011, **284**, 138.
67 W. C. Ketchie, M. Murayama and R. J. Davis, *Top. Catal.*, 2007, **44**, 307.
68 A. Corma and P. Serna, *Science*, 2006, **313**, 332.
69 K. Shimizu, Y. Miyamoto, T. Kawasaki, T. Tnji, Y. Tai and A. Satsuma, *J. Phys. Chem. C*, 2009, **113**, 17803.
70 F. Cardenas-Lizana, S. Gómez-Quero, N. Perret and M. Keane, *Gold Bull.*, 2009, **42**, 124.
71 J. E. Bailie and G. J. Hutchings, *Chem. Commun.*, 1999, 2151.
72 C. Milone, M. L. Tropeano, G. Guline, G. Neri, R. Ingoglia and S. Galvano, *Chem. Commun.*, 2002, 868.
73 C. Milone, R. Ingoglia, M. Tropeano, G. Neri and S. Galvagno, *Chem. Commun.*, 2003, 1359.
74 C. Mohr, H. Hofmeister, J. Radnik and P. Claus, *J. Am. Chem. Soc.*, 2003, **125**, 1905.
75 P. Claus, *Appl. Catal. A: General*, 2005, **291**, 222.
76 Y. Segura, N. López and J. Pérez-Ramírez, *J. Catal.*, 2007, **247**, 383.
77 A. Hugon, L. Delannoy and C. Louis, *Gold Bull.*, 2008, **41**, 127.
78 A. Hugon, L. Delannoy, J. M. Krafft and C. Louis, *J. Phys. Chem. C.*, 2010, **114**, 10823.
79 X. F. Yang, A. Q. Wang, Y. L. Wang, T. Zhang and J. Li, *J. Phys. Chem. C*, 2010, **114**, 3131.
80 L. Barrio, P. Liu, J. A. Rodriguez, J. M. Campos-Martıín and J. L. G. Fierro, *J. Chem. Phys.*, 2006, **125**, 164715.
81 A. Corma, M. Boronat, S. González and F. Illas, *Chem. Commun.*, 2007, 3371.
82 M. Boronat, F. Illas and A. Corma, *J. Phys. Chem. A*, 2009, **113**, 3750.
83 B. Yang, X. M. Cao, X. Q. Gong and P. Hu, *Phys. Chem. Chem. Phys.*, 2012, **14**, 3741.
84 R. Zanella, C. Louis, S. Giorgio and R. Touroude, *J. Catal.*, 2004, **223**, 328.
85 E. Bus, J. T. Miller and J. A. van Bokhoven, *J. Phys. Chem. B*, 2005, **109**, 14581.
86 E. Bus, R. Prins and J. A. van Bokhoven, *Catal. Commun.*, 2007, **8**, 1397.
87 T. Fujitani, I. Nakamura, T. Akita, M. Okumura and M. Haruta, *Angew. Chem. Int. Ed.*, 2009, **48**, 9515.
88 C. Mohr, H. Hofmeister and P. Claus, *J. Catal.*, 2003, **213**, 86.

89 C. Lemire, R. Meyer, Sh. Shaikhutdinov and H. J. Freund, *Angew. Chem. Int. Ed.*, 2004, **43**, 118.
90 J. D. Grunwaldt, M. Maciejewski, O. S. Becker, P. Fabrizioli and A. Baiker, *J. Catal.*, 1999, **186**, 458.
91 F. Bocuzzi and A. Chiorino, *J. Phys. Chem. B*, 2000, **104**, 5414.
92 M. Maciejewski, P. Fabrizioli, J. D. Grunwaldt, O. S. Becker and A. Baiker, *Phys. Chem. Chem. Phys.*, 2001, **3**, 3846.
93 T. Diemant, Z. Zhao, H. Rauscher, J. Bansmann and R. J. Behm, *R. J. Topics in Catal.*, 2007, **44**, 83.
94 A. Chiorino, M. Manzoli, F. Menegazzo, M. Signoretto, F. Vindigni, F. Pinna and F. Bocuzzi, *J. Catal.*, 2009, **262**, 169.
95 M. Boronat, P. Concepción and A. Corma, *J. Phys. Chem C*, 2009, **113**, 16772.
96 H. U. Blaser, H. Steiner and M. Studer, *ChemCatChem*, 2009, **1**, 210.
97 H. U. Blaser, U. Siegrist and H. Steiner, In *Fine Chemicals through Heterogeneous Catalysis: Aromatic Nitro Compounds*, R. A. Sheldon and H. van Bekkum, Eds., Wiley: New York, 2001, p. 389.
98 R. F. Kovar and F. E. Armond (U.S. Air Force), U.S. Patent 3,975,444, 1976.
99 M. Suchy, P. Winternitz and M. Zeller (Ciba-Geigy), WO Patent 91/02278, 1991.
100 A. Burawoy and J. P. Critchley, *Tetrahedron*, 1959, **5**, 340.
101 U. Siegrist, P. Baumeister and H. U. Blaser, *Chem. Ind.*, 1998, **75**, 207.
102 M. Boronat, P. Concepción, A. Corma, S. González, F. Illas and P. Serna, *J. Am. Chem. Soc.*, 2007, **129**, 16230.
103 H. U. Blaser, *Science*, 2006, **313**, 312.
104 A. Grirrane, A. Corma and H. García, *Science*, 2008, **322**, 1661.
105 A. Corma, P. Concepción and P. Serna, *Angew. Chem. Int. Ed.*, 2007, **46**, 7266.
106 P. Serna, P. Concepción and A. Corma, *J. Catal.*, 2009, **265**, 19.
107 A. Sarkany, P. Hargittai and A. Horvath, *Top. Catal.*, 2007, **46**, 121.
108 F. Gao, Y. L. Wang and D. W. Goodman, *J. Am. Chem. Soc.*, 2009, **131**, 5734.
109 E. Bus and J. A. van Bokhoven, *Phys. Chem. Chem. Phys.*, 2007, **9**, 2894.
110 F. Tao, S. Dag, L. W. Wang, Z. Liu, D. R. Butcher, H. Bluhm, M. Salmeron and G. A. Somorjai, *Science*, 2010, **327**, 850.
111 C. A. Witham, W. Y. Huang, C. K. Tsung, J. N. Kuhn, G. A. Somorjai and F. D. Toste, *Nature Chemistry*, 2010, **2**, 36.
112 G. A. Somorjai and J. Y. Park, *Topics in Catal.*, 2008, **49**, 126.
113 A. Corma, P. Serna, P. Concepción and J. J. Calvino, *J. Am. Chem. Soc.*, 2008, **130**, 8748.
114 M. Boronat and A. Corma, *Langmuir*, 2010, **26**, 16607.

PdZn based catalysts: connecting electronic and geometric structure with catalytic performance

Karin Föttinger
DOI: 10.1039/9781849737203-00077

In the recent years, the potential of PdZn intermetallic compounds and related compositions for improving and consequently replacing conventionally used catalysts has been explored for a range of diverse processes, such as selective hydrogenation reactions, methanol synthesis and steam reforming. PdZn has similar electronic properties and reactivity as Cu, a widely used metal catalyst, *e.g.* Cu is industrially applied in the low temperature water gas shift reaction and methanol synthesis. The higher stability of PdZn makes it an attractive alternative for certain applications. This review will give an overview over selected important potential applications and the correlation of the catalytic performance with properties, such as the electronic structure. A broad range of materials from oxide supported nanoparticles to single crystal based model systems is covered.

1 Introduction

Alloys offer a wide range of desirable and tuneable properties, which are often different from the individual metals.[1–2] For application in catalysis enhancement of activity, selectivity and stability are the main goals that make alloys favourable for many applications. Also intermetallic compounds (IMC) have attracted attention as catalysts. In contrast to substitutional alloys, IMCs are per definition single-phase compounds consisting of metals, with the crystal structure of an IMC being different from the structure of the constituting metals.[3] Due to a partly covalent or ionic bond character a higher structural stability is often observed, especially against segregation, which is a frequently observed phenomenon and major drawback when using (substitutional) alloys as catalysts. In terms of composition, IMCs do not necessarily have to possess a certain stoichiometric composition, but can have a broader stability region. In the following, the terms "alloy" and "intermetallic compound" will not be distinguished in each single case, but will be used equivalently, resembling their somewhat "mixed" usage in the largest part of catalysis-related literature. In general, the so-called ensemble effect, a geometric effect (*e.g.* site isolation), and/or the ligand effect, an alteration of the electronic properties, can be relevant when utilizing alloys as catalysts for achieving the desired catalytic properties.[1–2,4–7]

In the recent years, PdZn has emerged as a promising catalyst for a range of applications. One of the most widely used applications is methanol steam reforming (MSR). Iwasa and co-workers[8–14] were the first to discover the excellent activity and selectivity of Pd/ZnO in MSR exhibiting similar

Vienna University of Technology, Institute of Materials Chemistry, 1060 Vienna, Austria. E-mail: karin.foettinger@tuwien.ac.at

reactivity as Cu, but higher long-term stability. Stability is the major drawback of the commercially used Cu-based catalyst due to sintering already at moderate temperatures (>280 °C) and pyrophoricity.[15–17] Nowadays, there is general agreement that the promising catalytic properties of Pd/ZnO are attributed to PdZn alloy formation upon reduction.

Tsai *et al.*[18] introduced the concept of an IMC replacing an element that is not one of the constituting elements on the basis of similar valence electron densities of states. This concept opens up a wide field of designing catalysts for certain reactions based on the requirements in terms of electronic properties, which can overcome specific drawbacks of conventional catalysts, such as stability issues. They demonstrated this concept for PdZn substituting Cu in MSR.[18]

In the following, an overview over synthesis procedures, the physicochemical, structural and electronic properties (section 2) as well as some examples of catalytic applications (section 3) will be given. Specifically, two examples will be discussed in more detail where PdZn has important advantages compared to the commercially used catalysts: the main focus will be on the application of PdZn in selective hydrogenation reactions and in methanol steam reforming (MSR). The effect of site isolation and alteration of the electronic structure on the selectivity and stability will be discussed.

2 Preparation, structural and electronic properties

2.1 Preparation and formation of Pd-Zn compounds

Different classes of PdZn-based materials have been prepared and studied in the literature, ranging from high surface area oxide supported nanoparticles, metallurgically prepared bulk IMCs to model systems including single crystal-based surface alloys. An overview over (some of) the most prominent and common approaches will be given in this section, covering synthesis approaches for design of PdZn particles as well as for formation of the bimetallic particles upon exposure of Pd supported on ZnO to a reducing atmosphere. In addition, use of different oxides *vs.* ZnO as support will be covered.

In general, the challenge and difficulty in preparation of alloys is to achieve control over composition, particle size, homogeneity, while at the same time obtaining a high surface area, which is an important requirement for the application of such materials in catalysis. Especially for electropositive elements or metals with precursors that are not easy to reduce, such as Zn, controlled incorporation into alloys/IMCs is difficult. Besides, inhomogeneous particle size distribution can significantly affect the catalytic properties (particle size effects, structure sensitivity) but also the IMCs formed (incomplete formation, different phases present).

In general agreement, PdZn forms *in situ* under reducing conditions starting from Pd particles supported on ZnO at temperatures of ~300 °C. Spillover of hydrogen activated on the Pd onto the ZnO leads to the formation of reduced Zn species, which then migrate into the Pd. Iwasa and coworkers[8–14] noticed formation of an alloy by *ex situ* characterizing reduced and used catalysts by XRD, Auger electron spectroscopy (AES) and XPS. XRD showed that PdZn alloy formation starts at about 200 °C and goes further on with increasing temperature.[8] Similarly, Tew *et al.*[19] reported appearance of PdZn reflexes in XRD patterns at 250 °C. However, in X-ray absorption near edge spectra (XANES) indications of alloy formation were already observed upon reduction at 100 °C. Therefore, surface changes probably start at lower temperatures, while the formation of larger crystalline domains occurs at higher temperatures, as detected by XRD. However, Chin *et al.*[20] found that Pd/ZnO was less selective in MSR when reduced at 125 °C than when reduced at 350 °C, indicating presence of metallic Pd and incomplete alloy formation. Overall, the exact temperature of ZnO reduction in presence of Pd and PdZn formation may depend on various parameters, such as the particle size, the synthesis, details of reaction conditions like H_2 pressure, heating rate or dwell time. Föttinger *et al.*[21] showed that alloy formation starts at the surface and then grows further "inwards" by Zn diffusion into the metallic core which gets smaller with time. Diffusion of Zn overlayers, grown on Pd(111), into the subsurface of the Pd single crystal starts already at about 25 to 80 °C,[22–24] while diffusion of Zn from the PdZn surface/subsurface layers into the Pd bulk sets in at around 350 °C.[25] A surface monolayer of PdZn exhibited higher thermal stability than "deeper" (2nd, 3rd, etc) PdZn layers,[25] which may explain the formation of a (nearly) complete alloy shell on the surface of the Pd nanoparticles, prior to alloy formation towards the particle "bulk". In good agreement with the high surface area real world catalysts, Penner *et al.*[26] proposed that PdZn alloy formation proceeded *via* topotactic formation of PdZn on top of the Pd particles supported on ZnO thin films, which resulted in PdZn-encapsulated Pd cores.

Most commonly in literature, Pd on ZnO is prepared by impregnation or by coprecipitation. Karim *et al.* studied the influence of the PdZn particle size[27] and of the morphology of the ZnO crystallites[28] on the MSR activity and selectivity. Because ZnO dissolves under acidic conditions, Pd/ZnO catalysts were prepared by coprecipitation.[27] First, ZnO was dissolved in in nitric acid Pd-containing solution. In a second step, the pH was adjusted to 9.5 using NaOH. Particle sizes and the extent of alloy formation were varied independently in this work by different catalyst pretreatments. Karim *et al.*[28] also studied the influence of the ZnO morphology on MSR. The often applied impregnation of Pd from aqueous acidic solution was observed to affect the morphology of the ZnO crystals due to partial dissolution. Upon utilization of organic Pd precursors, such as Pd acetate dissolved in acetone, the faceted ZnO particles could be preserved which had a beneficial effect on MSR performance. Also Chin *et al.*[29] found that impregnation of ZnO from a highly acidic Pd nitrate aqueous solution strongly alters the textural properties of the ZnO, such as porosities and crystalline structure, by a partial dissolution of ZnO. They suggested that the extent of mixing between Pd^{2+} and Zn^{2+} during preparation may have an impact on the PdZn alloy formation and consequently on the catalytic properties. An opposite trend of selectivity dependence on particle size than Karim and co-workers was found by Agrell *et al.*[30] They varied the particle size by utilizing a microemulsion technique.[30] In this approach a metal salt is reduced in the water droplets of a water-in-oil microemulsion, which is stabilized by surfactants. In[30] Pd was prepared in a microemulsion obtained by adding an aqueous solution of $Pd(NH_3)_4Cl_2$ to a solution of nonyl-phenolethoxylate in cyclohexane and reduced with hydrazine, followed by deposition of Pd onto ZnO by chemical destabilisation of the suspension. This synthesis route allows for the preparation of fine and monodispersed particles and thereby control over particle size.

In order to achieve better dispersion of the metal particles and higher stability against sintering, PdZn was supported on other oxides such as silica[31] and alumina.[32] $Pd/Zn/Al_2O_3$ catalysts showed smaller particles with a higher stability against sintering,[32] however, these catalysts also contained ZnO and fcc Pd in addition to PdZn, resulting in a lower selectivity for MSR. Bollmann *et al.*[33] added different amounts of Zn to a Pd/Al_2O_3 catalyst. They found increasing Pd-Zn coordination with increasing Zn loading, as determined by EXAFS. However, always lower average Pd-Zn coordination numbers were observed on these samples than on the ordered 1:1 PdZn/ZnO, and more than 85% of the Zn species reacted with alumina to form zinc aluminate. Green *et al.*[31] did an extensive study on the effect of synthesis parameters like Pd:Zn ratio, addition of Pd by impregnation or ion exchange, different treatments like reduction prior to the subsequent Zn addition, *etc.* They determined an optimum PdZn atomic ratio of 1:2 in the hydrogenation of methyl acetate. The exact composition was difficult to determine, but Pd-Zn distribution was likely between PdZn and Pd_2Zn_3 with excess Zn^{2+} on silica, according to XRD and EDX-TEM.[31] Arana *et al.*[34] prepared silica supported PdZn catalysts with different Pd:Zn ratios (Pd:Zn 3:1, Pd:Zn 2:3) starting from Pd/SiO_2 followed by free adsorption of $ZnCl_2$ dissolved in acetone. The surface composition after reduction was determined by XPS and by CO and H_2 chemisorption combined with XRD bulk structure determination and Fourier transformed Infrared (FTIR) spectroscopy of CO adsorption. While the Zn rich composition contained mostly alloyed Pd according to XRD and XPS (in the bulk and at the surface; the state of Zn was difficult to determine, the presence of some Zn as ZnO can thus not be excluded), the Pd rich material was proposed to consist of PdZn particles decorated with metallic Pd. This picture emerged from XRD patterns of PdZn, but absence of Zn at the surface according to XPS.[34] Generally, controlling the composition and homogeneity seems to be a difficult task when supporting Pd-Zn on other oxides. Often, separate Pd and bimetallic particles of different compositions as well as excess oxidized Zn species have been observed to be simultaneously present.

In order to improve the required intimate contact between Pd and Zn supported on alumina, Lenarda *et al.*[35] reported a preparation protocol where Pd acetate was deposited on alumina followed by a pre-reduction with a zinc borohydride solution at room temperature. In addition, Lenarda[36] prepared PdZn catalysts by impregnating the spinel-oxide mixture $ZnAl_2O_4/ZnO$ obtained by the thermal decomposition of a Zn-Al hydrotalcite with Pd. In another approach, supported PdZn was synthesized by reductive decomposition of a ternary hydrotalcite-like compound obtained by co-precipitation from aqueous solution.[37] Hydrotalcite-like precursors offer the advantage of a homogeneous distribution and an intimate contact of the compounds. The Pd^{2+} cations were fully incorporated into the cationic slabs of the PdZnAl-hydrotalcite-like precursor. After reduction, the nanoparticulate intermetallic PdZn was present. Upon reduction in hydrogen, Ota *et al.*[37] found metallic Pd segregating from the precursors at low temperature and formation of the IMC PdZn at 250 °C, probably *via* a Pd_2Zn intermediate stage.

The major disadvantage of *in situ* formation of PdZn by reduction of Pd/ZnO as starting material is the very limited control over composition, structure and phase-purity, which can be adjusted only to a limited extent.

Preparation of alloyed PdZn nanoparticles in a controlled manner with a desired composition and particle size is far from straightforward. Approaches that were used to obtain some control on different aspects include colloidal synthesis, and use of organometallic precursors. Bimetallic Pd-Zn colloidal nanoparticles were prepared by Bronstein *et al.*[38] in block copolymer micelles derived from polystyrene-block-poly-4-vinylpyridine. Also in[39] colloidal synthesis was applied. Pd nanoparticles were synthesized *via* solution-precipitation and deposited on ZnO by decomposing a Pd precursor in a solution of trioctylphosphine or in a solution of octylamine at 300 °C. Tkachenko *et al.*[40] reported an advanced synthesis approach using mixed metal complexes. In[40] they used a heterobimetallic Pd-Zn carboxylate-bridged complex ($Pd(\mu\text{–}OOCMe)_4Zn(OH_2)$, Fig. 1) as precursor for the preparation of alloyed nanoparticles of controlled composition. Upon mild reduction in H_2/He at around 150–250 °C alloyed Pd-Zn nanoparticles were formed in the unsupported crystalline state as well as when supported on carbon. This was evidenced by elemental analysis, XANES and EXAFS, where the local structure of Pd and Zn clearly changed upon reduction (Fig. 1). However, as a major disadvantage this approach requires the synthesis of complex molecular precursors.

Cable *et al.*[41] developed a synthesis route in solution for preparing Pd-Zn intermetallic nanocrystals using zero-valent organometallic Zn precursors, which allows for a certain control of the nanoparticle morphology. Their approach consists of two steps: a preparation of Pd nanocrystals of a certain shape and size in solution, followed by addition of diethylzinc Et_2Zn into the hot organoamine solution. By this, a diffusion of Zn into the pre-formed Pd nanoparticle was achieved, while maintaining the pre-determined shape of the particles. Thus, the morphology of the pre-formed Pd particles, which is much easier to control since it is a monometallic system, then acts as a template. Beside Pd-Zn, this synthesis strategy was successfully applied to the preparation of Au-Zn and Cu-Zn IMCs and was proposed to be extendable to other systems offering potential "generalizability".[41]

Metallurgical preparation of PdZn by high temperature melting and long annealing times under inert atmosphere leads to phase pure bulk IMCs. Unfortunately, such materials typically have low specific surface areas, because they are synthesized in thermodynamic equilibrium. For the preparation of bulk PdZn Friedrich *et al.*[42] reported a procedure consisting of melting respective amounts of high purity elemental Zn and Pd in evacuated

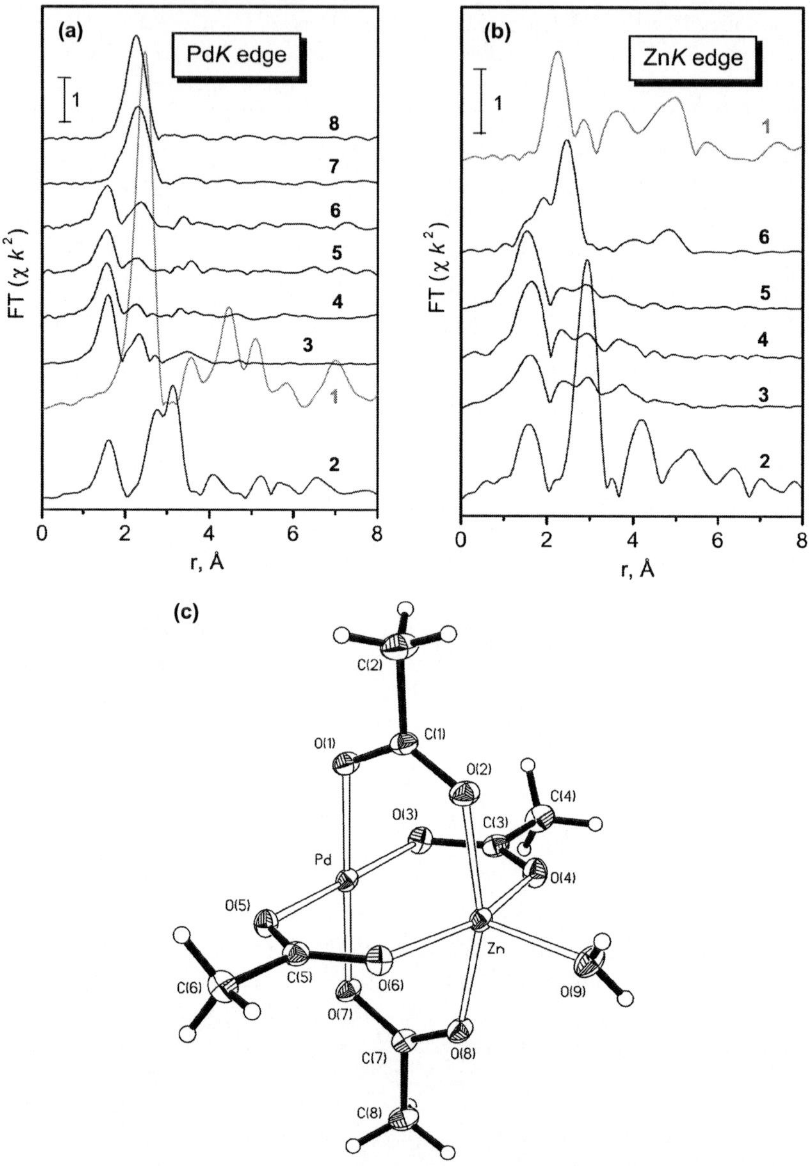

Fig. 1 EXAFS spectra of carbon-supported complex after different treatments (Fourier transform of k^2-weighted EXAFS function): (a) PdK edge, 1-Pd foil; 2-PdO; 3- unsupported complex; 4-supported sample as prepared; 5- supported sample after heating to 120 °C under He; 6- after treatment with 5% H_2/He at 50 °C; 7- with 5% H_2/He at 150 °C; and 8- with 5% H_2/He at 250 °C. (b) ZnK edge, 1-Zn foil; 2-ZnO; 3-unsupported complex; 4-supported sample as prepared; 5-supported sample after heating to 120 °C under He; 6-supported sample after treatment with 5% H_2/He at 250 °C. (c) Molecular structure of the complex PdZn(μ-OOCMe)$_4$(OH$_2$). Reproduced from ref. 40 with kind permission from Springer Science and Business Media.

and sealed quartz ampoules. The tubes were heated with 5 °C/min first to the melting point of Zn (420 °C) and then further with 1 °C/min to 850 °C for 6 d, followed by grinding under inert atmosphere in a glove box and another heat treatment in vacuum at 900 °C for 7 d. This recipe resulted in phase pure compounds of the desired 1:1 composition. Even though the practical application of bulk PdZn IMCs is limited by their low surface area, they can be considered as interesting and useful model systems.

Halevi and coworkers[43–45] obtained unsupported bimetallic alloy powders with high surface areas in the range of 1–10 m^2/g by an aerosol-based method. Such materials can be considered as in between complex conventional catalysts and well-defined single crystal-based model systems due to the fact that they have a lower complexity and heterogeneity than conventional catalysts and can be studied under realistic conditions due to reasonable surface areas, as compared to the latter systems. In this synthesis approach Pd and Zn nitrate were dissolved in nitric acid, the solution was dispersed into droplets, which dried upon passing through a furnace and were collected in a filter followed by reduction. The obtained powder was bulk phase pure $L1_0$ tetragonal PdZn[44] with a BET surface area of about 6 m^2/g. Some ZnO was present on the surface as determined by TEM and XPS, which was proposed to be a consequence of the high reactivity of Zn toward oxygen. Recently, the group prepared aerosol-derived phase-pure $PdZn_\alpha$ phase, a solid solution of Zn in fcc Pd, by the same synthesis approach.[43] Such materials have the potential to single out the role of the respective intermetallic phases and that of the support, provided that ZnO is absent and all Zn species is reducible.

Another approach for preparing PdZn nanoparticles in a controlled way was followed by Engels *et al.*[46] They applied the liquid phase polyol reduction using PVP (poly-N-vinylpyrrolidone) as protecting agent and obtained particles with very narrow particle size distributions centred around 3 nm mean diameter. Luo *et al.* explored different liquid phase reduction procedures to prepare unsupported PdZn nanoparticles, including coreduction of Pd^{2+} and Zn^{2+} by Superhydride, utilization of ZnH_2 as reduction agent and as Zn source and electroless plating of Zn nanoparticles with Pd^{2+}.[47] With all routes nanoparticulate PdZn was obtained.

In addition, different types of PdZn model systems have been investigated from Pd nanoparticles on ZnO thin films[26] or ZnO(0001) single crystals[48–49] to PdZn surface alloys prepared by depositing evaporated Zn in UHV on Pd(111) single crystals.[22–24,50–52]

Penner *et al.*[26] prepared thin film model catalysts by epitaxial growth of Pd particles embedded in a layer of amorphous ZnO and mechanically stabilized by SiO_2. The structural changes were studied by TEM and SAED after subjecting the Pd/ZnO/SiO_2 catalyst to hydrogen and annealing treatments. Ordered PdZn nanoparticles were formed upon reduction at 250–325 °C and were structurally stable in a wide temperature range of 200–600 °C.

Bayer *et al.*[22] prepared PdZn/Pd(111) surface alloys by depositing Zn *via* physical vapor deposition on a Pd(111) single crystal followed by annealing. Their recipe was adapted and further developed by several groups.[25,50,52–53] Well-defined PdZn(111)/Pd(111) surface alloys with a 1:1 Pd:Zn composition were obtained upon annealing between 225 and 325 °C or upon deposition in this temperature range. Variation of the amount of deposited Zn between 1 monolayer (ML) and 3 ML resulted in both cases in a Pd:Zn surface composition of 1:1.

Another approach was applied by Rodriguez.[54] Pd_xZn_y alloys were obtained by evaporation of both Pd and Zn and deposition onto a Ru(001) single crystal substrate at 77 °C. Neither Zn nor Pd is soluble in the substrate Ru. However, the composition of the PdZn films was non-homogeneous.[54] Kovacs *et al.*[55] prepared Pd_xZn_y alloys on Pd(100) by decomposition of diethyl zinc. However, the alloys obtained by the latter two approaches were not as well-defined in terms of composition, structure and homogeneity, as the alloys prepared by following the recipe of Bayer.

Bera *et al.*[48–49] prepared Pd/ZnO(0001) and PdZn/ZnO(0001) by vapour deposition of Pd and characterized the materials by HREELS and XPS. Heating to 77 °C was found to be sufficient to induce partial agglomeration of as-deposited 2-dimensional Pd films into three-dimensional particles. Upon further heating to 425 °C partial reduction of the ZnO surface occurred leading to the formation of a PdZn alloy. Hyman *et al.*[56] compared Pd particles and films deposited on ZnO(10–10) and ZnO(0001) single crystals and found that PdZn alloy formation was more facile on ZnO(0001).

2.2 Bulk structure

Pd and Zn are able to form alloys with a wide range of compositions, as well as ordered IMCs, as can be seen in the phase diagram[57] (Fig. 2). The most important bimetallic phase is the thermodynamically stable tetragonal PdZn phase with a 1:1 stoichiometry. It has a broad composition range from 37 to 56% Zn and crystallizes in the AuCu type structure. The 1:1 composition is most often obtained upon reduction of Pd/ZnO in a temperature range around 300 °C, which is relevant for many catalytic applications (such as for MSR, see section 3.2).

The tetragonal CuAu crystal structure ($L1_0$ type, space group P4/mmm) of PdZn with the 1:1 composition[58] is derived from an fcc structure by shortening the c axis ($c/a = 0.82$) and is illustrated in Fig. 2(b).[18] According to a recent experimental and quantum chemical study[42] the cubic modification displayed in the phase diagram does not exist at high temperatures.

The respective X-ray diffractograms (measured with Cu Kα radiation, $\lambda = 0.154$ nm) of PdZn formation upon reduction of Pd/ZnO at different temperatures is shown in Fig. 3(A).[13] The catalyst reduced at room temperature exhibits reflections of Pd metal ($2\theta = 40.1°$) and ZnO, which gives rise to characteristic reflections at 2θ values above 47° and below 38°. Upon reduction at 623 K characteristic reflections of tetragonal PdZn emerge at $2\theta = 41.2°$ (111) and 44.1° (200).

In the ordered 1:1 IMC each atom has eight nearest-neighbour heteronuclear bonds and four next-nearest neighbour homonuclear bonds. Coordination numbers and bond distances were also determined by Extended X-ray Absorption Fine Structure (EXAFS).[19,21,33] EXAFS and XANES were utilized to investigate the formation process of the PdZn alloy with temperature upon reduction of Pd/ZnO in H_2 by following the structural and electronic changes visible in the spectra.[19,21] The spectral changes in the near-edge region are illustrated in Fig. 3(B).[19] The amplitude of the oscillations present in metallic Pd decreases strongly upon PdZn alloying to a rather flat region after the edge.

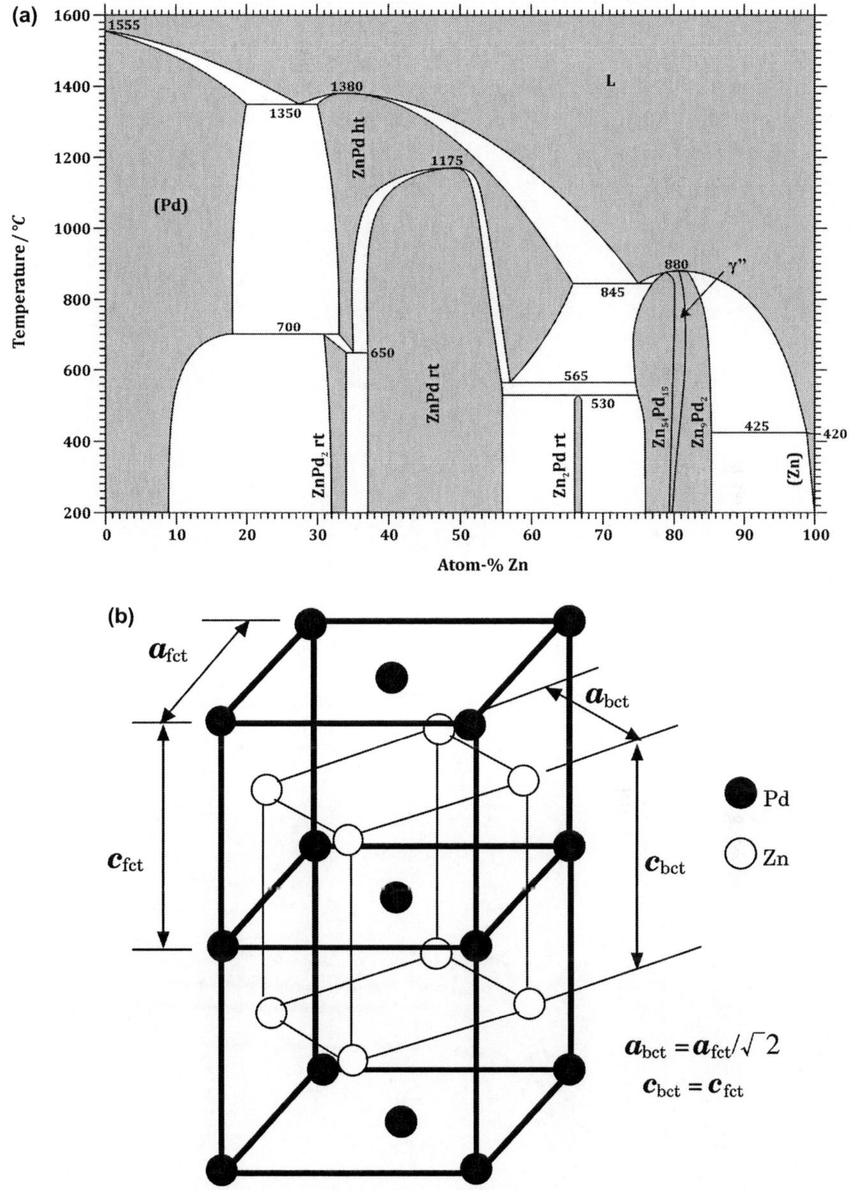

Fig. 2 (a) Phase diagram of the binary system Pd-Zn. Reproduced from ref. 57 with permission of ASM International. (b) Illustration of the tetragonal $L1_0$ crystal structure of PdZn. Circles in black are Pd atoms and those in white are Zn atoms. Reproduced from ref. 18 with permission of The Physical Society of Japan.

2.3 Surface structure and composition

DFT calculations by Chen *et al.*[59] showed that on the PdZn 1:1 alloy the (111) surface is thermodynamically the most stable surface but with only a slightly lower energy than the (100) surface. Both exhibit stoichiometric composition. Several groups studying PdZn/Pd(111) surface alloys[22–24,50–52] detected experimentally a 1:1 surface composition with a p(2 × 1) structure, which has been explained by the surface consisting of alternating rows of Pd

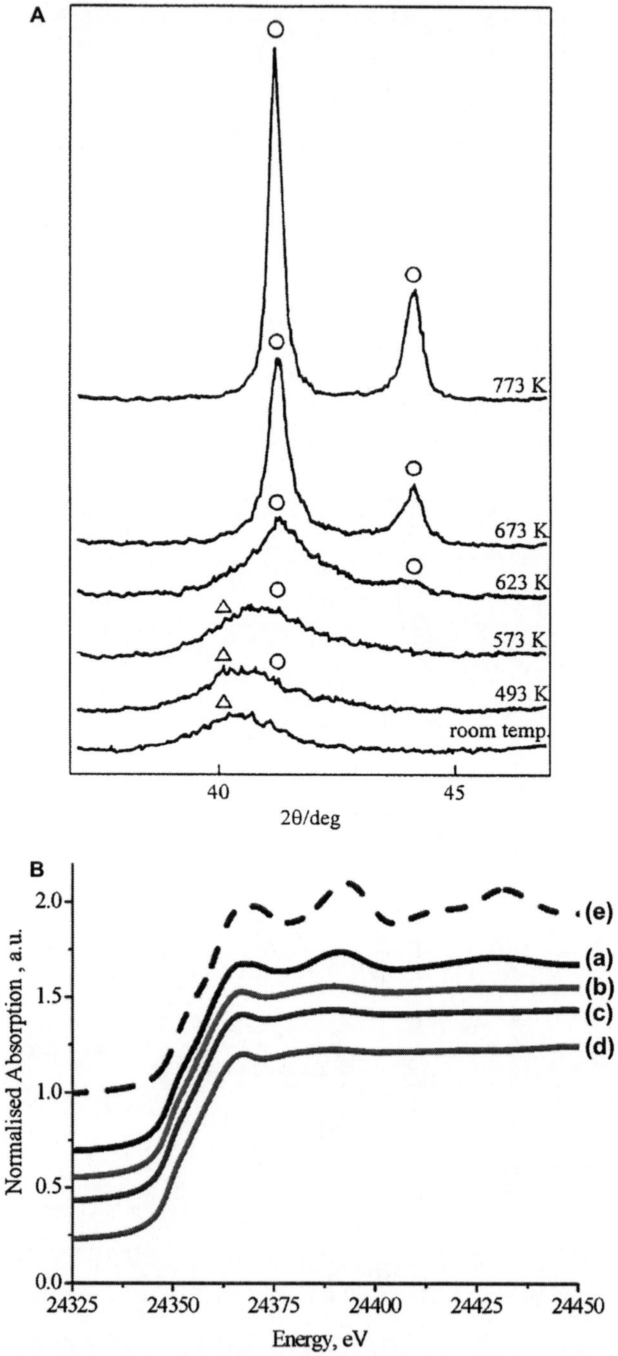

Fig. 3 (A) XRD patterns for 10 wt% Pd/ZnO previously reduced at different temperatures in H_2. Reduction temperatures of the catalysts are indicated. Pd (Δ), PdZn (O). Adapted from ref. 13, Copyright (1997), with permission from Elsevier. (B) Pd K edge XANES spectra of Pd/ZnO before and after reduction at high temperature: Pd/ZnO in He at 100 °C after reduction at 100 °C (a); in 4%H_2/He: Pd/ZnO at 250 °C (b), at 300 °C (c), and at 350 °C (d). Pd K edge XANES of foil (e) is shown for comparison. Reprinted with permission from ref. 19. Copyright (2011) American Chemical Society.

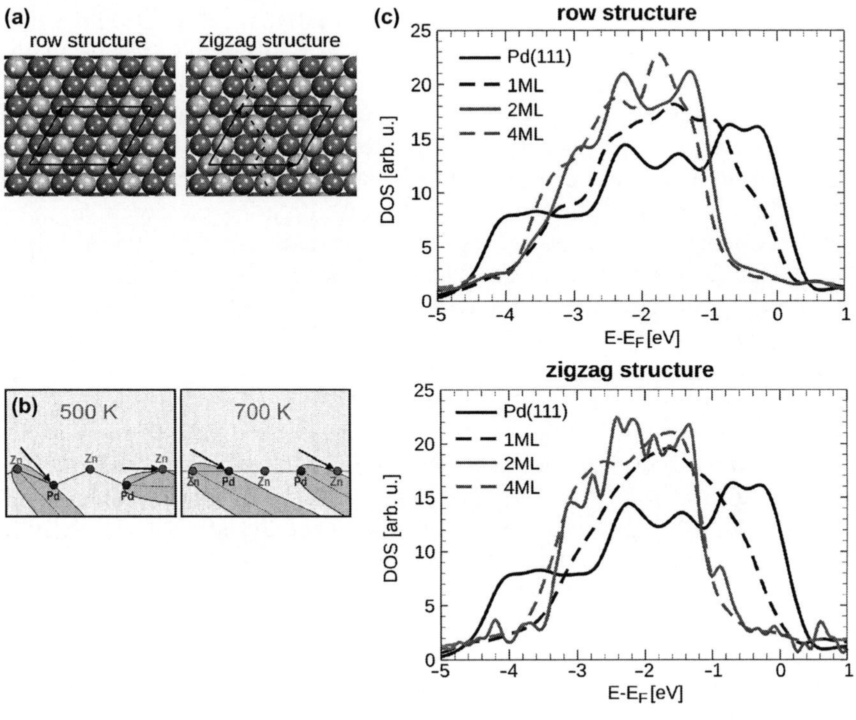

Fig. 4 (a) PdZn surface alloy on Pd(111) with bulk terminated row structure and with reconstructed zigzag structure. Pd atoms are displayed in bright colour, Zn atoms are displayed in dark. Lateral dimensions of calculated supercells are marked with solid lines, the zigzag arrangement formed by Pd atoms is marked with a dashed line. (b) Schematic view of the surface corrugation: 500 K corresponds to the corrugation of the multilayer alloy, while 700 K treatment results in a monolayer of PdZn. (c) Calculated DOS of 1, 2, and 4 layers thick PdZn surface alloys on Pd(111) in the absence of CO projected on Pd atoms located on the surface. Energies of electronic states are given with respect to the Fermi energy. (a) and (c) Reprinted with permission from ref. 60. Copyright (2012) American Chemical Society. (b) Adapted with permission from ref. 61. Copyright (2010) American Chemical Society.

and Zn (Fig. 4(a) "row structure"). This structure and composition forms after annealing at around 500 K on surface alloys grown by depositing evaporated Zn in UHV on Pd(111) single crystals.

An important aspect of PdZn/Pd(111) surface alloys is surface corrugation of the topmost layer. DFT calculations predicted a corrugation of the surface, with the Pd atoms moving inwards, while the Zn atoms relax outwards to the surface with respect to the bulk terminated geometry.[59] This Zn-out/Pd-in geometry was experimentally confirmed by Stadlmayr *et al.*,[23] who detected by low energy ion scattering (LEIS), which is sensitive to the outermost surface layer, that Zn atoms are located 0.25 Å above the Pd (Fig. 4(b)). The surface of PdZn thus consists of alternating rows of Zn and Pd atoms, with the Zn atoms lifted about 0.25 Å above the Pd atoms. However, the corrugation depends on the number of alloy layers, as found by DFT calculations[24] as well as by experimental measurements.[61] Stadlmayr *et al.*[61] studied the influence of temperature treatment on the geometric structure and corrugation in detail by combining low-energy ion scattering (LEIS), XPS, AES and Polarization Modulated Infrared

Reflection Absorption Spectroscopy (PM-IRAS). Heating Zn films deposited on the Pd(111) in the range of 125 °C to 225 °C results in the formation of multilayer PdZn alloy films with a composition close to 1:1. In the temperature regime above 275 °C, however, the subsurface layers deplete quickly in Zn due to Zn diffusion into the Pd bulk, while the composition of the surface layer changes only slightly, indicating formation of a monolayer PdZn film. Interestingly, an alteration of the corrugation was observed from a Zn-out/Pd-in surface corrugation on multilayer alloys to a Pd-out/Zn-in structure on a PdZn monolayer.[61] Since the Pd-Zn bond is stronger than the Pd-Pd or Zn-Zn bond, an increase of the Pd content in the second layer due to Zn depletion pulls the Zn atoms of the top layer inward, triggering the observed change of the corrugation from Zn-out/Pd-in toward a Zn-in/Pd-out structure. Such structures were also corroborated by DFT calculations[24] which predict – in full agreement with experiments – a Zn-up/Pd-down structure for 2 ML and Pd-up/Zn-down for 1 ML PdZn films. The effect of thick *vs.* thin alloys and the observed differences in corrugation on the electronic properties and the reactivity in MSR was studied by Rameshan *et al.*[25] and is described in section 3.2.

Very recently, Weilach *et al.*[60] presented evidence for a geometric rearrangement of the PdZn surface structure induced by CO. This is of relevance for the utilization of CO to probe available surface sites and for reactions where CO is involved either as reactant or as a product and is thus present under actual reaction conditions, *e.g.* in MSR and water-gas shift (WGS). CO adsorbs only on the Pd.[2] Bridge and hollow sites, which are typically observed on Pd metal, are absent on the intermetallic PdZn supported nanoparticles[21,32,33,62] as well as on PdZn/Pd(111) surface alloys.[25,60] By combining TPD and PM-IRAS measurements of adsorbed CO with DFT calculations, the surface structure was proposed to change from the row structure to a zig-zag-like arrangement, while maintaining the Pd:Zn 1:1 composition and keeping the electronic structure unchanged.[60] Note that the zig-zag was suggested as a model for the geometric re-arrangement, no indication for an ordered zig-zag arrangement was claimed by the authors. The reconstructed surface was calculated to be more stable in the presence of CO and capable of adsorbing CO on top of Pd atoms up to a saturation coverage of 1/2 ML. In contrast, for the bulk-like geometric arrangement of Pd and Zn in rows CO was predicted to adsorb on bridge sites with a saturation coverage of ~1/3 ML according to DFT calculations.[60] Thus, the experimentally observed on-top adsorption of CO as confirmed with PM-IRAS and the saturation coverage of 1/2 ML obtained from TPD measurements, strongly support the suggested reconstruction. The proposed different surface structures of the PdZn alloys arranged in rows and zigzags (Fig. 4(a)) eventually lead to significant differences in interatomic distances at the surface and in surface corrugation, which most likely affect the adsorption and consequently also the reaction properties of the system, despite the very similar electronic structure (Fig. 4 (c)).

Another important aspect of the surface structure and composition of PdZn concerns potential segregation of one component to the surface. Consulting the surface free energies of the Pd and Zn metals the very large difference of $0.936\,J\,m^{-2}$ for Zn *vs.* $2.043\,J\,m^{-2}$ for Pd suggests Zn

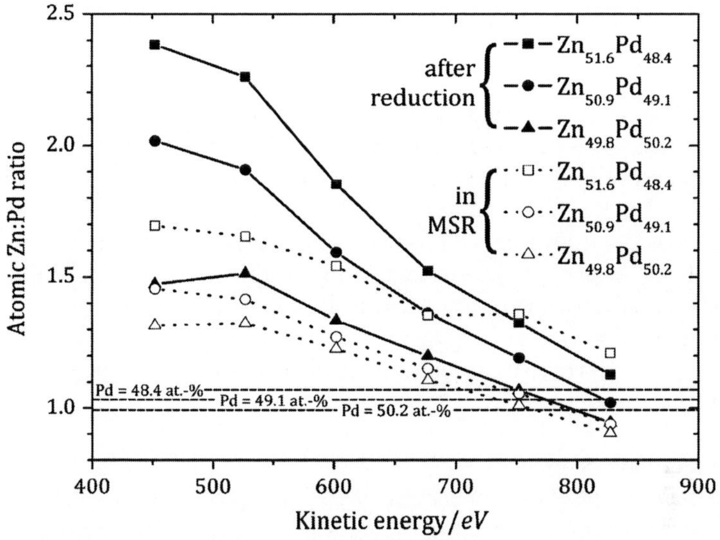

Fig. 5 Depth-depending atomic Zn/Pd ratio of $Zn_{51.6}Pd_{48.4}$, $Zn_{50.9}Pd_{49.1}$ and $Zn_{49.8}Pd_{50.2}$ after hydrogen reduction (420 °C, 0.5 mbar) and during methanol steam reforming (360 °C, 0.2 mbar, H_2O:MeOH = 2:1) derived from XPS measurements. Horizontal dashed lines represent the Pd bulk content of the investigated samples. Reprinted from ref. 64, Copyright (2012), with permission from Elsevier.

segregation to the surface.[54] However, due to the strong exothermicity of Pd-Zn bond formation, the 1:1 PdZn intermetallic is stable with a stoichiometric surface. Chen *et al.*[63] showed by DFT calculations that, due to stronger Pd-Zn bonding compared to the homonuclear bonds Pd-Pd and Zn-Zn, surface segregation of neither Zn nor Pd is favored in the alloy with the tetragonal structure. However, the situation may change when considering other compositions, and it may also be different under reaction conditions of surrounding gas atmosphere and higher temperatures where segregation could occur.

Friedrich *et al.*[64] investigated the surface composition of single phase intermetallic PdZn compounds with different bulk compositions (Pd_xZn_{100-x}, x = 46.8–59.1). *In situ* XPS spectra were recorded at mbar pressures in an as prepared state, upon reduction in H_2 and in MSR reaction atmosphere. Apparently, Zn is enriched at the surface (Fig. 5). The surface of the as prepared samples was enriched in mostly oxidized Zn due to grinding of the samples before the measurements leading to Zn surface segregation and oxidation. Upon reduction in H_2 at elevated temperatures ZnO was reduced and an increasing intensity of the signal of intermetallic Zn species was observed in the Zn Auger LMM spectra. On the Zn rich compositions Zn reduction was not complete, and oxidized Zn was still present after reduction, while on the Pd-rich samples all oxidized Zn was reduced at 420 °C. Still, the ZnPd ratio was higher than the bulk composition in all samples after the reduction treatment.

Also on high surface area supported PdZn nanoparticles oxygen induced segregation of Zn to the surface followed by oxidation to ZnO and consequently degradation of the PdZn bimetallic surface due to the strong interaction of Zn with oxygen was reported.[21,31,32,62,65] This is described in more detail in section 3.4 on the application of PdZn for oxidation reactions.

In summary, the presented results on segregation and reconstruction effects ultimately stress the significance of determination of not only the bulk but most importantly also the surface composition in order to be able to deduce assumptions concerning the nature of active sites of PdZn catalysts and to interpret the role and interplay of the PdZn intermetallic/alloy, of the PdZn-ZnO interface and of the presence of oxidized Zn species, as is currently intensively discussed for example in MSR. Obviously, even for small variations in bulk composition significant changes in the surface composition can occur and can strongly affect the catalytic properties.

2.4 Electronic properties

Alteration of the electronic structure of Pd upon alloy/IMC formation was proposed to be responsible for changes in catalytic performance in many applications. The electronic properties of PdZn were studied by XPS and UPS measurements of core level binding energies and valence band spectra. Additional information can be deduced indirectly from studies on the adsorption properties, such as determination of the strength of interaction by temperature programmed desorption (TPD) and from frequency shifts of adsorbed probe molecules like CO, as detected by FTIR spectroscopy. A change in electron density on the Pd can have a strong influence on the strength of adsorption of reactants, intermediates and products and thereby significantly alter the catalytic properties.

Upon Pd-Zn alloy formation, a large positive shift in the Pd binding energies was reported in good agreement by several authors.[18,22,25,54,65–69] The Pd $3d_{5/2}$ peak was found to shift from *ca.* 335.0 eV in metallic Pd by about 0.6–0.8 eV to higher values upon alloying (Fig. 6). That is in accordance for different types of systems from PdZn nanoparticles up to single crystal based model systems, which all show a similar shift.

Rodriguez[54] studied in detail the effect of alloy formation on the electronic properties of PdZn films supported on Ru(001) by valence and core level photoemission spectroscopy and TPD. The formation of Pd-Zn bonds increases the binding energy of the core and valence levels of Pd and decreases the binding energy of the core and valence levels of Zn due to a Pd(4d) → Zn(4p) charge transfer and a Pd(4d) → Pd(5s,5p) rehybridization. As a consequence the Pd-CO Pd(4d)-CO(2π) interaction was found to be weakened resulting in an up to 50–70 kJ/mol weaker Pd-CO bond on Zn rich alloy surfaces compared to Pd. A charge transfer from Zn to Pd was also predicted by DFT calculations.[70] Analysis of the chemical bonding revealed considerable charges of $Pd^{-0.4}Zn^{+0.4}$ for bulk PdZn.[42]

The weaker bonding of adsorbed CO on PdZn as observed by Rodriguez was confirmed by further CO-TPD studies *e.g.* in.[43,50,52,60] Jeroro *et al.*[52] studied the effect of Zn in PdZn model systems on the CO adsorption. CO did not adsorb on Zn-containing sites above −170 °C (ensemble effect of Zn), and additionally Zn also had a strong electronic effect destabilizing the bonding of CO on nearby Pd sites. The heat of adsorption of CO was

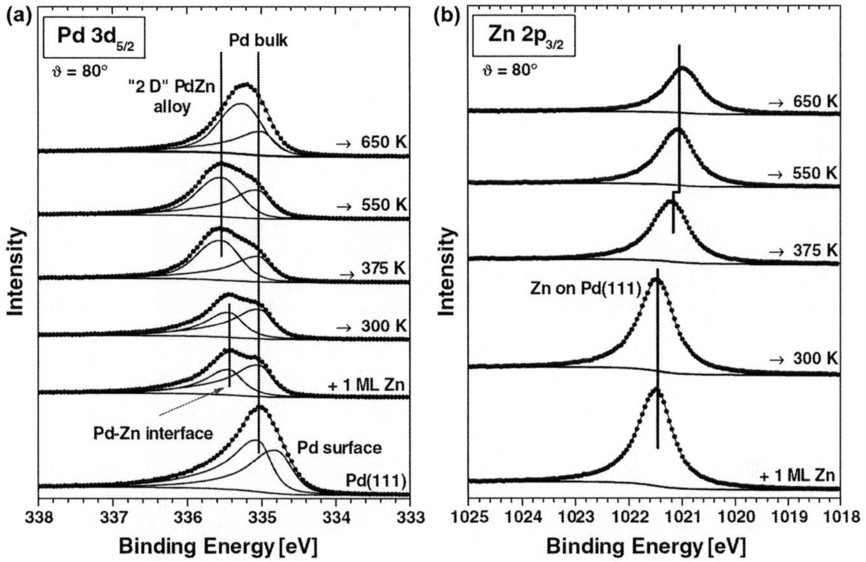

Fig. 6 (a) Pd $3d_{5/2}$ and (b) Zn $2p_{3/2}$ XP spectra of 1ML Zn on Pd(111), after deposition at 105 K and after annealing (for 30 s) at various temperatures up to 650 K. For comparison, also the spectrum of Pd(111) (in (a)) and the fitting results are shown. The spectra are measured at grazing emission ($\vartheta = 80°$) to enhance the surface sensitivity. Reprinted from ref. 22, Copyright (2006), with permission from Elsevier.

lowered by 16 kJ/mol on threefold Pd sites with Zn next nearest neighbors compared to threefold sites on the clean Pd(111) surface.[52] The CO desorption temperature was thus lowered to 25 °C. CO TPD spectra obtained upon dosing 0.5 L CO on Zn/Pd(111) are shown in Fig. 7. Weilach *et al.*[60] reported two main desorption features on multilayer 1:1 surface alloys after 0.5 L CO dosing: a shoulder at around −100 °C and the peak maximum at −60 °C.

A modification in electronic properties can also be deduced from the shift in the frequency of the stretch vibration of adsorbed CO, as observed by FTIR spectroscopy. CO was shown not to adsorb on sites including Zn-atoms at temperatures above around −120 to −60 °C,[2,25,61,68] which means that the types of adsorption sites available at the surface are on top of Pd and at Pd-Pd bridge sites.[60] Very similar frequencies of CO adsorbed on top of Pd in PdZn were observed at around 2070 cm^{-1} for ZnO supported PdZn nanoparticles[21,32,62] well as for single crystal based surface alloys[25,60] (Fig. 8), which corroborates the good comparability of model and real catalysts for the PdZn system that can also be deduced from the XPS results. The frequency of on-top CO on PdZn is shifted to lower wavenumber by about 20–30 cm^{-1} compared to purely metallic Pd nanoparticles. The electron transfer from the Zn to Pd induces this red-shift by the increased backdonation from Pd d-bands to CO, which increases the on-top CO adsorption energy and concomitantly weakens the internal C-O bond.[71] In addition, bridge and hollow bonded CO are absent on PdZn surfaces, again for supported nanoparticles as well as model systems. The destabilisation of bridging adsorption sites may

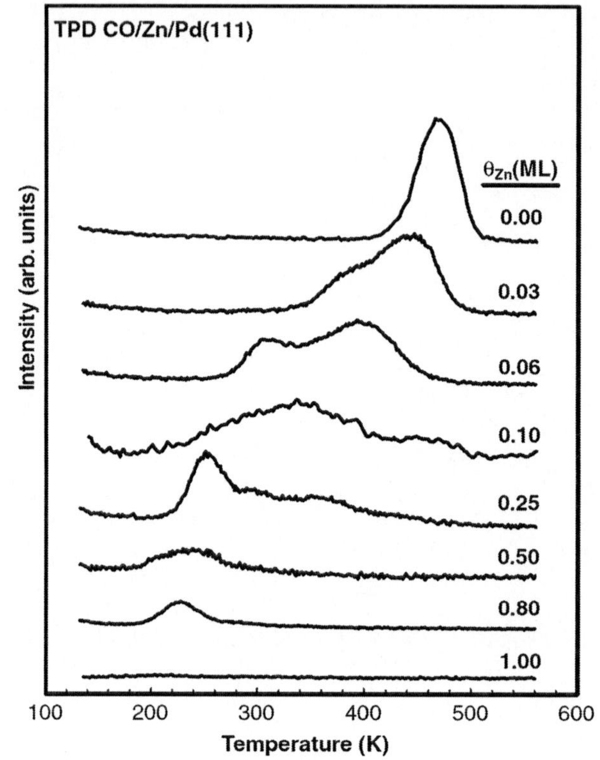

Fig. 7 CO TPD from Zn/Pd(111) that had been annealed at 550 K as a function of the Zn coverage. A 0.5 L CO dose was used in each TPD run. Reprinted from ref. 52, Copyright (2007), with permission from Elsevier.

originate from the increased Pd-Pd interatomic distances in the respective intermetallic compound as compared to metallic Pd. A recently proposed alternative explanation is a CO-induced surface reconstruction. A geometric rearrangement to a "zig-zag"-like structure was suggested, which leads to the absence of multiply bonded CO. This was described in section 2.3.

Chen *et al.*[59] performed a computational study on the PdZn system and showed that the local electronic structure of Pd in PdZn alloy is different from Pd but resembles that of Cu, with respect to the density of states at the Fermi level, band width and energetic position. The calculated valence electron density of states (DOS) is shown in Fig. 9. Bayer[22] measured valence band spectra of PdZn/Pd(111) by UPS. The valence band spectrum of PdZn showed a low density of states at the Fermi level and the measured spectrum closely resembles the spectrum of Cu(111), in good agreement with the calculated density of states for the PdZn alloy of 1:1 stoichiometry.[59] The similarity between the UPS spectra for the PdZn alloy and for Cu(111) is proposed as explanation for the similar reactivity of both systems in methanol steam reforming (see section 3.2). These findings on electronic properties are in agreement with Tsai. Also Tsai *et al.*[18] measured valence band spectra of unsupported bulk alloys and performed energy band calculations on PdZn. These results were combined with catalytic

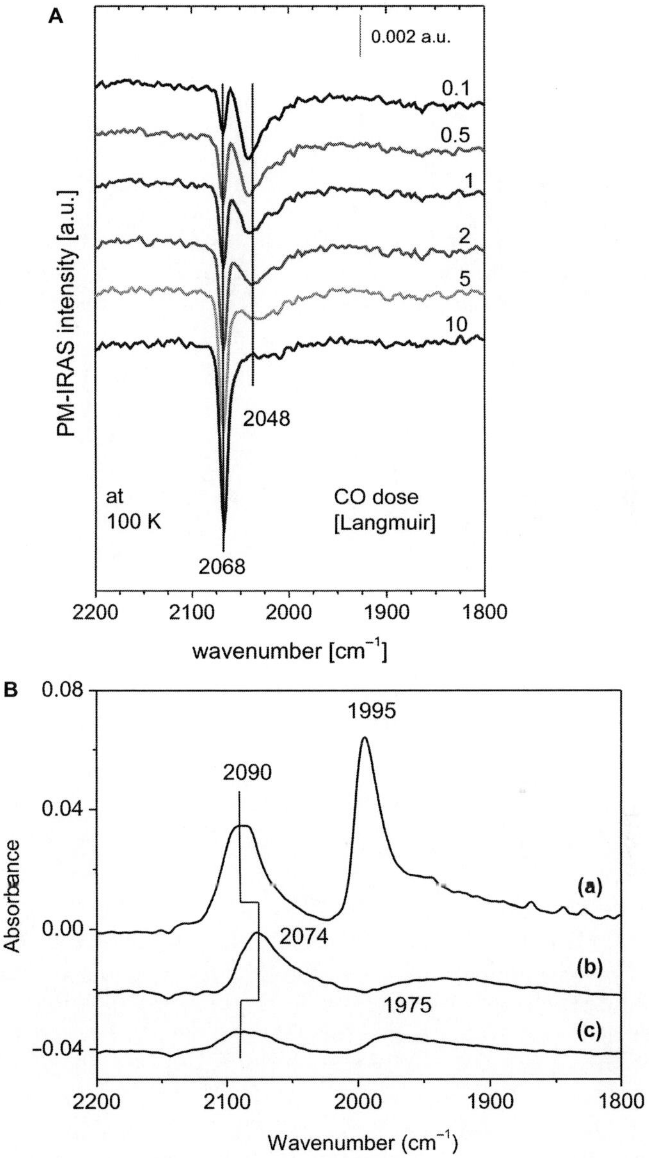

Fig. 8 (A) Coverage dependent IR spectra of CO adsorption on >4 ML PdZn/Pd(111) at 100 K. Reprinted with permission from ref. 60. Copyright (2012) American Chemical Society. (B) FTIR spectra of CO adsorption (303 K, 5 mbar CO) on Pd/ZnO after (a) reduction at 303 K; (b) after reduction at 623 K; and (c) after oxidation at 573 K and re-reduction at 303 following (b). Difference spectra are shown, obtained by subtraction of the corresponding spectra before CO exposure. Reprinted with permission from ref. 21. Copyright (2011) American Chemical Society.

measurements of MSR. They found, in excellent agreement with Bayer and the work of Rösch and Neyman, that PdZn exhibits a similar valence electron density as Cu and is catalytically equivalent to Cu in MSR. Although PtZn possesses the same crystal structure as PdZn, the catalytic performance was worse than PdZn presumably due to the different valence

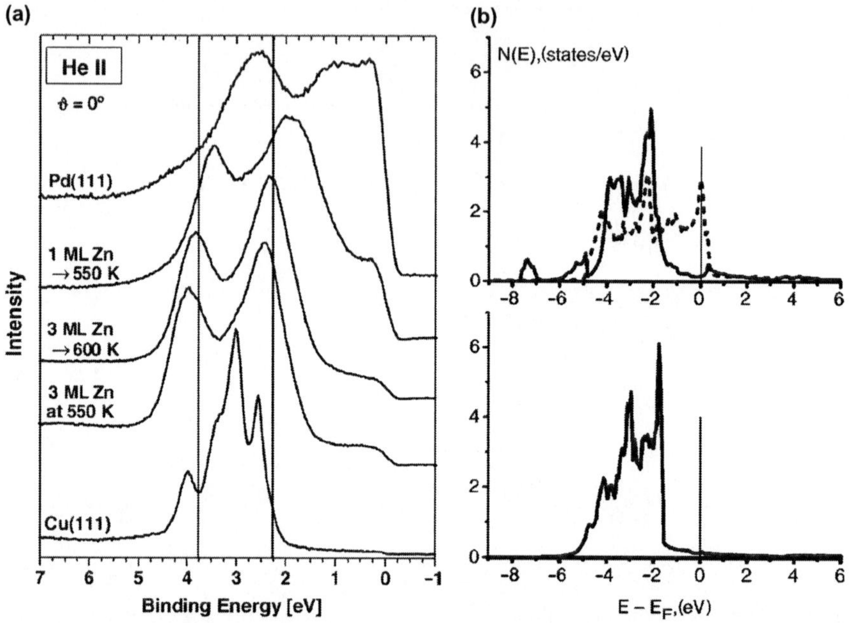

Fig. 9 (a) Comparison of the "Pd 4d" region of the He II UP spectra of differently prepared PdZn alloys on Pd(111). The spectra of Cu(111) and Pd(111) are shown for comparison. The spectra were recorded with a pass energy of 40 eV, only for Cu(111) a pass energy of 20 eV was used. Reprinted from ref. 22, Copyright (2006), with permission from Elsevier. (b) Local density of states: upper panel: Pd 4d band of bulk Pd (dashed) and PdZn (solid); lower panel: Cu 3d-band of bulk Cu. Reprinted with permission from ref. 59. Copyright (2003) by the American Physical Society.

band structures. From this, Tsai and co-workers concluded that the electronic structure governs the selectivity at least for MSR.

Also Rameshan characterized the electronic structure by measuring core level and valence band spectra.[25,68] They found a striking difference in the valence band structure on monolayer *vs.* multilayer PdZn/Pd(111) surface alloys, which exhibit valence electron DOS similar to Pd and Cu, respectively. The strong implications on the selectivity in MSR will be discussed in section 3.2.

In general, alteration of the electronic properties upon alloying can have crucial implications on the reactivity by modifying reaction barriers and different stabilization of intermediates. This will be shown and discussed in more detail in section 3.

3 Catalytic properties in selected applications

3.1 Selective hydrogenation reactions

Pd catalysts have been extensively used and studied for hydrogenation reactions for conversion of organic molecules, but also for CO and CO_2 hydrogenation, *e.g.* to methanol. Using ZnO as support was found to strongly alter the reactivity of the Pd in hydrogenation reactions.[11,19,31,72,73]

The origin of this strong influence is the generally agreed PdZn alloy formation that occurs upon reduction and can lead to the modification of catalytic properties due to the well-known ensemble and ligand effect. As described in section 2.4 a strong modification of the electronic properties of the Pd upon alloy/intermetallic formation and charge transfer from neighboring Zn atoms was clearly observed (ligand effect), which can influence the strength of adsorption of reactants, products and intermediates. On the other hand, the dilution of the Pd atoms at the surface (ensemble effect) and thus the lower concentration of hydrogen at and near the surface can also considerably affect the selectivity in selective hydrogenation reactions. Depending on the reaction system either the ligand or the ensemble effect or both may be prevailing. In addition, the absence of hydride formation in PdZn reduces the hydrogen supply which is beneficial for achieving high selectivity.

Sarkany *et al.*[72] applied Pd/ZnO catalysts in the selective hydrogenation of small amounts of 1,3-butadiene in butene. Reduction at $>150\,^{\circ}C$ caused a pronounced decrease in catalytic activity and an increase in selectivity for the hydrogenation of butadiene to olefins due to formation of the PdZn intermetallic phase, as detected by XRD. Based on XPS data, the authors suggested that the Pd crystallites were "decorated" with reduced Zn metal which significantly decreased the number of available adsorption sites, as determined by hydrogen and CO chemisorption. Although the overall activity decreased after high-temperature reduction, the turnover frequency (TOF) values remained virtually unchanged, as well as the activation energy and reaction orders. Iwasa *et al.*[11] found 99% selectivity to ethylamine in the hydrogenation of acetonitrile over alloyed PdZn/ZnO compared to $<5\%$ for Pd black. While Pd/Al_2O_3 and Pd/SiO_2 produce only little crotyl alcohol in the hydrogenation of crotonaldehyde, similar amounts of crotyl alcohol and butyraldehyde were obtained over PdZn.[74] Similar explanations are suggested as origin of the change in the catalytic properties in butadiene and acetonitrile hydrogenation on PdZn/ZnO. The pronounced decrease in the hydrogen adsorption would be a major cause for the lower activity, and the Zn-modified surface sites would contribute to a weaker interaction with the reactants and/or intermediates and thereby affect activity and selectivity.

Another important application where Pd-based intermetallic catalysts have successfully been applied[19,75–78] is the hydrogenation of alkynes to alkenes in an excess of alkene. Acetylene selective hydrogenation to ethylene is of industrial importance for cleaning up the ethylene feed from the steam cracking unit. Acetylene is problematic even in small amounts acting as a poison for the polymerization catalyst in polyethylene production. Typically about 1% acetylene has to be removed to the low ppm level without hydrogenation of the desired ethylene and avoiding total hydrogenation to undesired ethane. Pd is active but not very selective in this reaction. Besides, it deactivates due to coking. To improve selectivity and long-term stability, the Pd catalyst is selectively poisoned by addition of Pb (Lindlar catalyst)[79] or co-feeding CO[80] in order to reduce the active ensembles size by surface blocking. Furthermore, the lower supply of hydrogen due to the absence of neighbouring sites is beneficial for the selectivity and stability by suppressing full hydrogenation[81] and oligomerization,[82] respectively. Similar effects explain the superior performance of the industrially employed Pd-Ag alloys.[83] Again, surface Pd atoms are diluted, which disfavours di-σ-adsorbed ethylene on neighbouring sites, identified as a likely precursor to carbonaceous deposits.[84–85] In addition, the concentration and supply of hydrogen is strongly lowered due to absence of H on neighbouring sites and of bulk dissolved H (no hydride formation), which influences selectivity and stability. Studt *et al.*[86] performed DFT calculations and found that Ag addition lowers the energy barrier of desorption of the intermediately formed ethylene. However, a major disadvantage of the PdAg catalyst is occurrence of segregation. Comprehensive studies covering different types of materials ranging from model systems and metallurgically prepared IMCs up to supported nanoparticles have been performed on related Pd-Ga IMCs.[75–78] PdZn has also shown promising catalytic properties for this highly relevant reaction. Tew *et al.*[19] studied pentyne hydrogenation over PdZn/ZnO and compared the catalytic performance to Pd/SiO_2. By *in situ* high-resolution XRD and X-ray absorption spectroscopy (XAS) they determined complete alloy formation after reduction at 300 °C and above, with the structure of the catalyst being close to the 1:1 PdZn $L1_0$ structure. By *in situ* XAS, the presence of the PdZn alloy under reaction conditions of pentyne hydrogenation was confirmed. The kinetic data clearly showed that there is a pronounced effect of the extent of alloy formation on the selectivity. Total hydrogenation to pentane was completely absent on the PdZn catalyst, which was attributed to the electronic structure of the PdZn alloy and the site isolation. The selectivity is strongly influenced by the strength of adsorption of alkyne *vs.* alkene. The heat of adsorption is higher for alkynes than for alkenes on Pd.[87–88] On Cu, the heat of adsorption of alkenes is lower than on Pd according to DFT calculations,[86] and total hydrogenation of butyne to butane was absent on Cu/SiO_2.[89] Therefore, it was proposed that the changes in electronic properties from Pd to PdZn affect alkene adsorption more strongly resulting in faster desorption.[19] The similar electronic properties of PdZn to Cu, as described in detail in section 2.4, is a likely explanation for the absence of total hydrogenation due to faster desorption of the alkenes, concomitant with the site isolation in PdZn.

PdZn catalysts were also applied for the hydrogenation of esters. Wehner *et al.*[73] studied the hydrogenation of methyl acetate to methanol and ethanol in gas phase at low pressures over Pd/ZnO and found good activity and selectivity, in contrast to Pd on other supports, due to PdZn alloy formation. The major side reaction was secondary transesterification on the $Zn^{\delta+}$ species, which could be significantly reduced and nearly suppressed by adding small amounts of water to the reaction feed. The water was assumed to lead to a controlled *in situ* re-oxidation, thereby lowering acidity by reducing the amount of oxygen vacancies at the ZnO surface. However, too much water had a negative effect on the activity, presumably by disrupting the Pd-Zn interaction. Also Green *et al.*[31] applied $Pd\text{-}Zn/SiO_2$ for methyl acetate hydrogenation. They found a strongly enhanced activity on bimetallic Pd-Zn, as compared to monometallic Pd, with an optimum Pd:Zn ratio of 2:1. At higher Zn:Pd atomic ratios a decrease in selectivity occurred due to transesterification on the excess Zn^{2+} on SiO_2.

PdZn catalysts were tested in methanol synthesis by CO and/or CO_2 hydrogenation. Promotion by alkali metals was found to increase the rate of methanol formation even at low quantities,[90] which was attributed to a modification of the electronic properties. Since upon PdZn alloy formation strong alterations of electronic properties have been found in addition to changes in geometric properties, strong changes in catalytic properties can be expected for the PdZn system. Especially the similarity in the electronic structure to the methanol synthesis catalyst Cu makes PdZn a promising candidate.

Fujitani *et al.*[91] investigated Pd-based catalysts supported on a range of different oxides. When Pd was supported on ZnO and Ga_2O_3 strongly enhanced methanol formation rates from CO_2 hydrogenation were detected. They reported TOF values for the CO_2 hydrogenation to methanol which were nearly 5 times larger for Pd/ZnO and even 7 times larger for Pd/Ga_2O_3 than on Cu/ZnO and Cu/ZnO/Al_2O_3 catalysts under the same conditions. Based on *in situ* FTIR experiments a different reaction mechanism was proposed depending on the support oxide.

Also Ota *et al.*[37] applied PdZn and P_2Ga derived from a hydrotalcite-like precursor material for methanol synthesis (Fig. 10). Strongly improved activity compared to Pd on hydrotalcite-derived $MgAlO_x$ was achieved on PdZn and on P_2Ga catalysts, which were 100 and 200 times, respectively, more active for methanol synthesis than the monometallic Pd catalyst. However, in contrast to Fujitani, inferior activity of PdZn compared to Cu was observed. PdZnAl showed a methanol selectivity of 60% as compared to 82% on the CuZnAl sample prepared by the same procedure. The methanol TOF of CuZnAl was about 10 times higher than that of the PdZn.

Arana *et al.*[34] studied CO and CO_2 hydrogenation on silica-supported Pd-Zn catalysts. While a Zn-rich catalyst (Pd:Zn ratio 2:3) with alloyed PdZn at the surface showed low hydrogenation activity and produced mostly higher hydrocarbons (therefore acting as a Fischer Tropsch catalyst), a Pd rich catalyst (Pd:Zn ratio 3:1) with mostly metallic Pd at the surface showed the highest activity for methanol and higher oxygenates production. This work is obviously in contradiction to the studies described above.

Iwasa *et al.*[92] tested Pd/ZnO catalysts in methanol synthesis by CO_2 hydrogenation. In agreement with Fujitani, the turnover frequency and selectivity of methanol formation were markedly larger than on the Cu/ZnO control catalyst. No activity was observed, however, for CO hydrogenation on Pd/ZnO.

As described in the next section, most likely similar effects are relevant in methanol synthesis from CO_2 and in MSR, the reverse reaction, which explain the different reactivity of Pd in intimate contact with Zn, *i.e.* the electronic properties similar to Cu and different species present at the surface under reaction conditions.

3.2 Methanol steam reforming (MSR)

PdZn-catalysts have been most widely used as catalysts in MSR. In the recent years methanol has emerged as a promising compound to chemically store hydrogen. It has numerous advantages such as it is a liquid under ambient conditions, thus easy to store, transport and handle, it has a high H:C ratio and requires only mild reaction conditions. Since the reaction does not require cleavage of C–C bonds, it is less prone to coking, in

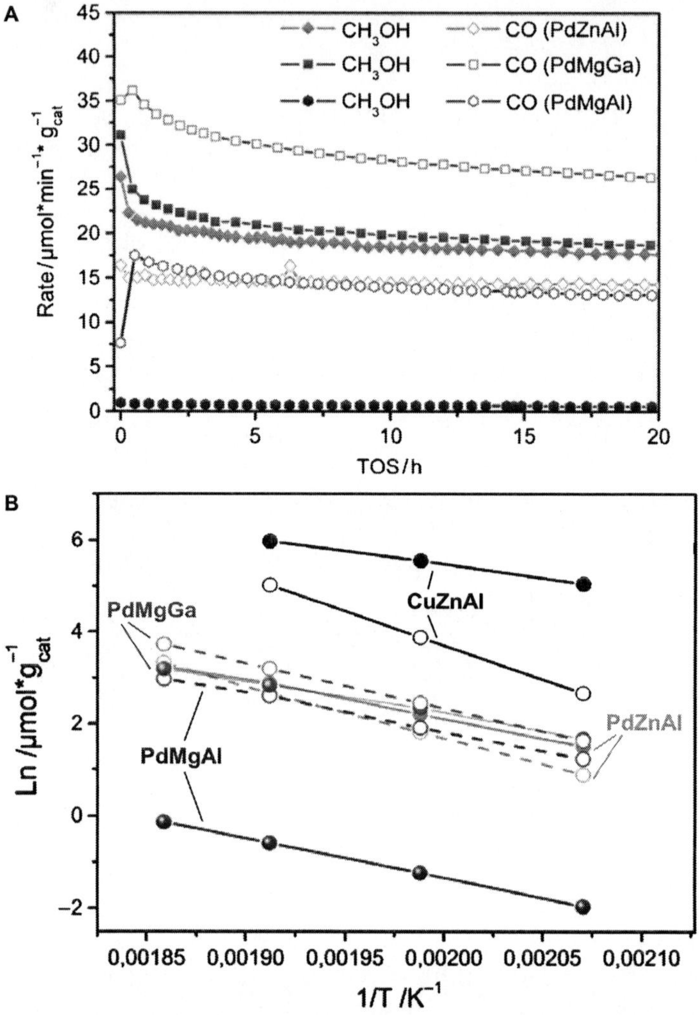

Fig. 10 (A) Rate of CO_2 hydrogenation over PdZnAl, PdMgGa and PdMgAl at 523 K (30 bar, 400 mg catalyst, 100 ml/min CO_2/H_2). (B) Arrhenius plot for methanol synthesis from CO_2 (closed symbols: MeOH, open symbol: CO). Reprinted from ref. 37, Copyright (2012), with permission from Elsevier.

comparison, *e.g.* to ethanol as feedstock, and reaction temperatures required are low (typically 200–300 °C). Therefore, it is especially interesting for decentralized and mobile applications.[93] A further advantage is that methanol is already produced in large quantities worldwide, as it is one of the major base chemicals, and the process of industrial methanol synthesis is well established. The syngas needed for methanol synthesis can be produced from fossil, but also from renewable sources and from CO_2 hydrogenation.

$$CH_3OH + H_2O \leftrightarrow CO_2 + 3\,H_2 \quad \text{methanol steam reforming (MSR)}$$

$$CH_3OH \leftrightarrow CO + 2\,H_2 \quad \text{methanol decomposition (MDC)}$$

Similar catalysts are active and selective in MSR like in methanol synthesis. Mostly, Cu-based catalysts have been used due to their superior performance.[16,93] Analogies to methanol synthesis *via* CO_2 hydrogenation are obvious, since MSR is the reverse reaction, thus likely involving similar or the same reaction intermediates and active sites. Therefore, PdZn has appeared as a promising candidate, potentially able to substitute Cu, which has the drawback of low long-term stability due to sintering and besides, Cu is pyrophoric once it is reduced.[15–17] Pd/ZnO has also shown interesting catalytic properties for steam reforming of ethanol.[94]

Iwasa and coworkers were the first to report that Pd/ZnO has exceptionally high activity and selectivity for MSR[8–10,12,13,67,95] and connected the catalytic properties to PdZn alloy formation upon reduction at $>300\,°C$, as observed by *ex situ* XRD and XPS. The conversion and MSR selectivity *vs.* reduction temperature are shown in Fig. 11.[67] They tested a series of Pd catalysts supported on a range of different oxides (Table 1).[67] Typically, Pd catalyzes methanol decomposition to CO and H_2 (MDC). However, when supported on ZnO, Ga_2O_3 and In_2O_3 a completely different selectivity resembling that on Cu was observed. In all the three materials, formation of an alloy with reduced Zn, Ga and In was observed upon reduction. When comparing the catalytic performance in MSR for Pd on different supports (as shown in Table 1) and over supported group 8–10 metal catalysts, such as Pd, Pt, Co, Ni, Rh, Ir and Ru catalysts,[12,67,95] Pd/ZnO appeared as the most promising candidate for replacing Cu in MSR with a selectivity of more than 99% to the desired steam reforming reaction. A completely different reactivity was also observed in the absence of water. While Pd still catalyzes exclusively methanol decomposition yielding CO and H_2, methyl formate was the main product over PdZn.[67]

While it has often been considered that PdZn alloy formation is responsible for the excellent MSR selectivity of Pd/ZnO, a direct evidence for IMC formation under MSR conditions and the effect on the selectivity was recently presented in.[21] By *in situ* Quick-EXAFS, the formation of the PdZn intermetallic phase upon exposure of Pd/ZnO (without previous reduction) to the methanol steam reforming reaction atmosphere was monitored in real time. Simultaneously with the structural and electronic changes occurring with time, the reaction selectivity changed from CO/H_2 to CO_2/H_2 formation (Fig. 12).

Iwasa and coworkers[13,67] proposed a different reaction pathway on PdZn compared to Pd to explain the different selectivity (Fig. 13). While methanol is rapidly dehydrogenated to CO and H_2 *via* formaldehyde (HCHO) on Pd, on PdZn the intermediate HCHO reacts with water to CO_2 and H_2, most likely *via* HCOOH, in a similar pathway to that probably proceeding over Cu catalysts. Bera *et al.*[49] studied the adsorption and reaction of CH_3OH on Pd and PdZn films and particles supported on a ZnO(0001) single crystal using temperature-programmed desorption (TPD). They observed complete dehydrogenation of methanol to CO and H_2 on Pd and PdZn for both films and particles. Upon alloy formation CO adsorption was weakened resulting in lower desorption temperature, as observed by TPD. They identified a low temperature oxidation path of CO to CO_2 with oxygen provided by the ZnO(0001). However, due to the high quantity of formed CO_2, which is

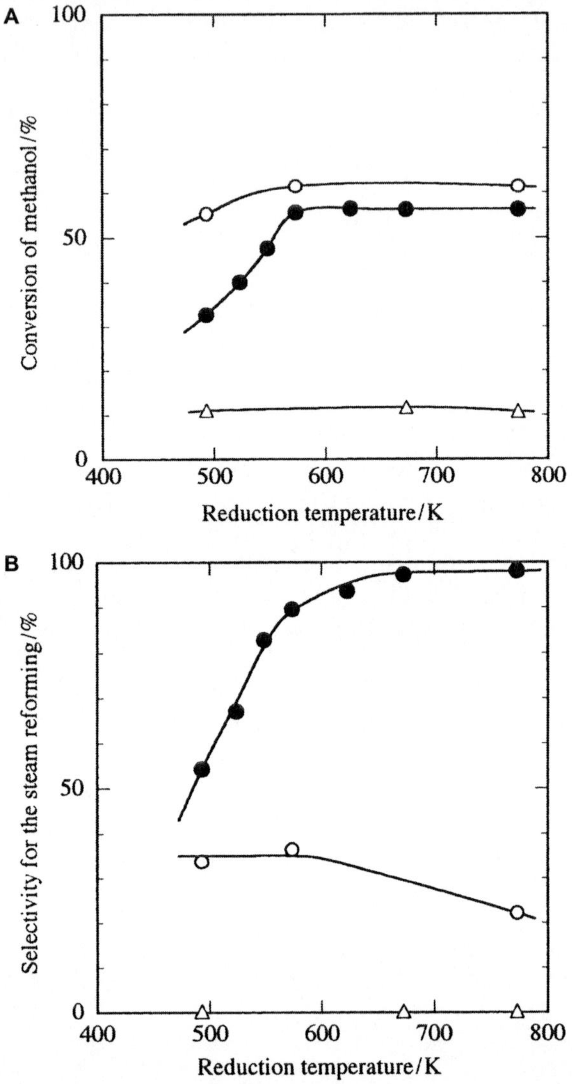

Fig. 11 (A) Conversion of methanol and (B) selectivity for steam reforming over Pd/ZnO, Pd/SiO_2 and Pd/ZrO_2 previously reduced at various temperatures. (●) Pd/ZnO, (Δ) Pd/SiO_2, (O) Pd/ZrO_2. Reproduced from ref. 67 with kind permission from Springer Science and Business Media.

higher than the thermodynamic equilibrium concentration that can be reached by WGS, a reaction sequence of methanol decomposition to CO and H_2 followed by water-gas shift reaction of CO with water to CO_2 and H_2 was mostly excluded as prevailing reaction pathway in MSR over PdZn.[13]

From surface science it is known that HCHO adsorbs in a different geometry on Cu as compared to Pd and other group 8–10 metals. Formaldehyde adsorbs preferentially in an on top η^1 (O) configuration on Cu,[96–97] whereas it adsorbs in a bridging η^2 (C, O) geometry on Pd, Pt and Ni.[98–104] The latter species undergoes rapid dehydrogenation to CO and H_2 already below −70 °C.[104] Iwasa and coworkers[13,67] suggested an η^1

Table 1 Steam reforming of methanol over various supported Pd catalysts. Reaction conditions: reaction temperature 220 °C, inlet partial pressures of methanol 10.1 kPa, inlet partial pressures of water 10.1 kPa. Reproduced from ref. 67 with kind permission from Springer Science and Business Media.

		Selectivity (%)				
Catalyst[a]	Conversion (%)	CO_2	CO	DME	Dispersion (%)	TOF[b] (s^{-1})
Pd/ZnO	54.2	99.2	0.8	0	2.2	0.497
Pd/In_2O_3[c]	28.3	95.5	4.5	0	2.1	0.291
Pd/Ga_2O_3	21.2	94.6	5.4	0	2.6	0.177
Pd/SiO_2	15.7	0	100	0	9.0	0.021
Pd/Al_2O_3	58.9	0	69.6	30.4	14.0	0.036
Pd/MgO	41.0	6.6	93.4	0	10.4	0.053
Pd/ZrO_2	64.3	18.4	81.6	0	6.6	0.139
Pd/CeO_2	62.4	22.7	77.3	0	36.2	0.025
Pd/A.C.	2.3	0	100	0	–	–
Pd black	10.9	0	100	0	2.1	0.0071
Cu/ZnO	19.3	100	0	0	1.0	0.300

[a] Pd loading 10 wt%.
[b] Turnover frequency for H_2 formation.
[c] Reduction temperature is 250 °C.

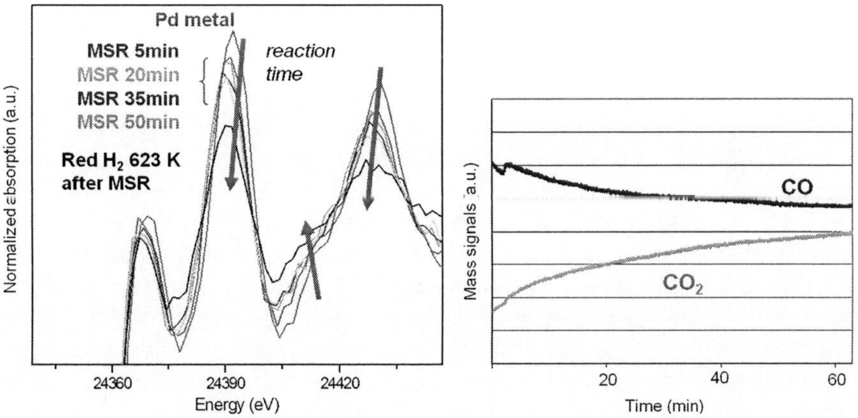

Fig. 12 Pd K edge XANES spectra and MS traces during MSR over Pd/ZnO at 623 K without pre-reduction of the catalyst (methanol : water = 1:1). Adapted with permission from ref. 21. Copyright (2011) American Chemical Society.

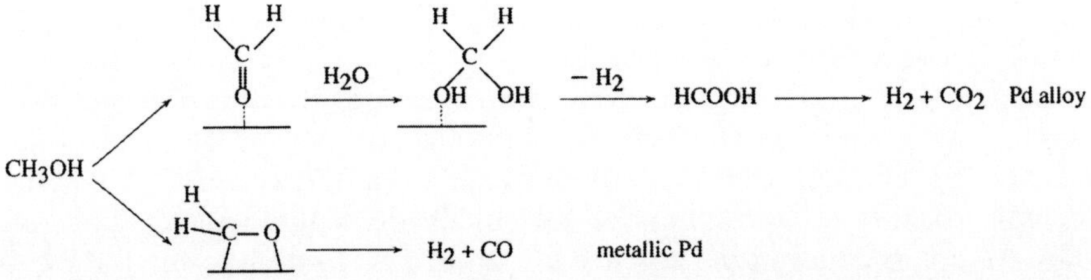

Fig. 13 Reaction mechanisms for the steam reforming of methanol over Pd alloys and metallic Pd. Reproduced from ref. 67 with kind permission from Springer Science and Business Media.

adsorption of HCHO on PdZn similar to Cu, which is less prone to C–H (and also C–O) bond cleavage than the bridging η^2 species on Pd. The proposed different adsorption geometry of intermediate formaldehyde on PdZn would thus be a plausible cause of the different reactivity on Pd and PdZn due to the differences in stabilization and further conversion of HCHO. However, according to DFT calculations carried out by Chen *et al.*[87] and Lim *et al.*[105] the η^2 (C, O) geometry is the most stable adsorption configuration also on PdZn with the C in formaldehyde bonded to Pd and O interacting with Zn. Experimental proof of η^1 adsorption of HCHO on PdZn high surface area catalysts under actual reaction conditions is lacking up to now. According to the theoretical calculations of Neyman, Rösch and coworkers[105–107] as well as Huang *et al.*,[108] instead of altering the bonding configuration of intermediates, the role of Zn may be the increase of the reaction barriers for C–H bond cleavage in adsorbed methoxy and formaldehyde. This would lead to a stabilization of intermediate formaldehyde at the surface and thus to a higher probability of further reaction with hydroxyls to formate, which can then decompose to CO_2 and H_2. The fast dehydrogenation of formaldehyde on Pd is considerably slowed down by the increased reaction barriers. The activation energies for C–H bond breaking in methoxy and formaldehyde were calculated to be only 33 and 38 kJ/mol for Pd(111) but 113 and 64 kJ/mol on PdZn(111), respectively.[105,107] In good agreement with these conclusions based on DFT, Jeroro and Vohs[51] came to similar conclusions on the role of the Zn in PdZn surface alloys. They investigated the surface species formed upon dosing HCHO and CH_3OH on different amounts of Zn/Pd(111) by high-resolution electron energy loss spectroscopy (HREELS) and TPD. Most of the formaldehyde was adsorbed in an η^2 configuration characterized by a ν_{CO} vibration at 1440–1460 cm^{-1}, while η^1 adsorbed HCHO, although observed in low amounts at low temperatures by the vibrational ν_{CO} band at 1630 cm^{-1}, was excluded as reaction intermediate. Already small amounts of Zn (as low as 0.03 ML) had a significant effect on the stability of surface species and especially on the dehydrogenation reactions of formaldehyde and methoxy, which were strongly decreased. This indicates an electronic effect playing the major role. Similar results were reported for ethanol and acetaldehyde.[109]

In this line, Tsai *et al.*[18] suggested that the reactivity of PdZn is governed by its strongly modified electronic structure, being similar to Cu, which is described in more detail in section 2.4. According to DFT calculations, the alteration of the electronic structure considerably modifies the reaction barriers for methanol decomposition, as shown in Fig. 14.[70] The activation barriers for C–H bond cleavage down to CO were found to be high on PdZn and Cu, and notably different on Pd.

Lin *et al.*[110] evaluated an alternative suggested reaction pathway proceeding *via* methyl fomate (MFO) as intermediate. Such a mechanism has also been proposed over Cu catalysts by several authors (*e.g.*[111–112]). MFO can be formed in a reaction of formaldehyde with methoxy and was observed as product in the absence of steam.[13] According to the DFT calculations of Lin[110] the activation barrier of the formaldehyde reaction with methoxy to MFO was considerably higher (by 32 kJ/mol) than that of the

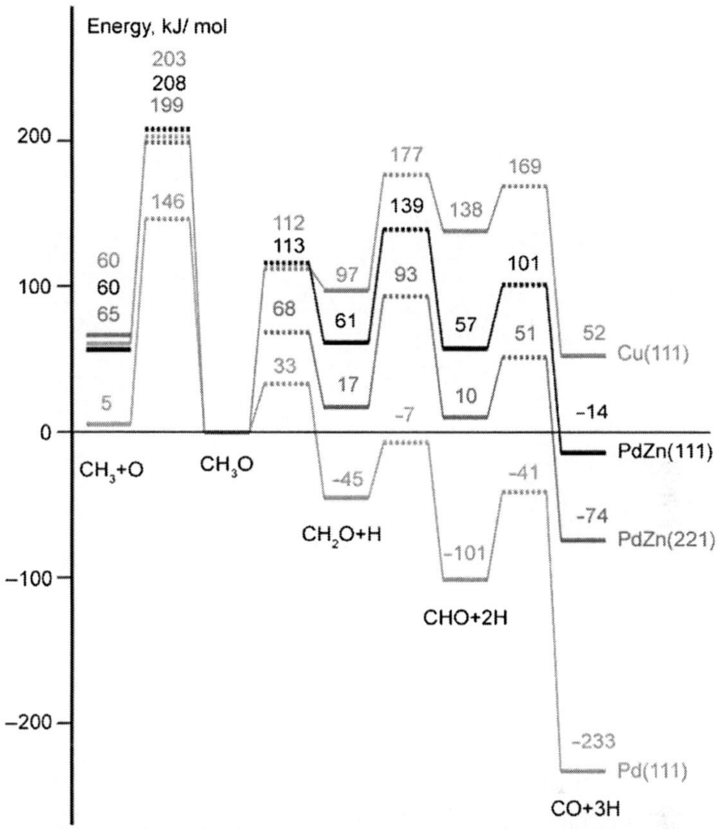

Fig. 14 Summary of calculated energies (without zero-point energy corrections) of methoxide decomposition on the planar surface PdZn(111) and the stepped surface PdZn(221) compared to the corresponding results for Cu(111) and Pd(111). Adsorbed CH_3O on a clean substrate is chosen as energy reference. Dashed lines mark transition states. For each configuration, the energy values are listed in the order of the substrates given at the right edge, *e.g.* the top-most values correspond to Cu(111) and the values at the bottom to Pd(111). Reproduced from ref. 70 with permission of the PCCP Owner Societies.

reaction with hydroxyls. Therefore, this pathway would only be relevant in the absence or with low amounts of water and thus OH. Recent mechanistic studies on the related Pd_2Ga/Ga_2O_3 system came to a similar conclusion excluding MFO as an intermediate in the selective MSR pathway to H_2 and CO_2.[113]

By utilizing *in situ* XPS at mbar pressures of methanol/water Rameshan *et al.*[25,68] showed that the Cu-like electronic structure is also present under conditions of methanol steam reforming. In addition, they compared the properties and reactivity of monolayer and multilayer PdZn/Pd(111) surface alloys.[25,68] These surface alloys show considerable differences in the surface structure (different corrugation), as described in section 2.3, but also in the reaction selectivity (CO *vs.* CO_2). While 1 ML PdZn is completely unselective to the desired MSR reaction and yields only CO and H_2, multilayer alloys are highly selective. This was connected with changes in the electron density of states by comparing core-level and valence band spectra (Fig. 15). Although the composition of the topmost layer is the same for

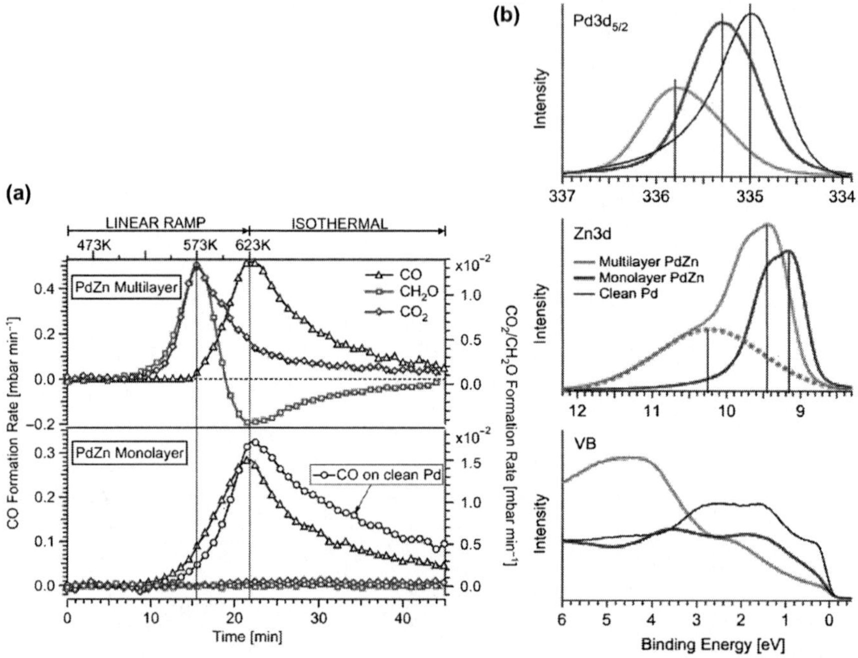

Fig. 15 (a) Temperature-programmed methanol steam reforming on the multilayer PdZn 1:1 alloy on Pd foil (upper panel) versus MSR reaction on "Zn-lean" monolayer PdZn surface and MSR reaction on clean Pd foil (lower panel). Reaction conditions: 12 mbar methanol, 24 mbar water, 977 mbar He; linear temperature ramp (9 °C/min) up to 350 °C, and subsequent isothermal reaction for 24 min. The decrease of the CO formation rate in the isothermal region is caused by progressive methanol consumption and carbon poisoning of the catalyst surface. Complete reaction mass balance involving stoichiometric hydrogen formation was verified by mass spectrometry analysis. (b) Ambient Pressure-XPS spectra (Pd 3d, Zn3d, and valence-band (VB) regions) acquired *in situ* during MSR on the PdZn 1:1 multilayer (light grey/red online curves) and monolayer alloy (dark grey/blue online curves). For comparison, the respective clean Pd spectra are added (black lines). The oxidized ZnOH component is highlighted by the dashed red line (middle panel). To obtain equal information depths for all spectra the Pd3d spectra were recorded with 650 eV photon energy, the Zn3d and valence-band regions at 120 eV. Reaction conditions: 0.12 mbar methanol, 0.24 mbar water, 280 °C. All spectra show raw unsmoothed data. Reproduced from ref. 25 with permission of John Wiley and Sons.

mono- and multilayer alloys, the subsurface composition obviously determines the electronic and geometric structure and thus the reactivity. When comparing valence band spectra (Fig. 15), the monolayer alloy shows a "Pd-like" valence band, whereas the selective multilayer PdZn alloy resembles the valence band of Cu. In addition, the position of the Pd3d signal is 335.3 eV in monolayer *vs.* 335.8 eV in multilayer alloys. Clearly, the electronic structure determines the selectivity in MSR. Beside the changes in electronic structure, which likely alter the reaction barriers, also a different surface chemistry of the Zn species was deduced by XPS at mbar pressures. From the Zn3d spectra acquired under MSR reaction conditions it was deduced that multilayer alloys are able to activate water as detected by the presence of ZnOH species (shoulder at 10.25 eV in the Zn3d, Fig. 15), while this species was not observed on the monolayer alloy under the same conditions. From that it can be concluded that the monolayer alloy does not activate water, which is required for the oxidation step of intermediate formaldehyde to form CO_2. Thus, the dehydrogenation of methanol to CO is more likely to occur on the unselective monolayer alloy. This result was supported by the corresponding C1s spectra, where signals of oxygenates such as CH_2O were observed on the multilayer alloy in the low temperature range, while CO was detected above 300 °C where the transformation to the monolayer alloy occurs.

While there is general agreement on the importance of PdZn alloy formation, open questions concern on the one hand the role of the support and on the other hand that of the nature of the intermetallic phase (ordered intermetallic *vs.* random alloy phases) and potential differences in reactivity of IMCs of different compositions. Halevi *et al.* compared the reactivity of unsupported PdZn α[43] and PdZn β particles[44] prepared by spray pyrolysis to figure out the importance of the nature of the IMC/alloy. While nearly 100% selectivity to MSR was found on the ordered 1:1 teragonal PdZn β phase, the PdZn α was 100% selective to CO, behaving similar to Pd. The so-called α phase is a solid solution of Zn in fcc Pd. In[43] the sample contains about 12 atom% Zn. While the TOF was improved as compared to pure Pd due to a weakening of the CO bond, a small amount of Zn was not enough to alter the selectivity in MSR. The authors thus concluded that the formation of the tetragonal 1:1 intermetallic compound is required to obtain a selective MSR catalyst.[43] The suggested reaction pathway on Pd, PdZn α and PdZn β and a structural scheme are shown in Fig. 16.[43]

In this line, Friedrich *et al.*[64] determined atomic Zn/Pd ratios of single phase intermetallic PdZn compounds with different bulk compositions (Pd_xZn_{100-x}, $x = 46.8$–59.1) by *in situ* XPS at mbar pressures and connected the composition with the catalytic performance in MSR. Fig. 5 in section 2.3 shows the detected surface Zn/Pd ratios after reduction in H_2 and under

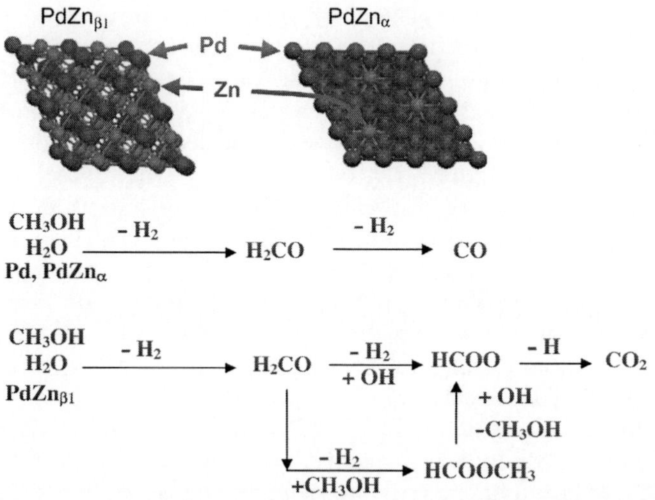

Fig. 16 Top: Schematic illustration of PdZn(111) beta and alpha phase structure, from a top-down perspective. Pd atoms are displayed in dark (blue), Zn are brighter (purple). Evident is the alternating Pd and Zn "stacked sheet" structure of the beta phase and random Zn-substitution of the Pd fcc lattice in the alpha phase. Bottom: Suggested reaction pathway diagram for MSR over Pd and PdZn catalysts. Reprinted from ref. 43, Copyright (2012), with permission from Elsevier.

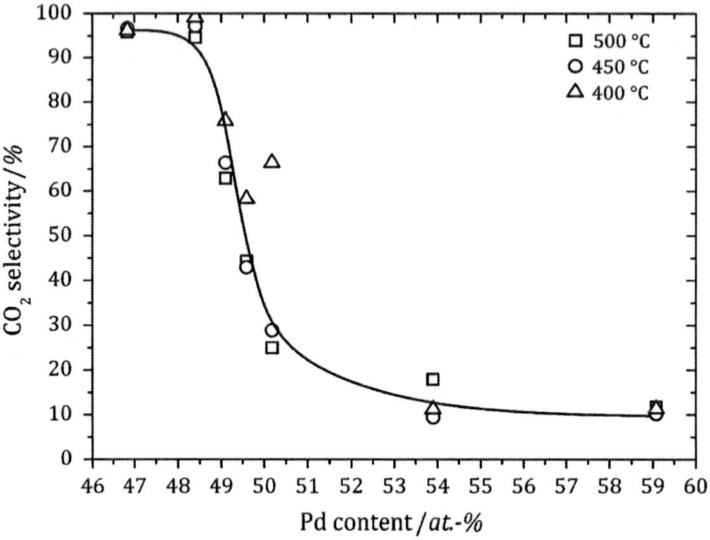

Fig. 17 CO_2 selectivity of Pd_xZn_{100-x} samples in methanol steam reforming depending on the reaction temperature (atmospheric pressure, H_2O:MeOH = 1:1). Reprinted from ref. 64, Copyright (2012), with permission from Elsevier.

MSR conditions and compares them to the bulk composition. Apparently, Zn is enriched at the surface. In addition, valence band spectra were recorded and revealed changes in the electronic structure. The observed variations in surface composition (*i.e.* the extent of Zn enrichment), Zn oxidation state and in electronic properties were connected with the catalytic properties for MSR, which revealed low selectivity to the desired CO_2 and H_2 formation on Pd-rich samples and higher activity and substantially higher selectivity on Zn-rich samples (Fig. 17). The much better steam reforming selectivity was explained by the existence of oxidized Zn species at the surface of the Zn-rich samples allowing for an easier adsorption and dissociation of methanol and water as on pure metallic surfaces. An important role of the oxide support was also suggested in[114] based on DFT calculations, which showed very low or practically no activation barriers of methanol and water dissociation on the polar Zn-terminated ZnO(0001) surface (Fig. 18). Contrary, dissociation of methanol and water were highly activated on flat defect-free PdZn surfaces, such as (111) and (100). Therefore, the reaction might occur at the metal-oxide interface, with the initial reaction steps of methanol and water dissociation occurring on the oxide and the metal being involved in further dehydrogenation of methoxy.[114] This could explain the dependence of the MSR activity on the ZnO morphology as observed by Karim *et al.*[28] Hyman *et al.*[56] observed large differences in methanol TPD on Pd/ZnO(0001) and Pd/ZnO(10–10). The structure sensitivity was explained by a synergism between Pd and the ZnO(0001) surface and the presence of highly active sites at the interface, which are absent on Pd/ZnO(10–10). Recently, for the related Pd/Ga_2O_3 catalysts the support oxide was found to play a main role.[113,115]

The open questions concerning the nature of the actual active sites and the suggested synergism between ZnO and the intermetallic PdZn

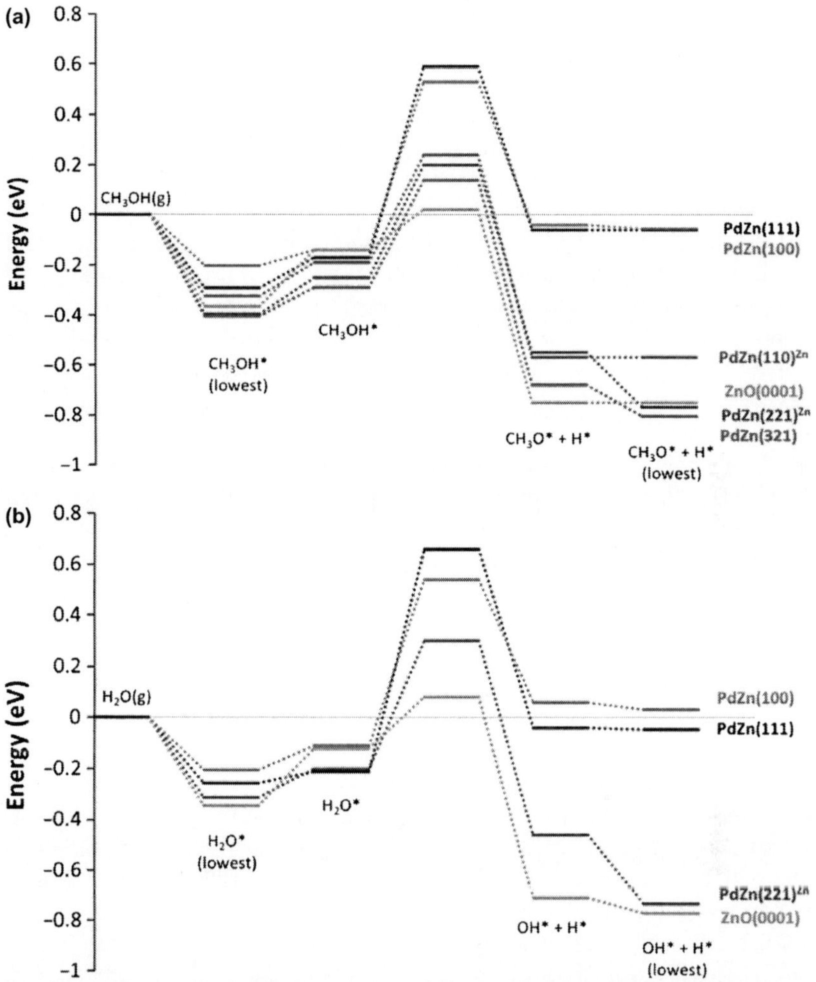

Fig. 18 (a) Potential energy surfaces for methanol O–H scission on PdZn (100), PdZn (111), PdZn $(110)^{Pd}$, PdZn $(110)^{Zn}$, PdZn $(221)^{Zn}$, PdZn (321) and ZnO (0001). (b) Potential energy surfaces for water O–H scission on PdZn (111), PdZn (100), PdZn $(221)^{Zn}$, and ZnO (0001). All ZPE-corrected energies (eV) are referenced against the sum of energies for methanol respectively water in the gas phase and either clean PdZn or clean ZnO surfaces. Reprinted from ref. 114, Copyright (2011), with permission from Elsevier.

complicate the determination of TOF values, which are frequently used to compare different catalyst systems. Therefore, reported TOFs need to be considered critically keeping this in mind. If the active sites are indeed located at the bimetal-oxide interface, TOF values based on the number of accessible Pd atoms (determined for instance by H_2 chemisorption measurements) would only represent the lower limit of the actual TOF values.

The uncertainties about the nature of the active sites and the detailed reaction mechanism may also be reflected in differing activation energies E_A reported in the literature. Experimentally determined activation energies can roughly be assembled into two groups: E_A values around 50[44] and around 100 kJ/mol.[20,68,116,117] The experimentally determined values can be compared to activation barriers obtained by quantum mechanical calculations. For the dehydrogenation of methoxy to formaldehyde, which is considered as the rate-determining step with the highest calculated barrier in the likely reaction sequence, E_A amounts to 113–120 kJ/mol on PdZn(111) according to Chen and coworkers.[70,107,118] Similar values were obtained for PdZn(100),[118] while lower activation energies were calculated when considering the stepped surfaces $PdZn(221)^{Pd}$ and $PdZn(221)^{Zn}$ with values of 50 and 108 kJ/mol, respectively.[106] Smith *et al.*[114] also calculated the initial steps in the methanol steam reforming reaction by DFT. The barriers for methanol and water dissociation as the very first reaction steps are displayed in Fig. 18 and were much lower on the polar Zn-terminated ZnO (0001) surface and on defective PdZn as compared to defect-free, flat PdZn surfaces. For methoxy dehydrogenation, presumably the rate-determining step, similar reaction barriers were reported as those of Chen *et al.* on flat surfaces. They calculated values of 86 kJ/mol on PdZn(111) and 102 kJ/mol on PdZn(100). Overall, there is good agreement between experimentally determined and calculated activation energies.

For potential application, the long-term stability is crucial and presents the major advantage of PdZn over Cu catalysts. Conant *et al.*[32] found a better long-term stability of $Pd/ZnO/Al_2O_3$ compared to a commercial $Cu/ZnO/Al_2O_3$. The conversion initially decreased by 17% and then stayed constant for long operation times (60 h) on the $Pd/ZnO/Al_2O_3$, while the conversion over the Cu catalyst dropped by 40% within 60 h under the same conditions with a fast initial drop followed by a continuous long-term deactivation. Moreover, the Pd catalyst could be fully regenerated by oxidation-reduction cycles, which was not the case for Cu. Also Suwa *et al.*[119] observed an initial deactivation of Pd/ZnO in a 50 h life test, which was explained by the formation of a zinc carbonate hydroxide destroying the ZnO crystal particles, as detected by XRD. Zhang and Farrauto[120] tested the long-term stability of a $PdZn/CeO_2$ catalyst up to 600 h run time. At 400 °C the MSR selectivity improved with time on stream from 70 to 90% because of an increasing extent of PdZn alloy formation, while the conversion decreased with operation time by about 25% after 600 h. Accumulation of carbonaceous deposits was identified as the reason for the deactivation, and regeneration could be achieved by oxidation at 400–600 °C in air.

Overall, the essential requirements determining the high activity and selectivity of PdZn are the "right" Cu-like electronic structure and the PdZn/ZnO interface. Obviously, formation of the intermetallic 1:1 PdZn component with a Cu-like electronic structure is crucial. An essential step is water activation, which likely occurs either on the ZnO support (nearby the intermetallic PdZn particles) or on Zn within the PdZn surface layer, that is arranged in an Zn-out/Pd-in corrugation.

3.3 Water-gas shift reaction (WGS)

Due to the very similar electronic properties of the PdZn to Cu, as described in section 2.4, PdZn is likely also a promising candidate for the WGS reaction, where Cu/ZnO is known to be a highly active catalyst which is applied for low temperature shift catalysis in industrial processes. This expectation is supported by the catalytic performance of PdZn in methanol synthesis (see section 3.1) and MSR (compare section 3.2) being completely different to Pd but very similar to Cu.

$$CO + H_2O \leftrightarrow CO_2 + H_2 \qquad \text{WGS}$$

Bollmann *et al.*[33] added different amounts of Zn (0–19 wt%) to a 2 wt% Pd/Al_2O_3 catalyst and studied the effect on the WGS reaction. In addition, structural properties such as Pd-Zn average coordination numbers were determined by means of EXAFS. Most of the Zn (over 85%) was found to react with alumina to form zinc aluminate. Only a fraction of the Zn reacted with Pd, with an increasing Pd-Zn coordination number with increasing Zn loading. On the PdZn samples, a strong increase of the turnover rate for the WGS reaction was observed. On Pd/ZnO, where an ordered 1:1 intermetallic compound was formed, with the largest Pd-Zn coordination number of all studied materials, turnover rates for WGS increased even more. A good correlation between the average Pd-Zn coordination numbers, the fraction of surface Pd that binds CO linearly and the WGS turnover rates was observed (Fig. 19). Clearly, a beneficial effect of Zn addition and a higher WGS activity on Pd/ZnO materials is demonstrated in this work. In comparison to the monometallic Pd catalyst, the WGS turnover rate on Pd/ZnO increased by a factor of 20 (per exposed Pd). Ota *et al.*[37] reported TOF values for reverse water-gas shift over PdZn and Pd catalysts derived from hydrotalcites as precursors. The activities of both catalysts expressed as TOF were very similar.

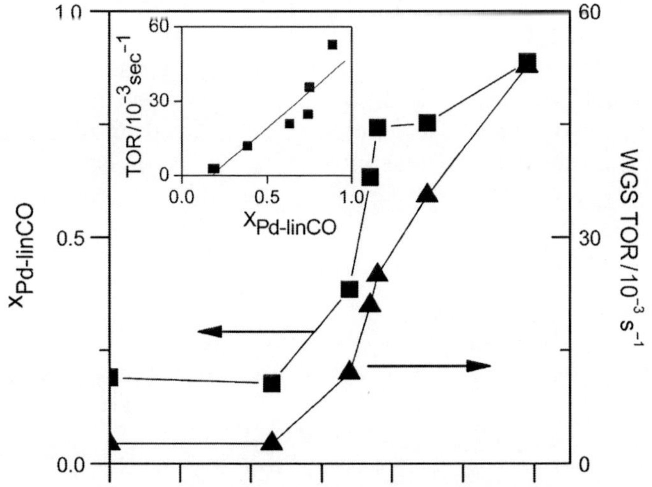

Fig. 19 (■) Fraction of surface Pd that binds CO linearly ($X_{Pd\text{-}linCO}$) under WGS conditions and (▲) WGS turnover rates (TOR) as a function of the average Pd–Zn coordination number N_{PdZn} measured by EXAFS. Rates are expressed per exposed mole of Pd and measured at 280 °C and 1 atm (6.8% CO, 8.5% CO_2, 37% H_2O, 37.4% H_2, balance Ar). Inset: correlation between the WGS TOR and $X_{Pd\text{-}linCO}$. Reprinted from ref. 33, Copyright (2008), with permission from Elsevier.

Dagle *et al.*[121] determined the reaction rates of WGS, RWGS, methanol steam reforming and methanol decomposition on $Pd/ZnO/Al_2O_3$. While the reaction rate constants were approximately of the same magnitude for WGS and MSR, RWGS rate constants were about 20 times lower than that of MSR and similar to MDC rate constants. Based on the detailed kinetic analysis it was suggested that WGS is not involved in MSR, but that RWGS could be one of the sources for the small amount of CO formed in MSR. In this line, Lebarbier *et al.*[122] identified RWGS as a likely potential reaction pathway to undesired CO formation in MSR on Pd/ZnO catalysts.

In summary, PdZn has shown promising activity for the WGS reaction with the advantage of higher thermal stability as compared to the commercially applied Cu/ZnO low temperature shift catalyst.

3.4 Oxidation reactions

Not many applications have been reported so far for PdZn as oxidation catalyst. Most likely, this is due to the instability of the PdZn (surface) against O_2 containing atmospheres.[21,31–32,62,65] Wehner *et al.*[65] observed by XPS that Zn in bimetallic PdZn/ZnO gets oxidized upon contact with air. Also Green *et al.*[31] found the bulk to be stable upon air exposure at room temperature according to XRD measurements, but the surface was affected and the Pd-Zn interaction disrupted. Similarly, Lebarbier *et al.*[62] found characteristic vibrational bands of CO on metallic Pd after an oxidation treatment of PdZn. Recently, Föttinger *et al.*[21] investigated in detail the detrimental effect of oxygen on PdZn/ZnO by combining XAS and FTIR spectroscopy. Upon O_2 exposure, the XANES spectra gradually changed back to the characteristic spectrum of metallic Pd with exposure time. FTIR spectroscopy of CO adsorption indicated metallic Pd or an alloy depleted in Zn by reappearance of bridge and hollow bonded CO, which are not available on the PdZn alloy, as described in section 2.3. Interestingly, the intensity of the CO bands after the oxidation treatment was strongly reduced, which was explained by a partial decoration of the (bi)metallic particles with oxidized ZnO species (Fig. 8(b), spectrum (c)). The effect of O_2 on PdZn is schematically illustrated in Fig. 20. A similar effect was found

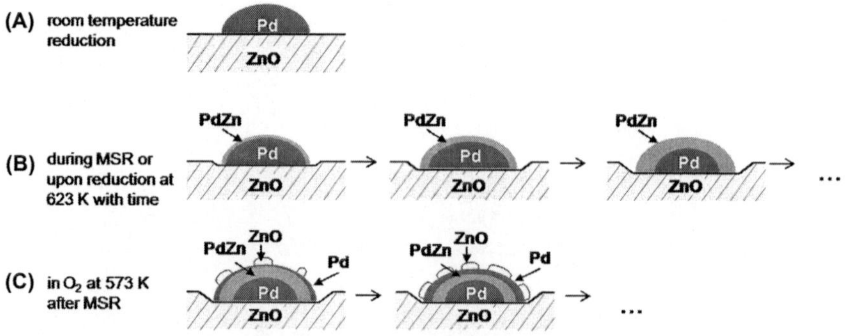

Fig. 20 Illustration of the suggested structural changes of Pd/ZnO in various environments. Reprinted with permission from ref. 21. Copyright (2011) American Chemical Society.

on related Pd_2Ga/Ga_2O_3 catalysts.[123–124] Therefore, a limited stability of PdZn surfaces in oxidation reactions can be expected.

PdZn was, however, used in oxidative methanol steam reforming (OMSR) and in the partial oxidation of methanol (POM) to hydrogen and CO_2. In this system, the poduced hydrogen can likely re-form the PdZn alloy and thereby balance or overcome the degradation by O_2. Under the POM and OMSR reaction conditions various reactions can occur, including methanol combustion, water-gas shift reaction, methanol steam reforming and methanol decomposition. Selectivity therefore depends strongly on the conditions such as temperature, conversion and methanl:O_2:H_2O ratio.

Cubeiro *et al.*[66,125] studied POM on Pd/ZnO and Pd/ZrO_2 catalysts. They performed a detailed analysis of reaction selectivity depending on a range of reaction parameters, such as temperature, feed composition and conversion (Fig. 21). While the reaction occurred through consecutive oxidation and reforming on Pd/ZnO,[66] the methanol decomposition reaction seemed to occur to a greater extent on Pd/ZrO_2, as expected for group 8 metals.[125] Agrell *et al.*[30] studied the partial oxidation of methanol over Pd/ZnO. They suggested a reaction network of sequential methanol combustion, steam reforming and decomposition. By XRD, they detected

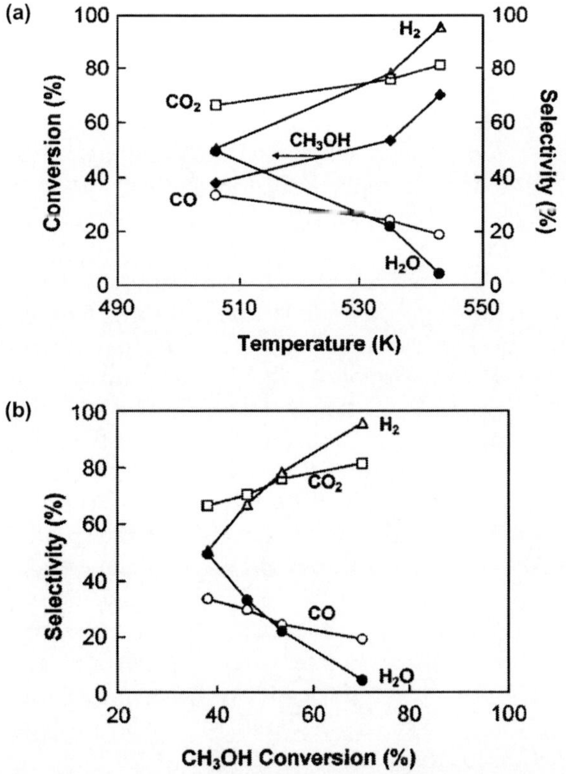

Fig. 21 (a) Dependence of methanol conversion and selectivities on the temperature and (b) selectivities as a function of methanol conversion for the partial oxidation of methanol over 1% Pd/ZnO using a feed ratio O_2/CH_3OH of 0.3. Reprinted from ref. 66, Copyright (1998), with permission from Elsevier.

Pd as Pd^0, PdZn and PdO on the used catalyst, which was explained by either considering that not all Pd forms an alloy or that the alloy obtained by H_2 reduction decomposes partially during the reaction in the presence of O_2.[30]

In contrast, Iwasa *et al.*[12] carried out OMSR and found that the PdZn alloy remains unchanged in the absence or presence of O_2. Conant *et al.*[32] observed formation of PdO and ZnO after oxidative treatment at 500 °C, which breaks up the PdZn alloy. However, during MSR reaction they observed identical catalytic properties after a certain reaction time independent of the applied pre-treatment (atmosphere, temperature). This behaviour was attributed to the establishment of a certain equilibrium state of the surface during the reaction, which depends on the reaction conditions and not on the pre-treatment. Thus, the detrimental effect of oxygen is likely "healed" by hydrogen or methanol.

Liu *et al.*[126] compared the long-term catalytic performance of Pd/ZnO and Cu/ZnO for MSR, OMSR and POM. Whereas Cu/ZnO lost activity at constant selectivity, Pd/ZnO exhibited a more stable activity. However, an increasing CO production was observed with time on stream. Based on characterization by XPS and XRD, the deactivation of Pd/ZnO was explained by carbon deposition and surface oxidation of the Pd-Zn alloy, breaking down the Pd-Zn alloy to elemental Pd on ZnO.[126]

4 Conclusions and outlook

The last section of the review is devoted to discuss some aspects of the PdZn system in view of its potential for the application in commercial processes. In spite of the higher costs of noble-metal containing catalysts, PdZn has shown a high potential to replace Cu for selected applications due to its considerably higher stability, which is an enormous advantage for commercial applications.

PdZn proved to be especially attractive in methanol steam reforming for hydrogen production, *e.g.* in mobile applications. This is reflected in a recent patent by Castellano *et al.*[127] An interesting approach in this direction is the successful application of PdZn catalysts in microreactor technology.[128–131] An optimization in decreasing the amount of Pd used will certainly help to boost its application. To achieve this, one has to know the essential parameters for high activity, selectivity and stability. In the past 15 years since the discovery of the advantageous catalytic performance of Pd/ZnO in MSR by the group of Iwasa, certain essential requirements have been identified. The enormous progress in understanding the PdZn system is based on the valuable combination of theoretical calculations and detailed *in situ* studies on different types of systems, covering conventional high surface area nanoparticulate catalysts, unsupported intermetallic compounds and UHV-grown single-crystal based model systems,

In summary, the parameters essential for achieving high selectivity to MSR identified up to now are:

1. the "right" Cu-like electronic structure, which determines the reaction barriers for intermediate formaldehyde conversion thus avoiding the undesired pathway to CO/H_2 but directing the reaction towards CO_2/H_2.

This requirement is fulfilled in the tetragonal PdZn with a 1:1 composition, which is anyway the thermodynamically most stable phase and typically forms upon reduction in a medium temperature range of around 300–500 °C.

2. the ability to activate water. The active sites for water activation are most likely Zn in PdZn multilayer alloys and oxidized Zn species at the PdZn/ZnO interface. On high surface area catalysts a PdZn monolayer is rather unlikely to be formed and PdZn multilayers are therefore intrinsically present. The importance of the interface opens up potential for an optimization of the PdZn-ZnO interaction.

For hydrogenation reactions, up to now considerably more work has been done on the related Pd-Ga system. PdZn has, however, shown promising potential to be explored further. The high stability of the PdZn IMC in reducing atmosphere is the major benefit as compared to conventional substitutional alloys. To achieve high selectivity the important factors are assumed to be the well-known ensemble and ligand effects on alloys:

1. the lower concentration of hydrogen available at the surface and the surface dilution of Pd (site isolation). In addition, hydride is absent in PdZn.
2. an alteration of the electronic properties and thereby a weaker interaction with certain products/intermediates, which for example slows down follow-up reactions.

The utilization of PdZn for oxidation reactions is, however, limited because of the instability of PdZn in oxidative environments. Overall, PdZn catalysts are therefore especially promising for applications in reducing atmosphere, as demonstrated in selective hydrogenation reactions and in MSR for H_2 production. In both, selectivity and stability are crucial, and the exceptional selectivity of PdZn makes it an attractive catalyst.

References

1 J. H. Sinfelt, *Acc. Chem. Res.*, 1977, **10**, 15.
2 J. Rodriguez, *Surf. Sci. Rep.*, 1996, **24**, 225.
3 G. E. R. Schulze, *Metallphysik*, Springer Verlag Wien New York, 1967.
4 W. M. H. Sachtler and R. A. van Santen, *Adv. Catal.*, 1977, **26**, 69.
5 J. H. Sinfelt, *Bimetallic Catalysts*, Wiley, New York, 1983.
6 V. Ponec, *Adv. Catal.*, 1983, **32**, 149.
7 W. H. M. Sachtler, *Faraday Discuss. Chem. Soc.*, 1981, **72**, 7.
8 N. Iwasa, S. Masuda, N. Ogawa and N. Takezawa, *Appl. Catal., A*, 1995, **125**, 145.
9 N. Iwasa, T. Mayanagi, S. Masuda and N. Takezawa, *React. Kinet. Catal. Lett.*, 2000, **69**, 355.
10 N. Iwasa, N. Ogawa, S. Masuda and N. Takezawa, *Bull. Chem. Soc. Jpn.*, 1998, **71**, 1451.
11 N. Iwasa, M. Yoshikawa and M. Arai, *Phys. Chem. Chem. Phys.*, 2002, **4**, 5414.
12 N. Iwasa, M. Yoshikawa, W. Nomura and M. Arai, *Appl. Catal., A*, 2005, **292**, 215.
13 N. Takezawa and N. Iwasa, *Catal. Today*, 1997, **36**, 45.
14 N. Iwasa, S. Masuda, N. Ogawa and N. Takezawa, *Appl. Catal., A*, 1995, **125**, 145.

15 K. D. Jung, O. S. Joe, S. H. Han, S. J. Uhm and I. J. Chung, *Catal. Lett.*, 1995, **35**, 303.
16 D. L. Trimm and Z. I. Onsan, *Catal. Rev.*, 2001, **43**, 31.
17 C. Zhang, Z. Yuan, N. Liu, S. Wang and S. Wang, *Fuel Cells*, 2006, **6**, 466.
18 A. P. Tsai, S. Kameoka and Y. Ishii, *J. Phys. Soc. Jpn.*, 2004, **73**, 3270.
19 M. W. Tew, H. Emerich and J. A. van Bokhoven, *J. Phys. Chem. C*, 2011, **115**, 8457.
20 Y. H. Chin, R. Dagle, J. L. Hu, A. C. Dohnalkova and Y. Wang, *Catal. Today*, 2002, **77**, 79.
21 K. Föttinger, J. A. van Bokhoven, M. Nachtegaal and G. Rupprechter, *J. Phys. Chem. Lett.*, 2011, **2**, 428.
22 A. Bayer, K. Flechtner, R. Denecke, H. P. Steinrück, K. M. Neyman and N. Rösch, *Surf. Sci.*, 2006, **600**, 78.
23 W. Stadlmayr, S. Penner, B. Klötzer and N. Memmel, *Surf. Sci.*, 2009, **603**, 251.
24 G. Weirum, M. Kratzer, H. P. Koch, A. Tamtögl, J. Killmann, I. Bako, A. Winkler, S. Surnev, F. P. Netzer and R. Schennach, *J. Phys. Chem. C*, 2009, **113**, 9788.
25 C. Rameshan, W. Stadlmayr, C. Weilach, S. Penner, H. Lorenz, M. Hävecker, R. Blume, T. Rocha, D. Teschner, A. Knop-Gericke, R. Schlögl, N. Memmel, D. Zemlyanow, G. Rupprechter and B. Klötzer, *Angew. Chem., Int. Ed.*, 2010, **49**, 3224.
26 S. Penner, B. Jenewein, H. Gabasch, B. Klötzer, D. Wang, A. Knop-Gericke, R. Schlögl and K. Hayek, *J. Catal.*, 2006, **241**, 14.
27 A. Karim, T. Conant and A. Datye, *J. Catal.*, 2006, **243**, 420.
28 A. M. Karim, T. Conant and A. K. Datye, *Phys. Chem. Chem. Phys.*, 2008, **10**, 5584.
29 Y. H. Chin, Y. Wang, R. A. Dagle and X. H. S. Li, *Fuel Process. Technol.*, 2003, **83**, 193.
30 J. Agrell, G. Germani, S. G. Jaras and M. Boutonnet, *Appl. Catal., A*, 2003, **242**, 233.
31 B. E. Green, C. S. Sass, L. T. Germinario, P. S. Wehner and B. L. Gustafson, *J. Catal.*, 1993, **140**, 406.
32 T. Conant, A. M. Karim, V. Lebarbier, Y. Wang, F. Girgsdies, R. Schlögl and A. Datye, *J. Catal.*, 2008, **257**, 64.
33 L. Bollmann, J. L. Ratts, A. M. Joshi, W. D. Williams, J. Pazmino, Y. V. Joshi, J. T. Miller, A. J. Kropf, W. N. Delgass and F. H. Ribeiro, *J. Catal.*, 2008, **257**, 43.
34 J. Arana, N. Homs, J. Sales, J. L. G. Fierro and P. R. de la Piscina, *Catal. Lett.*, 2001, **72**, 183.
35 M. Lenarda, E. Moretti, L. Storaro, P. Patrono, F. Pinzari, E. Rodriguez-Castellon, A. Jimenez-Lopez, G. Busca, E. Finocchio, T. Montanari and R. Frattini, *Appl. Catal., A*, 2006, **312**, 220.
36 M. Lenarda, L. Storaro, R. Frattini, M. Casagrande, M. Marchiori, G. Capannelli, C. Uliana, F. Ferrari and R. Ganzerla, *Catal. Commun.*, 2007, **8**, 467.
37 A. Ota, E. L. Kunkes, I. Kasatkin, E. Groppo, D. Ferri, B. Poceiro, R. M. Navarro Yerga and M. Behrens, *J. Catal.*, 2012, **293**, 27.
38 L. M. Bronstein, D. M. Chernyshov, I. O. Volkov, M. G. Ezernitskaya, P. M. Valetsky, V. G. Matveeva and E. M. Sulman, *J. Catal.*, 2000, **196**, 302.
39 P. D. Burton, D. Lavenson, M. Johnson, D. Gorm, A. M. Karim, T. Conant, A. K. Datye, B. A. Hernandez-Sanchez and T. J. Boyle, *Top. Catal.*, 2008, **49**, 227.

40 O. P. Tkachenko, A. Y. Stakheev, L. M. Kustov, I. V. Mashkovsky, M. van den Berg, W. Grünert, N. Y. Kozitsyna, Z. V. Dobrokhotova, V. I. Zhilov, S. E. Nefedov, M. N. Vargaftik and I. I. Moiseev, *Catal. Lett.*, 2006, **112**, 155.
41 R. E. Cable and R. E. Schaak, *Chem. Mater.*, 2007, **19**, 4098.
42 M. Friedrich, A. Ormeci, Y. Grin and M. Armbrüster, *Z. Anorg. Allg. Chem.*, 2010, **636**, 1735.
43 B. Halevi, E. J. Peterson, A. Roy, A. DeLariva, E. Jeroro, F. Gao, Y. Wang, J. M. Vohs, B. Kiefer, E. Kunkes, M. Hävecker, M. Behrens, R. Schlögl and A. K. Datye, *J. Catal.*, 2012, **291**, 44.
44 B. Halevi, E. J. Peterson, A. DeLaRiva, E. Jeroro, V. M. Lebarbier, Y. Wang, J. M. Vohs, B. Kiefer, E. Kunkes, M. Hävecker, M. Behrens, R. Schlögl and A. K. Datye, *J. Phys. Chem. C*, 2010, **114**, 17181.
45 E. J. Peterson, B. Halevi, B. Kiefer, M. N. Spilde, A. K. Datye, J. Peterson, L. Daemen, A. Llobet and H. Nakotte, *J. Alloys Comp.*, 2011, **509**, 1463.
46 V. Engels, D. A. Jefferson, F. Benaskar, P. C. Thüne, A. Berenguer-Murcia, B. F. G. Johnson and A. E. H. Wheatley, *Nanotechn.*, 2011, **22**, 205701.
47 Y. Luo, Y. Sun, U. Schwarz and M. Armbrüster, *Chem. Mater.*, 2012, **24**, 3094.
48 P. Bera and J. M. Vohs, *J. Chem. Phys.*, 2006, **125**, 64713.
49 P. Bera and J. M. Vohs, *J. Phys. Chem. C*, 2007, **111**, 7049.
50 H. Gabasch, A. Knop-Gericke, R. Schlögl, S. Penner, B. Jenewein, K. Hayek and B. Klötzer, *J. Phys. Chem. B*, 2006, **110**, 11391.
51 E. Jeroro and J. M. Vohs, *J. Am. Chem. Soc.*, 2008, **130**, 10199.
52 E. Jeroro, V. Lebarbler, A. Datye, Y. Wang and J. M. Vohs, *Surf. Sci.*, 2007, **601**, 5546.
53 M. Kratzer, A. Tamtögl, J. Killmann, R. Schennach and A. Winkler, *Appl. Surf. Sci.*, 2009, **255**, 5755.
54 J. A. Rodriguez, *J. Phys. Chem.*, 1994, **98**, 5758.
55 I. Kovacs, J. Kiss and F. Solymosi, *Vacuum*, 2007, **82**, 182.
56 M. P. Hyman, V. M. Lebarbier, Y. Wang, A. K. Datye and J. A. Vohs, *J. Phys. Chem. C*, 2009, **113**, 7251.
57 H. Okamoto, in *Binary Alloy Phase Diagrams*, ed. T. B. Massalski, Editon edn., 1990, p. 3068.
58 *PdZn, ICDD card # 06-0620.*
59 Z. X. Chen, K. M. Neyman, A. B. Gordienko and N. Rösch, *Phys. Rev. B*, 2003, **68**, 075417.
60 C. Weilach, S. M. Kozlov, H. H. Holzapfel, K. Föttinger, K. M. Neyman and G. Rupprechter, *J. Phys. Chem. C*, 2012, **116**, 18768.
61 W. Stadlmayr, C. Rameshan, C. Weilach, H. Lorenz, M. Hävecker, R. Blume, T. Rocha, D. Teschner, A. Knop-Gericke, D. Zemlyanov, S. Penner, R. Schlögl, G. Rupprechter, B. Klötzer and N. Memmel, *J. Phys. Chem. C*, 2010, **114**, 10850.
62 V. Lebarbier, R. Dagle, T. Conant, J. M. Vohs, A. K. Datye and Y. Wang, *Catal. Lett.*, 2008, **122**, 223.
63 Z. X. Chen, K. M. Neyman and N. Rösch, *Surf. Sci.*, 2004, **548**, 291.
64 M. Friedrich, D. Teschner, A. Knop-Gericke and M. Armbrüster, *J. Catal.*, 2012, **285**, 41.
65 P. S. Wehner, G. C. Tustin and B. L. Gustafson, *J. Catal.*, 1984, **88**, 246.
66 M. L. Cubeiro and J. L. G. Fierro, *J. Catal.*, 1998, **179**, 150.
67 N. Iwasa and N. Takezawa, *Top. Catal.*, 2003, **22**, 215.
68 C. Rameshan, C. Weilach, W. Stadlmayr, S. Penner, H. Lorenz, M. Hävecker, R. Blume, T. Rocha, D. Teschner, A. Knop-Gericke, R. Schlögl, D. Zemlyanow, N. Memmel, G. Rupprechter and B. Klötzer, *J. Catal.*, 2010, **276**, 101.

69 Z. Zsoldos, A. Sarkany and L. Guczi, *J. Catal.*, 1994, **145**, 235.
70 K. M. Neyman, K. H. Lim, Z. X. Chen, L. V. Moskaleva, A. Bayer, A. Reindl, D. Borgmann, R. Denecke, H. P. Steinrück and N. Rösch, *Phys. Chem. Chem. Phys.*, 2007, **9**, 3470.
71 I. Bako, R. Schennach and G. Palinkas, *J. Phys.: Conf. Ser.*., 2008, **100**, , 052067.
72 A. Sarkany, Z. Zsoldos, B. Furlong, J. W. Hightower and L. Guczi, *J. Catal.*, 1993, **141**, 566.
73 P. S. Wehner and B. L. Gustafson, *J. Catal.*, 1992, **135**, 420.
74 N. Iwasa, M. Takizawa and M. Arai, *Appl. Catal., A*, 2005, **283**, 255.
75 J. Osswald, R. Giedigkeit, R. E. Jentoft, M. Armbrüster, F. Girgsdies, K. Kovnir, T. Ressler, Y. Grin and R. Schlögl, *J. Catal.*, 2008, **258**, 210.
76 J. Osswald, K. Kovnir, M. Armbrüster, R. Giedigkeit, R. E. Jentoft, U. Wild, Y. Grin and R. Schlögl, *J. Catal.*, 2008, **258**, 219.
77 M. Armbrüster, M. Behrens, F. Cinquini, K. Föttinger, Y. Grin, A. Haghofer, B. Klötzer, A. Knop-Gericke, H. Lorenz, A. Ota, S. Penner, J. Prinz, C. Rameshan, Z. Révay, D. Rosenthal, G. Rupprechter, P. Sautet, R. Schlögl, L. Shao, L. Szentmiklósi, D. Teschner, D. Torres, R. Wagner, R. Widmer and G. Wowsnick, *ChemCatChem*, 2012, **4**, 1048.
78 A. Ota, M. Armbrüster, M. Behrens, D. Rosenthal, M. Friedrich, I. Kasatkin, F. Girgsdies, W. Zhang, R. Wagner and R. Schlögl, *J. Phys. Chem. C*, 2010, **115**, 1368.
79 H. Lindlar and R. Dubuis, *Org. Synth. Collect.*, 1973, **5**, 880.
80 M. L. Derrien, *Stud. Surf. Sci. Catal.*, 1986, **27**, 613.
81 G. C. Bond and P. B. Wells, *J. Catal.*, 1966, **5**, 65.
82 A. Borodzinski, *Catal. Lett.*, 1999, **63**, 35.
83 *US Pat.*, 6,054,409, 2000.
84 A. M. Doyle, S. K. Shaikhutdinov and H. J. Freund, *J. Catal.*, 2004, **223**, 444.
85 S. Shaikhutdinov, M. Heemeier, M. Bäumer, T. Lear, D. Lennon, R. J. Oldman, S. D. Jackson and H. J. Freund, *J. Catal.*, 2001, **200**, 330.
86 F. Studt, F. Abild-Pedersen, T. Bligaard, R. Z. Sørensen, C. H. Christensen and J. K. Nørskov, *Science*, 2008, **320**, 1320.
87 J. B. Conn, G. B. Kistiakowsky and E. A. Smith, *J. Am. Chem. Soc.*, 1939, **61**, 1868.
88 G. F. Taylor, S. J. Thomson and G. Webb, *J. Catal.*, 1968, **12**, 150.
89 R. A. Koeppel, J. T. Wehrli, M. S. Wainwright, D. L. Trimm and N. W. Cant, *Appl. Catal., A*, 1994, **120**, 163.
90 A. M. Kazi, B. Chen, J. G. Goodwin, J. G. Marcelin, N. Rodriguez and R. T. K. Barker, *J. Catal.*, 1995, **157**, 1.
91 T. Fujitani and I. Nakamura, *Bull. Chem. Soc. Jpn.*, 2002, **75**, 1393.
92 N. Iwasa, H. Suzuki, M. Terashita, M. Arai and N. Takezawa, *Catal. Lett.*, 2004, **96**, 75.
93 D. R. Palo, R. A. Dagle and J. D. Holladay, *Chem. Rev.*, 2007, **107**, 3992.
94 A. Casanovas, J. Llorca, N. Homs, J. L. G. Fierro and P. R. de la Piscina, *J. Mol. Catal. A: Chem.*, 2006, **250**, 44.
95 N. Iwasa, W. Nomura, T. Mayanagi, S. Fujita, M. Arai and N. Takezawa, *J. Chem. Eng. Jpn.*, 2004, **37**, 286.
96 B. A. Sexton, A. E. Hughes and N. R. Avert, *Surf. Sci.*, 1985, **155**, 366.
97 D. B. Clarke, D.-K. Lee, M. J. Sandoval and A. T. Bell, *J. Catal.*, 1994, **150**, 81.
98 J. L. Davis and M. A. Barteau, *Surf. Sci.*, 1990, **235**, 235.
99 R. W. McCabe, C. L. Dimaggio and R. J. Madix, *J. Phys. Chem.*, 1985, **89**, 854.
100 A. B. Anton, J. E. Parameter and W. H. Weinberg, *J. Am. Chem. Soc.*, 1986, **108**, 1823.

101 M. A. Henderson, Y. Zhou and J. M. White, *J. Am. Chem. Soc.*, 1989, **111**, 1185.
102 C. J. Houtman and M. A. Barteau, *J. Catal.*, 1991, **130**, 528.
103 R. J. Madix, T. Yamada and S. W. Johnson, *Appl. Surf. Sci.*, 1984, **19**, 43.
104 J. L. Davis and M. A. Barteau, *J. Am. Chem. Soc.*, 1989, **111**, 1782.
105 K. H. Lim, Z. X. Chen, K. M. Neyman and N. Rösch, *J. Phys. Chem. B*, 2006, **110**, 14890.
106 Z. X. Chen, K. H. Lim, K. M. Neyman and N. Rösch, *J. Phys. Chem. B*, 2005, **109**, 4568.
107 Z. X. Chen, K. M. Neyman, K. H. Lim and N. Rösch, *Langmuir*, 2004, **20**, 8068.
108 Y. Huang and Z.-X. Chen, *Langmuir*, 2010, **26**, 10796.
109 E. Jeroro and J. M. Vohs, *J. Phys. Chem. C*, 2009, **113**, 1486.
110 S. Lin, D. Xie and H. Guo, *J. Mol. Catal. A: Chem.*, 2012, **356**, 165.
111 B. Frank, F. C. Jentoft, H. Soerijanto, J. Kröhnert, R. Schlögl and R. Schomäcker, *J. Catal.*, 2007, **246**, 177.
112 B. A. Peppley, J. C. Amphlett, L. M. Kearns and R. F. Mann, *Appl. Catal., A*, 1999, **179**, 21.
113 A. Haghofer, D. Ferri, K. Föttinger and G. Rupprechter, *ACS Catal.*, 2012, **2**, 2305.
114 G. K. Smith, S. Lin, W. Lai, A. Datye, D. Xie and H. Guo, *Surf. Sci.*, 2011, **605**, 750.
115 S. E. Collins, J. J. Delgado, C. Mira, J. J. Calvino, S. Bernal, D. L. Chiavassa, M. A. Baltanás and A. L. Bonivardi, *J. Catal.*, 2012, **292**, 90.
116 P. Pfeifer, A. Kolbl and K. Schubert, *Catal. Today*, 2005, **110**, 76.
117 C. Cao, G. Xia, J. Holladay, E. Jones and Y. Wang, *Appl. Catal., A*, 2004, **262**, 19.
118 Z. X. Chen, K. H. Lim, K. M. Neyman and N. Rösch, *Phys. Chem. Chem. Phys.*, 2004, **6**, 4499.
119 Y. Suwa, S. I. Ito, S. Kameoka, K. Tomishige and K. Kunimori, *Appl. Catal., A*, 2004, **267**, 9.
120 Q. L. Zhang and R. J. Farrauto, *Appl. Catal., A*, 2011, **395**, 64.
121 R. A. Dagle, A. Platon, D. R. Palo, A. K. Datye, J. M. Vohs and Y. Wang, *Appl. Catal., A*, 2008, **342**, 63.
122 V. Lebarbier, R. Dagle, A. Datye and Y. Wang, *Appl. Catal., A*, 2010, **379**, 3.
123 A. Haghofer, K. Föttinger, F. Girgsdies, D. Teschner, A. Knop-Gericke, R. Schlögl and G. Rupprechter, *J. Catal.*, 2012, **286**, 13.
124 A. Haghofer, K. Föttinger, M. Nachtegaal, M. Armbrüster and G. Rupprechter, *J. Phys. Chem. C*, 2012, **116**, 21816.
125 M. L. Cubeiro and J. L. G. Fierro, *Appl. Catal., A*, 1998, **168**, 307.
126 S. T. Liu, K. Takahashi, K. Fuchigami and K. Uematsu, *Appl. Catal., A*, 2006, **299**, 58.
127 *US Pat.*, US2007258882 A1, 2007.
128 O. Ilinich, Y. Liu, C. Castellano, G. Koermer, A. Moini and R. Farrauto, *Plat. Met. Rev.*, 2008, **52**, 134.
129 J. D. Holladay, Y. Wang and E. Jones, *Chem. Rev.*, 2004, **104**, 4767.
130 J. D. Holladay, E. O. Jones, M. Phelps and J. Hu, *J. Power Sources*, 2002, **108**, 21.
131 G. Xia, J. D. Holladay, R. A. Dagle, E. O. Jones and Y. Wang, *Chem. Eng. Technol.*, 2005, **28**, 515.

Structural manipulation of the catalysts for ammonia decomposition

Xuezhi Duan,[a] Xinggui Zhou*[a] and De Chen[b]
DOI: 10.1039/9781849737203-00118

Ammonia decomposition is an important reaction in energy and environmental industries. The review is focused on the catalytic decomposition of ammonia as a key step. The performances of the often used catalysts (*i.e.*, Ru, Ni, Fe and bimetallic catalysts) are summarized and the effects of the size and shape of metal nanoparticles, promoters, supports and preparation techniques are reviewed.

1 Introduction

Ammonia decomposition has been widely applied for the removal of NH_3 arising from coal or biomass gasification, for the production of hydrogen as clean energy, or for the generation of reducing atmosphere in chemical, steel, glass and electronic industries [1–6]. Recently, it has attracted an increasing level of attention owing to its possible use as a CO_x-free hydrogen source for proton exchange membrane fuel cells (PEMFCs) [7–16]. At 293 K, ammonia is easily liquefied at 8 atm, and has a relatively high hydrogen storage (17.7 wt%) and a high energy density (3000 Wh/kg). For detailed information concerning the perspective of using ammonia as a possible energy carrier in the energy infrastructure, the readers are referred to the reviews and references therein [7,8,14,16].

Ammonia is a very stable molecule. Thermal ammonia decomposition is difficult to take place because of its extremely high activation energy (*ca.* 1172 kJ/mol). Various catalysts were developed to catalyze ammonia decomposition in the last century [17–25]. However, all these catalysts exhibited low activities. For example, the conversions of ammonia decomposition over commercial catalysts from Johnson Matthey, United Catalyst and Grace Davison were below 85.0% even at low gas hourly space velocity ($5\,h^{-1}$) and high temperature (700 °C) [17]. Hence, there is plenty of room for further improvements of the catalysts for ammonia decomposition.

Early in this century, Choudhary *et al.* [26] prepared a series of Ni-, Ir- and Ru-based catalysts by wet impregnation method with water as solvent, and the best catalyst was 10.0 wt% Ru/SiO_2, which had an ammonia conversion over 97% at 600 °C and *ca.* $30000\,h^{-1}$. More recently, Yin *et al.* [27] used carbon nanotubes (CNTs) to support Ru catalyst by incipient wetness impregnation with acetone as solvent. The 5.0 wt% Ru/CNTs catalyst showed much higher activity (*e.g.*, 84.7% NH_3 conversion at 500 °C) than the 10.0 wt% Ru/SiO_2 (*e.g.*, 64.0% NH_3 conversion at 500 °C)

[a]*State Key Laboratory of Chemical Engineering, East China University of Science and Technology, Shanghai 200237 China. E-mail: xgzhou@ecust.edu.cn*
[b]*Department of Chemical Engineering, Norwegian University of Science and Technology, Trondheim 7491, Norway*

[26]. These results indicate that preparing the catalyst in a controlled way could remarkably increase the activity.

Ammonia decomposition is known to proceed by stepwise dehydrogenation, followed by recombination of N and H to form N_2 and H_2, respectively. There is widespread consensus that the binding energy of N on a transition metal surface is a good descriptor for ammonia decomposition [10,13,28–32]. Too high a binding energy of the N atom to the metal surface is unfavorable for the associative desorption of strongly adsorbed N atoms, while too low a binding energy is unfavorable for the dehydrogenation of NH_x species [27,28,32]. This trade-off leads to a volcano-type relationship between the N binding energy and the rate of ammonia decomposition. Over the past a few decades, the catalytic properties of ammonia decomposition catalysts have been tuned either by modifying their compositions, which mediate the electronic structures, or by tailoring their sizes and shapes, which determine the surface atomic arrangements and coordinations [10,33–40]. It is expected that by manipulating the composition and structure of the catalysts, they have the optimal N binding energy and therefore high activities for ammonia decomposition.

In this chapter, we will review recent advances in the structural manipulation of Ru, Ni, Fe and bimetallic catalysts for ammonia decomposition, and explore the structure-activity relationship.

2 Ru-based catalysts

2.1 Size and shape effects

Experimental studies on single-metal catalysts have shown that Ru is the most active decomposition catalyst [13,16,36–40]. Ammonia decomposition over Ru-based catalysts is a structure-sensitive reaction. Rarog-Pilecka *et al.* [36] reported that the large nanoparticles of ruthenium (8.0 nm) have a TOF_{NH3} 8 times higher than the small nanoparticles (1.2 nm), in which the nanoparticles are assumed spherical shape independent of the size. However, the change of Ru particle size is usually accompanied by the change of Ru shape (Fig. 1a) [13,38–40]. Both the effects of Ru particle size and shape can be very difficult to understand, and to compound them at the same time is almost impossible to tackle. Therefore, the researchers use the number of Ru B_5-sites which are the active sites for ammonia decomposition to elucidate the difference in the catalytic activities [13,36–41]. As shown in Fig. 1b, the B_5-sites of Ru consist of an arrangement of three Ru atoms in one layer and two further Ru atoms in a lower layer. Jacobsen *et al.* [42] used first principles calculations to study the fraction of edge atoms and active sites relative to the total number of atoms as a function of the size of Ru particles, which were assumed hexagonal, and found that the maximum probability of finding B_5 sites is on Ru nanoparticles of *ca.* 2.0 nm (Fig. 1c). They further showed that the decreasing fraction of B_5 sites with increasing Ru crystal size was not due to a general decrease in the number of these sites but to the fast decrease in the fraction of edge atoms [42]. Recently, Karim *et al.* [13] showed that the maximum number of B_5 sites as a percent of surface atoms highly depends on Ru shape, and occurs at ~7.0 nm for elongated nanoparticles compared to that at ~1.8–3.0 nm for hemispherical

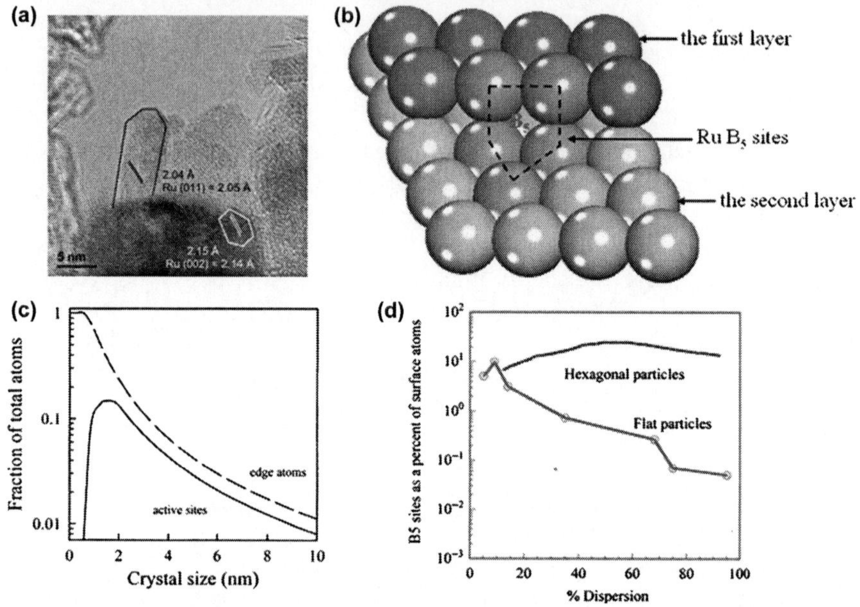

Fig. 1 (a) HRTEM image of Ru/γ-Al_2O_3 showing an elongated rectangular-shaped particle (left) and a more isotropic particle (right), reproduced from A. M. Karim, V. Prasad, G. Mpourmpakis, W. W. Lonergan, A. I. Frenkel, J. G. Chen and D. G. Vlachos, *J. Am. Chem. Soc.*, 2009, **131**, 12234 with permission of American Chemical Society; (b) schematic diagram of Ru B_5-sites; (c) fraction of edge atoms and active sites relative to the total number of atoms as a function of Ru crystal size, reproduced from C. H. Jacobsen, S. Dahl, P. L. Hansen, E. Tornqvist, L. Jensen, H. Topsoe, D. V. Prip, P. B. Moenshaug and I. Chorkendorff, *J. Mol. Catal. A: Chem.*, 2000, **163**, 25 with permission of Elsevier; (d) B_5 sites as a percent of total surface sites as a function of the Ru dispersion, reproduced from A. M. Karim, V. Prasad, G. Mpourmpakis, W. W. Lonergan, A. I. Frenkel, J. G. Chen and D. G. Vlachos, *J. Am. Chem. Soc.*, 2009, **131**, 12238 with permission of American Chemical Society.

nanoparticles (Fig. 1d). These findings indicate that instead of the size and shape effects of Ru nanoparticles, the number of Ru B_5 sites can serve as a criterion for a rational interpretation of the structure sensitivity of ammonia decomposition over Ru catalysts.

2.2 Support effects

The supports of Ru catalysts are mainly employed to enhance the dispersion, surface area and high-temperature stability of Ru nanoparticles. Au and co-workers studied the effect of the supports on Ru activity for ammonia decomposition, and indicated that the catalytic performance of Ru catalysts was support-dependent and the conductivity of support was a necessity for a highly active Ru catalyst [16,27,43–46]. On that basis, the researchers further correlated the relationship between the graphitization degree of carbon materials and the Ru activity (Fig. 2) [47,48]. They reported that the higher was the graphitization degree, the higher was the Ru activity. It must be noted that in these studies the size effects of Ru catalysts were not excluded, and the effects of the microstructures of different carbon materials were not revealed. Further investigations are therefore necessary for a deep understanding of the support effect of the catalyst.

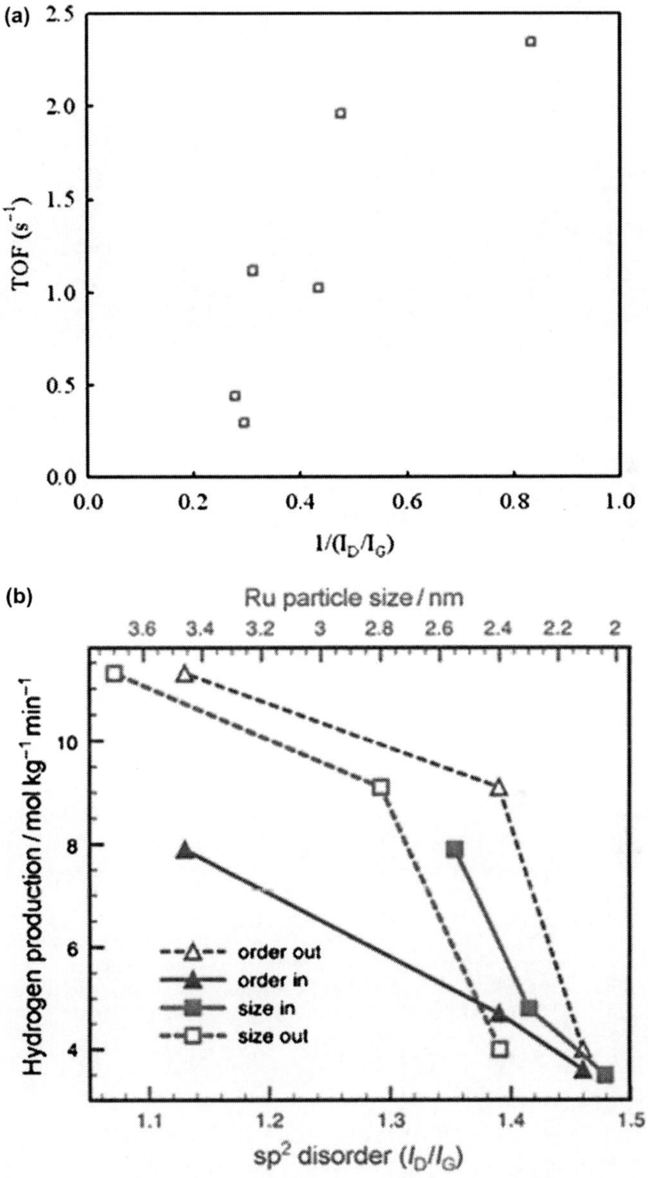

Fig. 2 (a) Turnover frequency for NH_3 decomposition as a function of $1/(I_D/I_G)$; (b) hydrogen production for NH_3 decomposition as a function of sp^2 disorder and Ru particle size, reproduced from W. Q. Zheng, J. Zhang, B. Zhu, R. Blume, Y. L. Zhang, K. Schlichte, R. Schlogl, F. Schuth and D. S. Su, *ChemSusChem*, 2010, **3**, 228 with permission of John Wiley and Sons.

Two nano-carbon materials with different graphitization degrees, *i.e.*, carbon nanofibers (CNFs) and CNTs (their HRTEM images shown in Fig. 3), were employed to support 3.2 wt% Ru catalysts, where CNTs had a graphitization degree higher than CNFs [37,41]. H_2-TPR was first used to characterize the interaction of Ru with carbon materials. The results showed that Ru particles had a stronger interaction with CNTs, which was beneficial for the transfer of electrons from support to Ru, and would

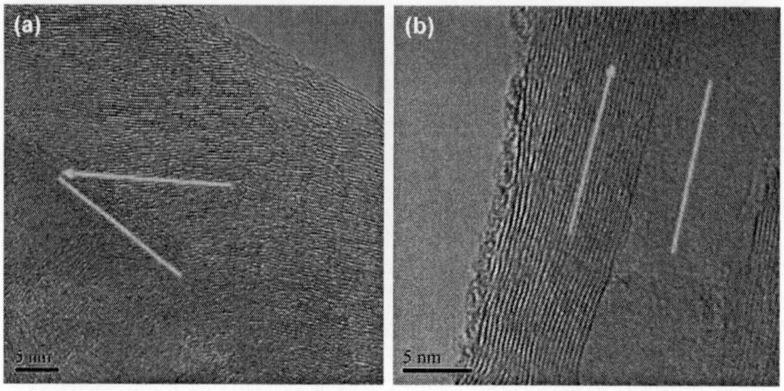

Fig. 3 HRTEM images of CNFs (a) and CNTs (b). Arrows and lines represent the direction of graphene layer and fiber axis of nanocarbon materials, respectively. Reproduced from X. Z. Duan, J. H. Zhou, G. Qian, P. Li, X. G. Zhou and D. Chen, *Chin. J. Catal.*, 2010, **31**, 981 with permission of Elsevier.

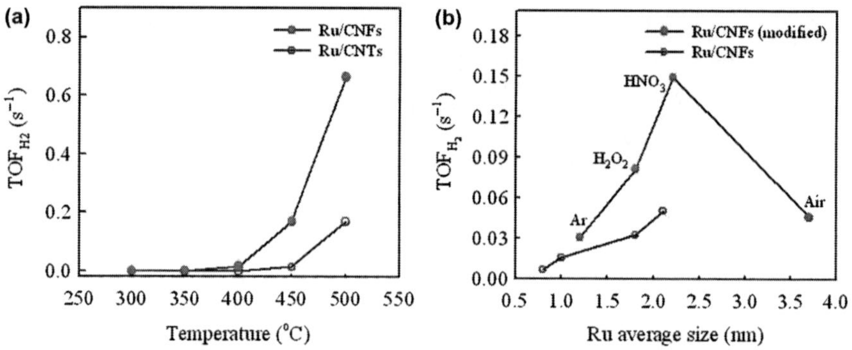

Fig. 4 (a) TOF_{H2} as a function of reaction temperature over Ru/CNFs and Ru/CNTs catalysts, reproduced from X. Z. Duan, J. H. Zhou, G. Qian, P. Li, X. G. Zhou and D. Chen, *Chin. J. Catal.*, 2010, **31**, 982 with permission of Elsevier; (b) TOF_{H2} as a function of Ru average size over Ru supported on CNFs with and without oxidation treatment, reproduced from X. Z. Duan, J. H. Zhou, G. Qian, P. Li, X. G. Zhou and D. Chen, *Chin. J. Catal.*, 2010, **31**, 985 with permission of Elsevier.

facilitate the associative desorption of N and consequently enhance the activity of the catalyst. Both Ru catalysts with similar average particle size were used to catalyze ammonia decomposition. Unexpectedly, Ru nanoparticles supported on CNFs were more active than those on CNTs (Fig. 4a), indicating that neither the graphitization degree alone is adequate to be directly correlated with the activity of Ru catalysts, nor the interaction between the Ru particle and the carbon support can account for the high activity of the Ru catalysts supported on CNFs. The differences in the microstructures of both nanocarbon materials (*e.g.*, surface defects, unsaturated carbon atoms and oxygen containing groups) may be responsible for the different activities of both Ru catalysts. In fact, there are two types of oxygen containing groups on carbon materials: (i) electron-withdrawing groups (such as carboxyl, hydroxyl, carbonyl, ester and anhydride) and (ii) electron-donating groups (such as phenolic hydroxyl and phenolic ether) [41]. Only the electron-donating groups facilitate the electron transfer from support to Ru, and hence enhance the Ru activity. Moreover, the surface defects and unsaturated carbon atoms on carbon materials are rich in electrons, and will facilitate ammonia decomposition on Ru particles. Compared to CNTs, CNFs have more surface defects, unsaturated carbon atoms and electron-donating groups than CNTs [49]. This provides a rational explanation why CNFs supported Ru catalysts have higher activities.

To explore the properties of the oxygen containing groups, the CNFs were further treated following the procedures reported earlier [37]. The TOFs of the Ru/CNFs catalysts were plotted in Fig. 4b to compare the activities of Ru particles supported on CNFs with and without oxidation treatment. The results showed that the Ru particles supported on the annealed but non-oxidized CNFs have similar activity to those supported on raw CNFs. It is well known that annealing the CNFs at high temperature in Ar atmosphere increases the graphitization degree and thus the electrical conductivity of the CNFs. This again demonstrated that the graphitization degree alone can not account for the similar activity of both Ru catalysts. More interestingly, Ru catalysts supported on CNFs oxidized by H_2O_2 and HNO_3 exhibited higher activities, which was because the treated CNFs have an abundance of electron donating groups, surface defects and unsaturated carbon atoms [41,50].

Moreover, the effects of the acidity-basicity properties of supports on the activities of Ru catalysts were studied by Au and co-workers [16,27]. They reported that the activity had a declining trend with the decrease of support basicity (Table 1), and hence suggested that the basicity of support was a necessity for a highly active Ru catalyst. However, the Ru nanoparticles supported on different supports had a large difference in Ru average particle size, which would also have a great impact on the activity. Therefore, further investigations are needed to disclose the effect of the acidity-basicity properties of supports on the Ru activity for ammonia decomposition.

2.3 Promoter effects

The addition of promoters (such as alkali, alkaline earth and rare earth metal ions) is routinely used to enhance the activities of catalysts [16,27,43–46,51–61]. Au and co-workers [45] systematically investigated the effects of the promoter cations (anion being NO_3^-) on the catalytic performance of

Table 1 Effects of supports on catalytic performance of Ru catalysts for NH_3 decomposition (Ref. 16 and 27).

Catalyst	$TOF_{H2}(s^{-1})$	NH_3 desorption peak (K)	CO_2 desorption peak (K)	Ru average particle size (nm)
Ru/CNTs	1.1	—	—	3.9
Ru/AC	1.2	—	—	4.1
Ru/TiO_2	0.5	415	408	—
Ru/Al_2O_3	1.5	369	440	8.7
Ru/MgO	2.0	—	375, 442 and 572	9.1
Ru/ZrO_2-BD	0.4	403 and 610	—	9.4

Table 2 NH_3 conversion and H_2 formation rate over Ru/CNTs modified by potassium with different anions at 643 and 673 K (Ref. 45).

	NH_3 conversion (%)		H_2 formation rate (mmol/(min · g_{cat}))	
Catalyst[a]	643	673	643	673
Ru	5.9	8.8	14.5	21.7
Ru-KNO_3	33.3	49.7	57.1	85.2
Ru-KOH	31.6	47.2	55.7	83.2
Ru-K_2CO_3	31.3	46.7	55.2	82.4
Ru-KF	14.8	22.1	28.9	43.1
Ru-KCl	11.3	16.8	23.0	34.3
Ru-KBr	5.7	8.5	12.9	19.3
Ru-K_2SO_4	0.07	0.1	0.8	1.1
Ru-K_3PO_4	0.1	0.2	1.2	1.8

[a] Atomic ratios of Ru/K are 1 in all the modified samples. Reaction conditions: $GHSV_{NH3} = 60000\,mL/(h \cdot g_{cat})$.

Ru/CNTs where the size and shape of Ru nanoparticles were not changed by the modification, and reported that the order of the promotional effect was K-Ru > Na-Ru > Li-Ru > Ce-Ru > Ba-Ru > La-Ru > Ca-Ru > Ru, suggesting that the higher was the electronegativity of the promoter, the lower was the activity of the catalyst. They further studied the effects of the promoter anions (cation being K^+). As shown in Table 2, of all the potassium salts adopted, KNO_3, KOH and K_2CO_3 show similar promotional effects. They suggested that KOH and/or K_2O was the active species of promoters because KNO_3 and K_2CO_3 were thermally decomposed into K_2O, which easily reacted with H_2O to form KOH. Moreover, a similar promotional effect of alkali metals (anion being NO_3^-) was also observed over Ru/MgO catalysts where Cs had a higher promotional effect than K [57].

Various experimental techniques such as nitrogen adsorption [55] or temporal analysis of products (TAP) [61] were carried out to determine the role of the promoters in different Ru catalysts for ammonia decomposition. Rarog-Pilecka *et al.* [55] performed N_2 adsorption experiments to determine the role of the promoters, and found that the role of the alkali (K and Cs) in Ru/C catalysts acting as an electronic promoter was different from that of Ba, which was considered to act as a structural promoter to modify the local arrangement of Ru surface atoms and thus create more B_5 active sites. Garcia-Garcia *et al.* [61] used TAP technique to study the effect of the Na promoter on the mechanism of ammonia decomposition over Ru catalysts. As shown in Fig. 5a, NH_3 conversion remains constant for all the catalysts investigated during the whole multi-pulse experiment because the whole number of NH_3 molecules is lower than that of the exposed Ru atoms. The evolution of the H_2 yield of each catalyst (Fig. 5b) shows that both the Na promoter and the high temperature treatment of the support prevent the diffusion of hydrogen from the metal to the support *via* spill-over. The evolution of the N_2 yield of each catalyst (Fig. 5c) suggests the formation of Na-N-Ru complex at the promoter-transition metal crystallite interface.

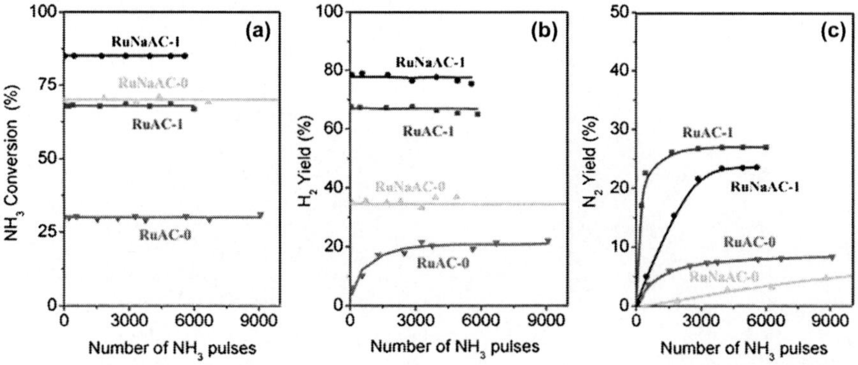

Fig. 5 (a) NH_3 conversion, (b) H_2 yields and (c) N_2 yields at 673 K obtained on RuAC-0, RuNaAC-0, RuAC-1 and RuNaAC-1 as a function of the NH_3/Ar pulse number. AC-0 and AC-1 donate a commercial activated carbon from ICASA-Spain with and without annealing treatment, respectively. Reproduced from Ref. 61 with permission from the PCCP Owner Societies.

Some interesting results have recently been reported concerning the promotion of Ru catalysts [58–59]. Pyrz *et al.* [58] used a hollandite (KRu_4O_8) catalyst supported on γ-Al_2O_3 to catalyze ammonia decomposition, and demonstrated a big enhancement of the NH_3 conversion by 30–50% at T = 350 °C. Klerke *et al.* [59] proposed a novel concept of doping by making use of the atomically dispersed alkali in the bulk titania support. They demonstrated that compared with the Ru/TiO_2 catalyst promoted with Cs, these novel nano-titanates supports have much higher promotional effect. In addition, the order of the promotional effect is Cs > K > Na. Unfortunately, both the hollandite phase and the nano-titanates were found to be unstable at high temperatures. Despite this, these concepts still provide a new perspective for rational design of Ru catalysts for ammonia decomposition.

2.4 High-temperature stability

For supported Ru catalysts, the metal nanoparticles tend to sinter during thermal treatment, leading to a decrease in the number of Ru active sites. In order to obtain active and stable Ru catalysts for ammonia decomposition, Ru nanoparticles were embedded into porous lanthanum-stabilized zirconia (LSZ) supports (Ru@LSZ) [62]. The results showed that the embedding strategy not only minimizes the undesirable sintering of the Ru nanoparticles, but also increases the activity compared to the reference impregnated Ru/LSZ catalyst.

Core-shell structured Ru@M (M = SiO_2, Al_2O_3 and MgO) catalysts were prepared by Au and co-workers [63–64], and were shown to be very active and stable even at high temperatures (up to 650 °C) thanks to the protection of the inert nano-shell. The high activity was accounted for by the enrichment of the reactants in the shells (Fig. 6). However, this interpretation was questioned by Schuth *et al.* [7], because it is contrast to the laws governing mass transfer in porous catalysts. Further investigations are needed to understand the enhanced activity of encapsulated Ru catalysts.

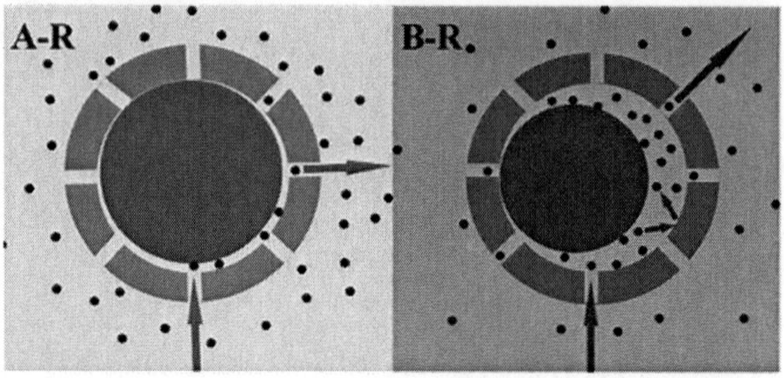

Fig. 6 Illustration of enhanced adsorption and reaction in a microcapsular-like reactor: (A-R) Dried-SMP-R@SiO_2; (B-R) Calcined-SMP-R@SiO_2. SMP and R donate submicroparticle and reduction, respectively. Reproduced from Ref. 63.

3 Ni-based catalysts

Since ammonia decomposition catalysts are aimed for general application such as producing H_2 for PEMFCs, the cost of the catalyst is crucial. To this end, there are two main lines of research and development which are of both fundamental and practical importance: (i) optimizing the noble Ru catalysts to reduce the loading of Ru, and (ii) searching for less expensive but still highly active alternatives.

Optimizing the Ru catalysts has been briefly discussed in Section 2. In the following sections, we will focus on the searching for non-noble metals catalysts, which include monometallic (such as Ni and Fe) and bimetallic catalysts. Ni-based, Fe-based and bimetallic catalysts for ammonia decomposition will be reviewed in Section 3, Section 4 and Section 5, respectively.

3.1 Structure sensitivity

Previous studies on ammonia decomposition over different Ni surfaces suggest that the reaction has a pronounced structure sensitivity [65–71]. Zhang *et al.* [72–73] studied the effect of Ni particle size and showed that the optimal size of Ni nanoparticles for ammonia decomposition appeared around 2.3 nm (Fig. 7). They attributed the reason to the higher concentration of Ni B_5 sites (Fig. 8) for this size, and suggested that the B_5 sites were the active sites for NH_3 decomposition.

Fully in line with these studies, Duan *et al.* [28–30] performed first-principles calculations to study the mechanism of ammonia decomposition on the close-packed Ni(111) and the stepped Ni(211) and Ni(110) surfaces. As shown in Fig. 9, the stepwise dehydrogenation of NH_3 preferentially proceeds on the stepped Ni(211) surface. However, because of the extreme barrier (2.96 eV) of the associative desorption of N, the Ni(211) surface will be blocked by strongly adsorbed N atoms. This indicates that the B_5 sites of Ni(211) surface are not the active sites for ammonia decomposition. A model for the two N atoms blocking Ni(211) surface was constructed and denoted as 2N-Ni(211). The potential energy diagram of ammonia decomposition on the 2N-Ni(211) surface has already included in Fig. 9. The first step of dehydrogenation of NH_3 preferentially proceeds on the stepped Ni(110) surface compared to the stepped 2N-Ni(211) surface. Then, NH_2 transfers to the close-packed Ni(111) surface and proceeds to complete

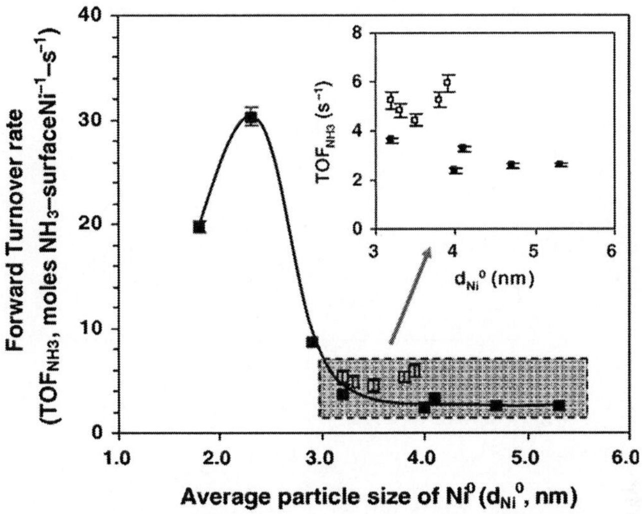

Fig. 7 TOF_{NH3} as a function of Ni particle size (solids: Ni/Al_2O_3; hollows: Ni/La-Al_2O_3). Reproduced from J. Zhang, H. Y. Xu and W. Z. Li, *Appl. Catal. A: Gen.*, 2005, **296**, 260 with permission of Elsevier.

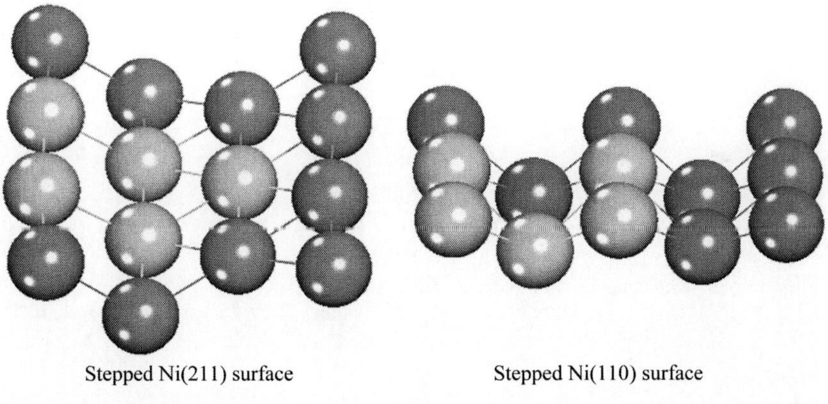

Fig. 8 Schematic diagram of the B_5-sites (labelled by five pale blue balls) of stepped Ni(211) and Ni(110) surfaces.

the catalytic cycle. This indicates that the B_5 sites of Ni(110) surface are also not the active sites for ammonia decomposition. Therefore, the explanation of the volcano-shaped curve of TOF_{NH3} versus Ni particle size (Fig. 7) with the number of B_5 sites is questionable. It is the appropriate ratio of Ni(111) to Ni(110) surface that is responsible for the size effects. These findings can not only reasonably interpret the size effects, but also provide the basis for the rational design of Ni catalysts.

3.2 Powdered Ni catalysts

In terms of cost, Ni is more attractive than Ru. A lot of studies have been performed to develop highly efficient Ni catalysts for ammonia decomposition [26–32,65–75]. Table 3 summarizes the H_2 formation rate over different Ni catalysts at 773 K. The H_2 formation rate was increased from

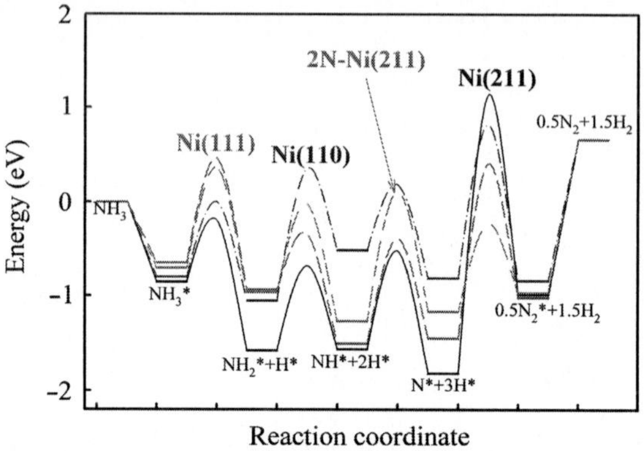

Fig. 9 Potential energy diagram of ammonia decomposition on Ni(111), Ni(211), Ni(110) and 2N-Ni(211) surfaces (Refs. 28–30). X* denotes an adsorbed species and all the energies include the ZPE corrected energies.

Table 3 H_2 formation rate (mmol/($g_{cat} \cdot$ min)) over the supported Ni catalysts at 773 K.

Catalyst	Ni loading (wt%)	H_2 formation rate	Preparation method[b]	Ni dispersion (%)	Ni average size (nm)	Ref.
Ni/CNTs	5.0	2.9	IWI	—	—	[27]
Ni/SiO_2	10.0	3.3	WI	0.9[c]	—	[26]
Ni/SiO_2-Al_2O_3[a]	65.0	6.8	—	7.7[c], 8.5[d]	—	[26]
Ni/Fumed SiO_2	5.0	6.5	IWI	—	—	[51]
Ni/MCM-41	5.0	9.0	IWI	—	—	[51]
Ni/MCM-41	5.0	8.9	TIE	—	—	[51]
Ni/SBA-15	5.0	7.3	IWI	—	—	[51]
Ni/Al_2O_3	58.0	12.8	CP	25.4[d]	3.9	[72]
Ni/SBA-15	80.8	19.1	DP	17.8[d]	5.7	[74]
Ni-CeO_2/SBA-15	13.3	21.7	DP	29.5[d]	3.4	[74]
Ni-CeO_2/Al_2O_3	50.8	24.1	CP	21.0[d]	—	[75]

[a] The catalyst was obtained from Aldrich chemicals.
[b] IWI, WI, TIE, CP and DP denote incipient wetness impregnation, wet impregnation, template ion-exchange, coprecipitation and deposition precipitation, respectively.
[c] From CO chemisorption.
[d] From H_2 chemisorption.

2.9 to 19.1 mmol/($g_{cat} \cdot$ min) by changing the method for catalyst preparation. When the Ni/Al_2O_3 and Ni/SBA-15 catalysts were promoted with ceria, the H_2 formation rates were further increased from 12.8 to 24.1 mmol/($g_{cat} \cdot$ min) and from 19.1 to 21.7 mmol/($g_{cat} \cdot$ min), repectively. The H_2 formation rate over Ni-CeO_2/Al_2O_3 was near to that over 5.0 wt% Ru/CNTs (27.9 mmol/($g_{cat} \cdot$ min)) [27]. Therefore, it is possible for Ni catalysts to have activities comparable to Ru catalysts by manipulating the size and morphology of the Ni particle. However, Table 3 also shows that the supported Ni catalysts are largely different in Ni dispersion and average particle size, thus leaving rooms for further improvement. Although the average size of Ni/SBA-15 catalyst reported by Liu *et al.* [74] was much larger than the optimal average size of Ni/Al_2O_3 catalysts (*ca.* 2.3 nm) reported by Zhang *et al.* [73], the Ni/SBA-15 catalyst exhibited higher activity. However, this has not been elucidated in the literature. Compared with that of ammonia decomposition over Ru catalysts, the reaction mechanism over Ni catalysts has not been well understood, and is in need of further investigations.

3.3 Structured Ni catalysts

In view of the practical applications of ammonia decomposition in energy and environmental industries, the conventional supported catalyst powders are not likely to withstand mechanical shaking, and may clump in a way that leads to fluid bypassing [76]. The powder packing usually has a large flow resistance, and the poor heat transfer through the powders will easily lead to unacceptably non-uniform temperature distributions. The catalysts in the form of structured reactor have smaller characteristic lengths, larger surface to volume ratio, and more compact size, and therefore have the advantages of intensified heat and mass transfer, improved reaction efficiency, and ease of use in portable applications [76–80]. Typical example includes micorfibrous composites bed reactor [81] and monolith structured reactor [12].

Lu and co-workers [81–86] developed a novel sintered microfibrous composite bed reactor for ammonia decomposition where the sintered microfibrous composites (their typical optical photograph shown in Figs. 10a, 10b and 10c) were prepared through wet layup papermaking method.

Fig. 10 Optical photographs of microfibrous structures prepared using 8 μm diameter Ni fibers (a), 6 μm diameter SS-316 fibers with 100–200 μm particulates (b) and 8 μm diameter Ni fibers with 100–200 μm particulates (c); profile of pressure drop (d) as a function of N_2 over per millimetre thickness of microbrous composite in (c). (a) reproduced from Y. Liu, H. Wang, J. F. Li, Y. Lu, H. H. Wu, Q. S. Xue and L. Chen, *Appl. Catal. A: Gen.*, 2007, **328**, 79 with permission of Elsevier; (b) reproduced from Y. Liu, H. Wang, J. F. Li, Y. Lu, Q. S. Xue and J. C. Chen, *AIChE J.*, 2007, **53**, 1845 with permission of John Wiley and Sons; (c) and (d) reproduced from Ref. 83.

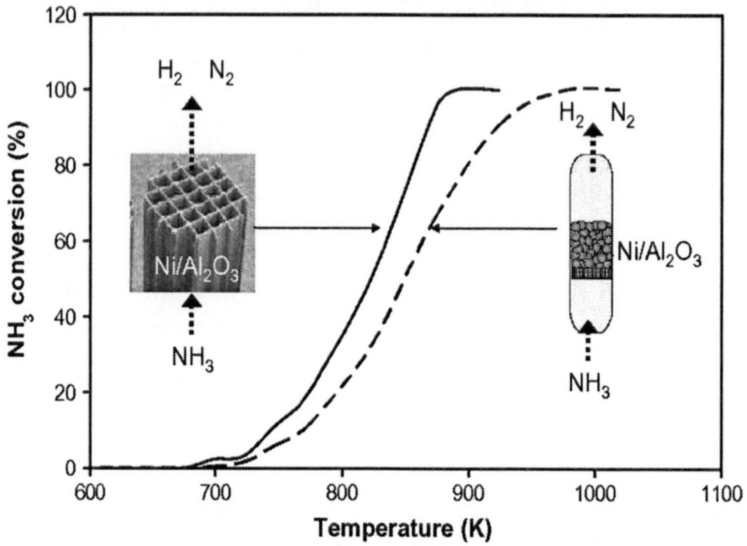

Fig. 11 NH_3 conversion as a function of reaction temperature over powdered Ni/Al_2O_3 and monolith-based Ni/Al_2O_3 catalysts. Schematic diagram of two kinds of Ni/Al_2O_3 catalyst is shown in the insets. Reproduced from Graphical Abstract by C. Plana, S. Armenise, A. Monzon and E. Garcia-Bordeje, *J. Catal.*, 2010, **275**, 228 with permission of Elsevier.

They systematically studied the effects of supports and rare earth oxide promoters on the performance of the microfibrous entrapped Ni catalysts. The results showed that Ni supported on Al_2O_3 was more active than those on SiO_2 or AC, and Ni/Al_2O_3 with 10 wt% Ni loading exhibited the best activity and stability. The order of the promotional effect was $Ni/CeO_2\text{-}Al_2O_3 > Ni/La_2O_3\text{-}Al_2O_3 \sim Ni/Pr_6O_{11}\text{-}Al_2O_3 > Ni/Sm_2O_3\text{-}Al_2O_3 > Ni/Y_2O_3\text{-}Al_2O_3$. At a face velocity of 12.8 cm/s (Fig. 10d), the pressure drop across the microfibrous composite bed was as low as 59 Pa per millimeter of the bed length. Moreover, compared to the packed bed with 2 mm catalyst pellets, the microfibrous composite bed provided a 4-fold reduction of catalytic bed volume and a 5-fold reduction of catalytic bed weight, and reduced the reaction temperature by up to 100 °C.

Recently, Plana *et al.* [12] employed a cordierite monolith reactor, which had a channel size of a few hundreds micrometers, for ammonia decomposition. The cordierite monolith was coated with mesoporous alumina and Ni catalysts were supported by eletrostatic adsorption. As shown in Fig. 11, the monolith-based Ni/Al_2O_3 catalyst achieved a complete conversion of pure NH_3 at a temperature as low as 873 K, which was more active than the powdered Ni/Al_2O_3 catalyst. Moreover, this monolith-based Ni catalyst had a lower pressure drop and exhibited good stability at high temperatures.

4 Fe-based catalysts

Although Fe-based catalysts have lower activities for ammonia synthesis than Ru-based catalysts, they still form the backbone of most catalysts used in ammonia plants because of the low cost and abundant resources. Hence, it comes as no surprise that Fe-based catalysts, with activities for ammonia decomposition lower than Ru- and Ni-based catalysts, have also been extensively studied.

Doubly (*i.e.*, Al_2O_3 and CaO) promoted fused iron catalysts were employed by Kowalczyk *et al.* [87] to catalyze ammonia decomposition. When the concentration of ammonia was less than *ca.* 25%, the rate of decomposition was insensitive to the change of ammonia concentration. However, when the concentration of ammonia exceeded 25–30%, the rate exhibited a systematic increase, presumably due to the formation of successive iron nitride phases. In fact, the catalytic decomposition of ammonia takes place on the surface of iron in parallel with the nitriding reaction. The nitriding of iron produces different iron nitrides, which include α-Fe(N), α''-$Fe_{16}N$, γ'-Fe_4N, ξ-Fe_xN and ζ-Fe_2N [88]. Arabczyk and co-workers [18,89–95] studied the two parallel reactions on Fe catalysts, and found that with the increase of nitriding potential ($P_{NH3}/P_{H2}^{1.5}$), the rate of ammonia decomposition increased on α-Fe(N), but decreased on the mixture of α-Fe(N) and γ'-Fe_4N. They suggested that α-Fe(N) was the active species for ammonia decomposition.

Recently, Lu *et al.* [96] prepared highly dispersed γ-Fe_2O_3 ($\sim$6 nm) on CMK-5 carbon and C/SBA-15 composites (Fig. 12) by using wet impregnation and thermal treatment. Owing to the confinement effect of the mesoporous ordered matrices, γ-Fe_2O_3 nanoparticles were fully immobilized within the channels of the supports. On CMK-5 carbon, even with very high Fe-loadings (up to *ca.* 12 wt%), no iron species were observed on the external surface of the support. At a space velocity of 7500 $cm^3/(g_{cat} \cdot h)$, a complete ammonia conversion over the catalysts was maintained at 600 °C for 20 h. At a higher space velocity of up to 60 000 $cm^3/(g_{cat} \cdot h)$, complete ammonia decomposition was achieved at 700 °C. However, some sintering and migration of large particles to the external surface of the CMK-5 support were observed.

In addition to the confinement effect of the mesoporous ordered matrices, the core-shell structured materials were also used to prevent the Fe nanoparticles from aggregation at high temperatures [64,97–99]. For example, the core-shell structured Fe catalysts (Fig. 13) prepared by Li *et al.* [98] were found to be more active and stable than the naked Fe catalysts, and the mesoporous silica shells encapsulated Fe catalysts exhibited higher activity than the microporous silica shells encapsulated Fe catalysts.

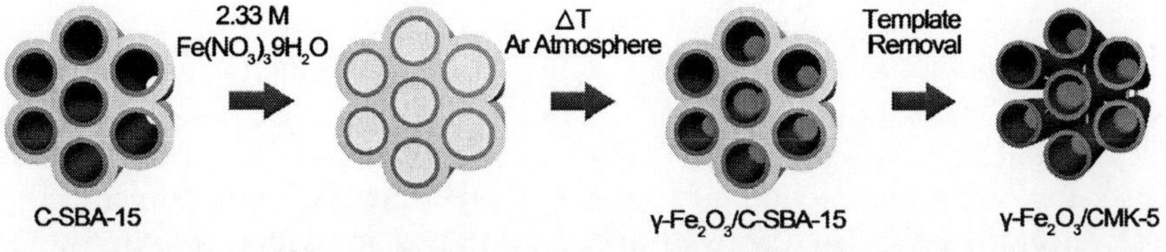

Fig. 12 Illustration of synthesis priciple of γ-Fe_2O_3 nanoparticles spatially confined within the tubes of CMK-5 carbon. Reproduced from A. H. Lu, J. J. Nitz, M. Comotti, C. Weidenthaler, K. Schlichte, C. W. Lehmann, O. Terasaki and F. Schuth, *J. Am. Chem. Soc.*, 2010, **132**, 14152 with permission of American Chemical Society.

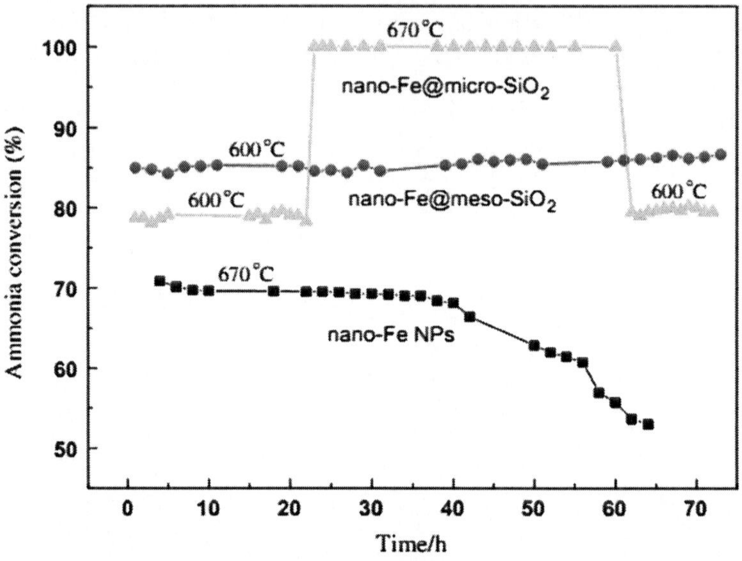

Fig. 13 Stability test of nano-Fe, nano-Fe@micro-SiO_2, and nano-Fe@meso-SiO_2 catalysts. Reproduced from Y. X. Li, S. Q. Liu, L. H. Yao, W. J. Ji and C. T. Au, *Catal. Commun.*, 2010, **11**, 371 with permission of Elsevier.

Fe is a catalyst for the formation of CNTs by catalytic chemical vapor deposition (CCVD), and the Fe particles will finally stay in the CNTs. These CNTs supported Fe particles were also found to be good catalysts for NH_3 decomposition [100]. However, the Fe particles are most likely to exist in the inner pores of CNTs, as a result of the specific growth mechanism of CNTs [101]. Stable as they are owing to the physical isolation by the CNTs, these Fe particles have less accessible surface for the reaction. In this regard, Duan *et al.* [9] proposed a method, which is schematically shown in Fig. 14 and compared with the traditional method for CNFs growth, to control the size and shape of Fe nanoparticles, and the way the Fe nanoparticles stay on the fibrous carbon materials.

As shown in Fig. 15, fishbone carbon nanofibers (FCNFs) were obtained on unsupported Fe catalysts, while platelet carbon nanofibers (PCNFs) were obtained on Fe/mica catalysts. All the Fe particles situated on the top of the CNFs, as a result of the top growth mechanism for the CNFs growth on the mica supported Fe catalysts. Although the shapes of unsupported and mica supported Fe particles were irregular, they ended up with nanoparticles of definite shapes and facets. The Fe particles on PCNFs had a short drum shape, in contrast with the Fe particles with a cone shape on FCNFs. Significant reconstruction of Fe particles had obviously taken place during the growth of CNFs.

The as-obtained Fe-FCNFs and Fe-PCNFs were directly used as catalysts for ammonia decomposition. Fe-PCNFs catalysts were much more active than Fe-FCNFs catalysts and commercial ammonia decomposition catalysts, which could be ascribed to the facts that the Fe particles in the Fe-PCNFs catalysts had more accessible surfaces, preferential surface orientation, and interstitial carbon atoms [9,41]. The reconstructing of the Fe

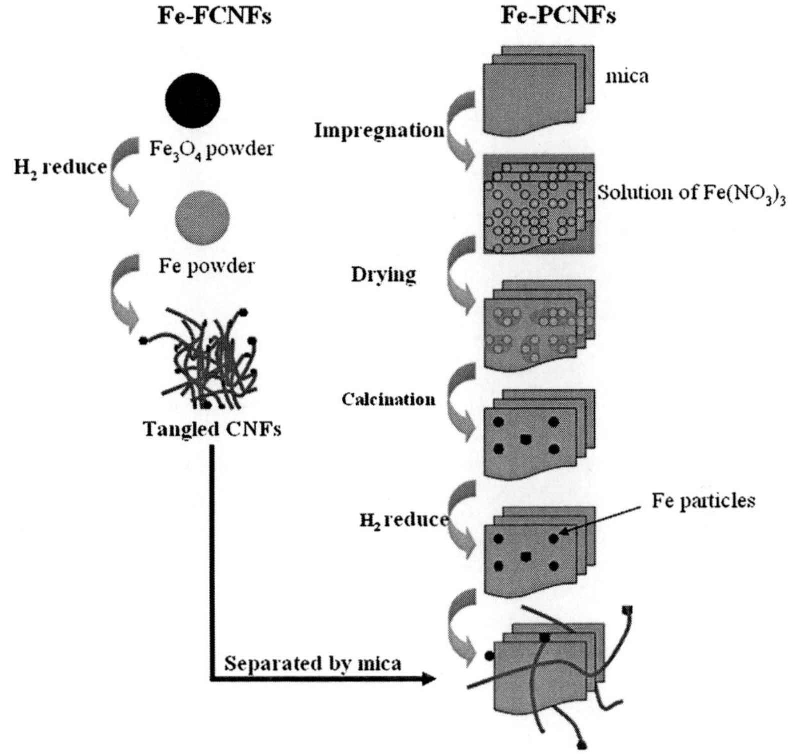

Fig. 14 Schematic diagram of synthesis routes to Fe-FCNFs and Fe-PCNFs catalysts. Reproduced from X. Z. Duan, G. Qian, X. G. Zhou, Z. J. Sui, D. Chen and W. K. Yuan, *Appl. Catal. B: Environ.*, 2011, **101**, 190 with permission of Elsevier.

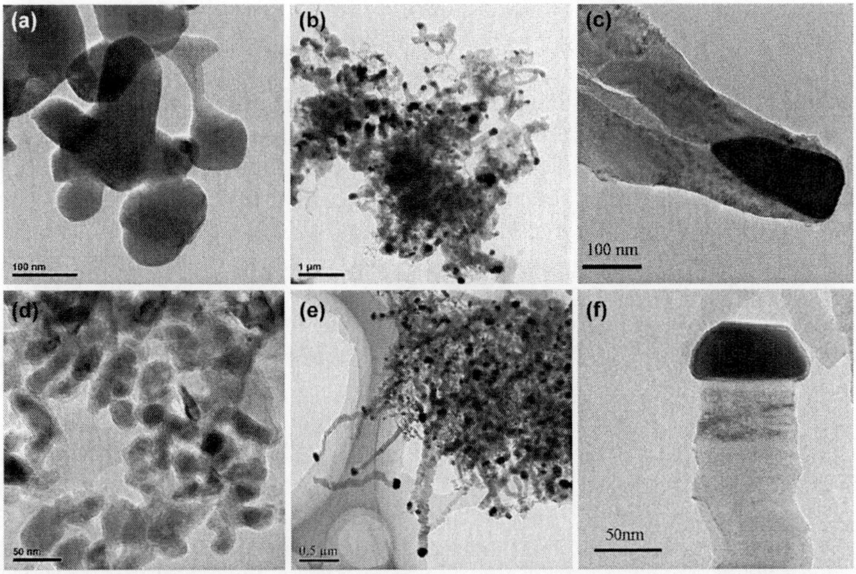

Fig. 15 HRTEM images of unsupported Fe particles (a), Fe-FCNFs (b and c), Fe/mica (d) and Fe-PCNFs (e and f). Reproduced from X. Z. Duan, G. Qian, X. G. Zhou, Z. J. Sui, D. Chen and W. K. Yuan, *Appl. Catal. B: Environ.*, 2011, **101**, 191–192 with permission of Elsevier.

particles by CNFs growth was further explored by using different supports, such as mica, SiO_2 and graphite flakes, and alumina powders [102]. They suggested that the flat surface of the supports favors the formation of platelet carbon nanofibers, thus leaving the Fe particles highly exposed for ammonia decomposition.

5 Bimetallic catalysts

As suggested by the volcano curve of the rate of ammonia synthesis against the N binding energy, there possibly exist some bimetallic catalysts as active as Ru catalysts. Encouraged by the finding of a bimetallic catalyst (Co-Mo) that is more active than Ru [103,104] in ammonia synthesis, great efforts have been made to develop highly efficient bimetallic catalysts for ammonia decomposition in the past decade [17,105–110]. For example, Zhang *et al.* [105] showed that the CNTs supported Fe-Co catalysts with varying Co/Fe ratio were as active as the CNTs supported Co catalysts, and the stability of Co catalysts was significantly improved by alloying with Fe. Liang *et al.* [106] prepared a $NiMoN_y/\alpha\text{-}Al_2O_3$ catalyst and evaluated its activity at 873–1023 K and GHSV = 1800–3600 h^{-1}. Compared to the pure Ni or Mo catalysts, this catalyst was more active and reached the maximum when the Ni/Mo atomic ratio was about 1.5. Lu *et al.* [107] studied the synergistic effects between Fe, Co, Ni and Mo supported on $\gamma\text{-}Al_2O_3$ for ammonia decomposition. They found that the Co promoted Mo catalyst had a higher activity, and the Co-Mo bimetallic catalyst with a Co/Mo atomic ratio of 0.8 exhibited the highest activity. Though the change of the atomic ratio of bimetallic catalysts was usually accompanied by the change of the bimetallic particle sizes, only the effects of the atomic ratio of bimetallic catalysts were concerned so far, while the size effects, which also significantly influenced their activities, were not studied.

MCM-41 with controlled mesoporosity was employed to support Co, Mo and Co-Mo bimetallic catalysts with different Co/Mo atomic ratios, and its channels served as nano-reactors to confine the metal nanoparticles so to have a uniform size distribution [33]. Figure 16 shows the NH_3 conversion as a function of the atomic ratio of Co/(Co + Mo) over MCM-41 supported Co, Mo and Co_xMo_y catalysts. The NH_3 conversion increases with the atomic ratio of Co/(Co + Mo) until 0.7, and decreases when the atomic ratio is further increased. The optimal Co/Mo atomic ratio, Co/Mo = 7/3, is different from that reported by Liu et al, *i.e.*, Co/Mo = 0.8 [107], which is because the size of the bimetallic particles also has great influence on the performance of the catalysts. As shown in Table 4, compared with other bimetallic catalysts for ammonia decomposition, the Co_7Mo_3/MCM-41 catalyst has a much higher activity, which may be due to a better synergistic effect of the bimetallic compound, smaller size of the bimetallic nanoparticles and better balance of the acidity-basicity of the supports.

Hansgen *et al.* [10,111] used density functional theory calculation and microkinetic modeling to identify good monolayer bimetallic (surface or subsurface) catalysts according to the N binding energies. The Ni-Pt-Pt(111) surface, with one monolayer of Ni atoms residing on a Pt(111) substrate,

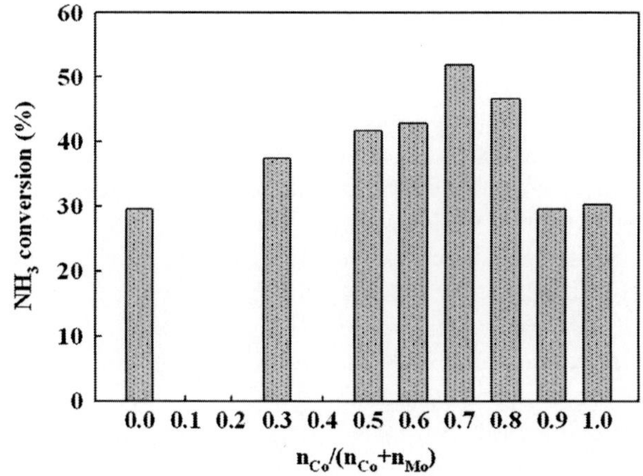

Fig. 16 NH_3 conversion as a function of the atomic ratio of Co/(Co + Mo) over MCM-41 supported Co, Mo and Co_xMo_y catalysts at 773 K (Ref. 33).

Table 4 Catalytic activity of various bimetallic catalysts for NH_3 decomposition at 873 K.

Catalyst	Metal content	$GHSV_{NH3}$ (mL/(h · g_{cat}))	NH_3 conversion (%)	Rate (mol/(g · h))	Ref.
Ni-Pt/Al_2O_3	1.0–5.0% Ni, <1% Pt	6600[a]	78.1	—	[17]
Co_7Mo_3/MCM-41	5.0%	36000	99.2	1.6	[33]
Fe_5Co/CNTs	5.0%	36000	48.0	0.8	[105]
$NiMoN_y/\alpha$-Al_2O_3	4.1% Ni, 6.7% Mo	3600[a]	78.9	—	[106]
Co_3Mo_3N/Mg(Al)O	10.0% Mo	7600[a]	—	0.9	[108]
Co_xMo_yN/γ-Al_2O_3	—	7600[a]	—	0.4	[109]

[a] The unit of ammonia space velocity is h^{-1}.

was predicted to be a catalytically active surface. This was verified by temperature-programmed desorption and high-resolution electron energy loss spectroscopy experiments. However, whether such results could be extrapolated to technical conditions remains to be seen, since the activity is dependent on the exact nature of the system [7].

6 Conclusions and perspectives

Ruthenium is the most active metal for ammonia decomposition, and B_5 sites have also been identified as the active sites. The appearance of the B_5 active sites does not only depend on the particle size, but also the particle shape. The electron donating groups, surface defects and unsaturated carbon atoms of carbon materials are beneficial for the transfer of electrons from support to Ru, and can facilitate the associative desorption of N and consequently enhance the activity of the catalyst. The higher is the electronegativity of the promoter, the lower is the NH_3 conversion. As for the Ni catalyst, the B_5 sites are not the active sites for ammonia decomposition. The stepped Ni(211) surface is easily blocked by strongly adsorbed N atoms. The first step dehydrogenation preferentially proceeds on the stepped Ni(110) surface, and the other steps preferentially proceed on the close-packed Ni(111) surface. An appropriate ratio of Ni(111) to Ni(110) surface is crucial for active Ni catalysts. For Fe catalyst, the active species for ammonia decomposition is α-Fe(N) formed by the nitriding reaction. It must be noted that the cost of the catalyst is crucial for the practical applications of ammonia decomposition. In this regard, Fe is more attractive than Ru and Ni. However, Fe is easy to excessively nitride into lower active iron nitride species. It is necessary to prevent Fe particles from the excessively nitriding process by manipulating their compositions and microstructures in furture study.

Significant advances have been witnessed in searching for more efficient catalysts for ammonia decomposition. The surface groups on carbon nanofibers have great influences on the activity of Ru catalysts. The core-shell structured and embedded Ru catalysts are more active and stable than the traditional impregnated Ru catalysts. Dispersing Ni particles in fibrous framework or on monolithic support produces more active and stable catalysts. Manipulating the size and surface orientation, and isolating the individual Fe nanoparticles by encapsulation or carbon nanofiber growth are very efficient ways to enhance their activities in ammonia decomposition. Bimetallic Co_7Mo_3/MCM-41 catalyst has been identified as very active catalyst. However, a detailed understanding of the structural effect on molecular or atomic level is still lacking, and will be the main research topic in this area. In addition, theoretical investigation also produces very promising results, which are to be verified by experiments.

Moreover, there is an additional research pathway for the practical application of ammonia decomposition, which is the combination of the catalysts with the hydrogen-selective membrane [112–115]. This technology can shift the reaction equilibrium and thus increase the hydrogen yield at a given temperature. Future work is needed to investigate the synergistic effect between the catalysts and the membrane.

Acknowledgements

This work is financially supported by the National Basic Research Program of China (2012CB720501), the China Postdoctoral Science Foundation (2012M520041), the Fundamental Research Funds for the Central Universities (WA1214020, WG1213011 and WA1114006), the Natural Science Foundation of China (21276077), the Shanghai Natural Science Foundation (12ZR1407300) and the 111 Project of Ministry of Education of China (B08021).

References

1 C. B. Xu, J. Donald, E. Byambajav and Y. Ohtsuka, *Fuel*, 2010, **89**, 1784.
2 W. Torres, S. S. Pansare and J. G. Goodwin, *Catal. Rev.*, 2007, **49**, 407.
3 N. Tsubouchi, H. Hashimoto and Y. Ohtsuka, *Energ. Fuel.*, 2007, **21**, 3063.
4 D. Sutton, B. Kelleher and J. R. H. Ross, *Fuel Process. Technol.*, 2001, **73**, 155.
5 Y. Ozawa and Y. Tochihara, *Chem. Eng. Sci.*, 2007, **62**, 5364.
6 O. Gorbenko, O. Stadel, G. Wahl and A. Kaul, *Chem. Vapor Depos.*, 2004, **10**, 109.

7 F. Schuth, R. Palkovits, R. Schlogl and D.S. Su, *Energy Environ. Sci.*, 2012, **5**, 6278.
8 R. Lan, J. T. S. Irvine and S. W. Tao, *Int. J. Hydrogen Energ.*, 2012, **37**, 1482.
9 X. Z. Duan, G. Qian, X. G. Zhou, Z. J. Sui, D. Chen and W. K. Yuan, *Appl. Catal. B: Environ.*, 2011, **101**, 189.
10 D. A. Hansgen, D. G. Vlachos and J. G. Chen, *Nature Chem.*, 2010 **2**, 484.
11 H. Muroyama, S. Saburi, T. Matsui and K. Eguchi, *Appl. Catal. A: Gen.*, 2012, **443–444**, 119.
12 C. Plana, S. Armenise, A. Monzon and E. Garcia-Bordeje, *J. Catal.*, 2010, **275**, 228.
13 A. M. Karim, V. Prasad, G. Mpourmpakis, W. W. Lonergan, A. I. Frenkel, J. G. Chen and D.G. Vlachos, *J. Am. Chem. Soc.*, 2009, **131**, 12230.
14 A. Klerke, C. H. Christensen, J. K. Norskov and T. Vegge, *J. Mater. Chem.*, 2008, **18**, 2304.
15 J. D. Holladay, Y. Wang and E. Jones, *Chem. Rev.*, 2004, **104**, 4767.
16 S. F. Yin, B. Q. Xu, X. P. Zhou and C. T. Au, *Appl. Catal. A: Gen.*, 2004, **277**, 1.
17 A. S. Chellappa, C. M. Fischer and W. J. Thomson, *Appl. Catal. A: Gen.*, 2002, **227**, 231.
18 W. Arabczyk and J. Zamlynny, *Catal. Lett.*, 1999, **60**, 167.
19 G. Papapolymerou and V. Bontozoglou, *J. Mol. Catal. A: Chem.*, 1997, **120**, 165.
20 K. Tamaru, *Acc. Chem. Res.*, 1988, **21**, 88.
21 J. P. McGeer and H. S. Taylor, *J. Am. Chem. Soc.*, 1951, **73**, 2743.
22 K. S. Love and P. H Emmett, *J. Am. Chem. Soc.*, 1941, **63**, 3297.
23 J. K. Dixon, *J. Am. Chem. Soc.*, 1931, **53**, 1763.
24 C. H. Kunsman, *J. Am. Chem. Soc.*, 1929, **51**, 688.
25 C. H. Kunsman, *J. Am. Chem. Soc.*, 1928, **50**, 2100.
26 T. V. Choudhary, C. Sivadinarayana and D.W. Goodman, *Catal. Lett.*, 2001, **72**, 197.
27 S. F. Yin, Q. H. Zhang, B. Q. Xu, W. X. Zhu, C. F. Ng and C. T. Au, *J. Catal.*, 2004, **224**, 384.
28 X. Z. Duan, G. Qian, Y. Liu, J. Ji, X. G. Zhou, D. Chen and W. K. Yuan, *Fuel Process. Technol.*, 2012, doi:10.1016/j.fuproc.2012.05.030.
29 X. Z. Duan, J. Ji, G. Qian, C. Fan, Y. A. Zhu, X. G. Zhou, D. Chen and W. K. Yuan, *J. Mol. Catal. A: Chem.*, 2012, **357**, 81.
30 X. Z. Duan, G. Qian, C. Fan, Y. A. Zhu, X. G. Zhou, D. Chen and W. K. Yuan, *Surf. Sci.*, 2012, **606**, 549.
31 A. Boisen, S. Dahl, J. K. Norskov and C. H. Christensen, *J. Catal.*, 2005, **230**, 309.
32 J. C. Ganley, F. S. Thomas, E. G. Seebauer and R. I. Masel, *Catal. Lett.*, 2004, **96**, 117.
33 X. Z. Duan, G. Qian, X. G. Zhou, D. Chen and W. K. Yuan, *Chem. Eng. J.*, 2012, **207–208**, 103.
34 D. A. Hansgen, D. G. Vlachos and J. G. Chen, *Surf. Sci.*, 2011, **23–24**, 2055.
35 K. Rasim, M. Bobeth, W. Pompe and N. Seriani, *J. Mol. Catal. A: Chem.*, 2010, **325**, 15.
36 W. Rarog-Pilecka, D. Szmigiel, A. Komornicki, J. Zielinski and Z. Kowalczyk, *Carbon*, 2003, **41**, 589.
37 X. Z. Duan, J. H. Zhou, G. Qian, P. Li, X. G. Zhou and D. Chen, *Chin. J. Catal.*, 2010, **31**, 979.
38 W. Q. Zheng, J. Zhang, H. Y. Xu and W. Z. Li, *Catal. Lett.*, 2007, **119**, 311.

39 F. R. Garcia-Garcia, A. Guerrero and I. Rodriguez-Ramos, *Top. Catal.*, 2009, **52**, 758.
40 L. Wang, J. L. Chen, L. Ge, Z. H. Zhu and V. Rudolph, *Energ. Fuel.*, 2011, **25**, 3408.
41 X. Z. Duan, *PhD Thesis*, East China University of Science and Technology, China, 2012.
42 C. H. Jacobsen, S. Dahl, P. L. Hansen, E. Tornqvist, L. Jensen, H. Topsoe, D. V. Prip, P. B. Moenshaug and I. Chorkendorff, *J. Mol. Catal. A: Chem.*, 2000, **163**, 19.
43 S. F. Yin, B. Q. Xu, W. X. Zhu, C. F. Ng, X. P. Zhou and C. T. Au, *Catal. Today*, 2004, **93–95**, 27.
44 S. F. Yin, B. Q. Xu, C. F. Ng and C. T. Au, *Appl. Catal. B: Environ.*, 2004, **48**, 237.
45 S. J. Wang, S. F. Yin, L. Li, B. Q. Xu, C. F. Ng and C. T. Au, *Appl. Catal. B: Environ.*, 2004, **52**, 287.
46 S. F. Yin, B. Q. Xu, S. J. Wang, C. F. Ng and C. T. Au, *Catal. Lett.*, 2004, **96**, 113.
47 L. Li, Z. H. Zhu, Z. F. Yan, G. Q. Lu and L. Rintoul, *Appl. Catal. A: Gen.*, 2007, **320**, 166.
48 W. Q. Zheng, J. Zhang, B. Zhu, R. Blume, Y. L. Zhang, K. Schlichte, R. Schlogl, F. Schuth and D. S. Su, *ChemSusChem*, 2010, **3**, 226.
49 J. H. Zhou, *PhD Thesis*, East China University of Science and Technology, China, 2007.
50 http://www.mhhe.com/physsci/chemistry/carey/student/olc/ch12substituenteffects.html.
51 X. K. Li, W. J. Ji, J. Zhao, S. J. Wang and C. T. Au, *J. Catal.*, 2005, **236**, 181.
52 S. F. Yin, B. Q. Xu, S. J. Wang and C. T. Au, *Appl. Catal. A: Gen.*, 2006, **301**, 202.
53 K. Hashimoto and N. Toukai, *J. Mol. Catal. A: Chem.*, 2000, **161**, 171.
54 W. Rarog, Z. Kowalczyk, J. Sentek, D. Skladanowski, D. Szmigiel and J. Zielinski, *Appl. Catal. A: Gen.*, 2001, **208**, 213.
55 W. Rarog-Pilecka, D. Szmigiel, Z. Kowalczyk, S. Jodzis and J. Zielinski, *J. Catal.*, 2003, **218**, 465.
56 D. Szmigiel, W. Rarog-Pilecka, E. Miskiewicz, Z. Kaszkur and Z. Kowalczyk, *Appl. Catal. A: Gen.*, 2004, **264**, 59.
57 J. Zhang, H. Y. Xu, Q.J. Ge and W. Z. Li, *Catal. Commun.*, 2006, **7**, 148.
58 W. Pyrz, R. Vijay, J. Binz, J. Lauterbach and D. J. Buttrey, *Top. Catal.*, 2008, **50**, 180.
59 A. Klerke, S. K. Klitgaard and R. Fehrmann, *Catal. Lett.*, 2009, **130**, 541.
60 K. Nagaoka, K. Honda, M. Ibuki, K. Sato and Y. Takita, *Chem. Lett.*, 2010, **39**, 918.
61 F. R. Garcia-Garcia, A. Guerrero-Ruiz, I. Rodriguez-Ramos, A. Goguet, S. O. Shektman and C. Hardacre, *Phys. Chem. Chem. Phys.*, 2011, **13**, 12892.
62 B. Lorenzut, T. Montini, C. C. Pavel, M. Comotti, F. Vizza, C. Bianchini and P. Fornasiero, *ChemCatChem*, 2010, **2**, 1096.
63 Y. X. Li, L. H. Yao, Y. Y. Song, S. Q. Liu, J. Zhao, W. J. Ji and C. T. Au, *Chem. Commun.*, 2010, **46**, 5298.
64 L. H. Yao, Y. X. Li, J. Zhao, W. J. Ji and C. T. Au, *Catal. Today*, 2010, **158**, 401.
65 P. M. Gundry, J. Haber and F. C. Tompkins, *J. Catal.*, 1962, **1**, 363.
66 C. W. Seabury, T. N. Rhodin, R. J. Purtell and R. P. Merrill, *Surf. Sci.*, 1980, **93**, 117.
67 M. Grunze, M. Golze, R. K. Driscoll and P. A. Dowben, *J. Vac. Sci. Technol.*, 1981, **18**, 611.

68 M. Huttinger and J. Kuppers, *Surf. Sci.*, 1983, **130**, L277.
69 I. C. Bassignana, K. Wagemann, J. Kuppers and G. Ertl, *Surf. Sci.*, 1986, **175**, 22.
70 G. Ertl and J. Rustig, *Surf. Sci.*, 1982, **119**, L314.
71 D. Chrysostomou, J. Flowers and F. Zaera, *Surf. Sci.*, 1999, **439**, 34.
72 J. Zhang, H. Y. Xu, X. L. Jin, Q. J. Ge and W. Z. Li, *Appl. Catal. A: Gen.*, 2005, **290**, 87.
73 J. Zhang, H. Y. Xu and W. Z. Li, *Appl. Catal. A: Gen.*, 2005, **296**, 257.
74 H. C. Liu, H. Wang, J. H. Shen, Y. Sun and Z. M. Liu, *Catal. Today*, 2008, **131**, 444.
75 W. Q. Zheng, J. Zhang, Q. J. Ge, H. Y. Xu and W. Z. Li, *Appl. Catal. B: Environ.*, 2008, **80**, 98.
76 J. C. Ganley, E. G. Seebauer and R. I. Masel, *AIChE J.*, 2004, **50**, 829.
77 J. C. Ganley, E. G. Seebauer and R. I. Masel, *J. Power Sources*, 2004, **137**, 53.
78 S. R. Deshmukh, A. B. Mhadeshwar and D. G. Vlachos, *Ind. Eng. Chem. Res.*, 2004, **43**, 2986.
79 R. Z. Sorensen, A. Klerke, U. Quaade, S. Jensen, O. Hansen and C. H. Christensen, *Catal. Lett.*, 2006, **112**, 77.
80 Christian, M. Mitchell, D. P. Kim and P. J. A. Kenis, *J. Catal.*, 2006, **241**, 235.
81 M. M. Wang, J. F. Li, L. Chen and Y. Lu, *Int. J. Hydrogen Energ.*, 2009, **34**, 1710.
82 Y. Liu, H. Wang, J. F. Li, Y. Lu, H. H. Wu, Q. S. Xue and L. Chen, *Appl. Catal. A: Gen.*, 2007, **328**, 77.
83 Y. Liu, H. Wang, J. F. Li, Y. Lu, Q. S. Xue and J. C. Chen, *AIChE J.*, 2007, **53**, 1845.
84 Y. Lu, H. Wang, Y. Liu, Q. S. Xue, L. Chen and M. Y. He, *Lab Chip*, 2007, **7**, 133.
85 J. F. Li, *PhD Thesis*, East China Normal University, China, 2008.
86 H. Wang, *Master Thesis*, East China Normal University, China, 2007.
87 Z. Kowalczyk, J. Sentek, S. Jodzis, M. Muhler and O. Hinrichseny, *J. Catal.*, 1997, **169**, 407.
88 J. Kunze, "Nitrogen and Carbon in Iron and Steel: Thermodynamics", *Akademie-Verlag in Berlin*, 1990.
89 W. Arabczyk and U. Narkiewicz, *Appl. Surf. Sci.*, 2002, **196**, 423.
90 R. Pelka, I. Moszynska and W. Arabczyk, *Catal. Lett.*, 2009, **128**, 72.
91 Z. Lendzion-Bielun, R. Pelka and W. Arabczyk, *Catal. Lett.*, 2009 **129**, 119.
92 R. Pelka and W. Arabczyk, *Top. Catal.*, 2009, **52**, 1506.
93 W. Arabczyk and R. Pelka, *J. Phys. Chem. A*, 2009, **113**, 411.
94 K. Kiełbasa, R. Pelka and W. Arabczyk, *J. Phys. Chem. A*, 2010, **114**, 4531.
95 R. Pelka, K. Kiełbasa and W. Arabczyk, *Cent. Eur. J. Chem.*, 2011, **9**, 240.
96 A. H. Lu, J. J. Nitz, M. Comotti, C. Weidenthaler, K. Schlichte, C. W. Lehmann, O. Terasaki and F. Schuth, *J. Am. Chem. Soc.*, 2010, **132**, 14152.
97 Y. X. Li, L. H. Yao, S. Q. Liu, J. Zhao, W. J. Ji and C. T. Au, *Catal. Today*, 2011, **160**, 79.
98 Y. X. Li, S. Q. Liu, L. H. Yao, W. J. Ji and C. T. Au, *Catal. Commun.*, 2010, **11**, 368.
99 M. Feyen, C. Weidenthaler, R. Guttel, K. Schlichte, U. Holle, A. H. Lu and F. Schuth, *Chem. Eur. J.*, 2011, **17**, 598.
100 J. Zhang, M. Comotti, F. Schuth, R. Schlogl and D. S. Su, *Chem. Commun.*, 2007, 1916.
101 S. Helveg, C. Lopez-Cartes, J. Sehested, P. L. Hansen, B. S. Clausen, J. R. Rostrup-Nielsen, F. Abild-Pedersen and J. K. Norskov, *Science*, 2004, **427**, 426.

102 X. Z. Duan, G. Qian, J. H. Zhou, X. G. Zhou, D. Chen and W. K. Yuan, *Catal. Today*, 2012, **186**, 48.
103 C. J. Jacobsen, S. Dahl, B. S. Clausen, A. Logadottir and J. K. Norskov, *J. Am. Chem. Soc.*, 2001, **123**, 8404.
104 A. Boisen, S. Dahl, J. K. Norskov and C. H. Christensen, *J. Catal.*, 2005, **230**, 309.
105 J. Zhang, J. O. Muller, W. Q. Zheng, D. Wang and D. S. Su, *Nano Lett.*, 2008, **8**, 2738.
106 C. H. Liang, W. Z. Li, Z. B. Wei, Q. Xin and C. Li, *Ind. Eng. Chem. Res.*, 2000, **39**, 3694.
107 C. S. Lu, X. N. Li, Y. F. Zhu and C. H. Zhou, *Chin. Chem. Lett.*, 2004, **15**, 105.
108 Y. Z. Xiang and X. N. Li, *Chin. J. Chem. Eng.*, 2005, **13**, 696.
109 X. B. Zhang, L. Y. Xu, J. F. Yuan, Y. J. Zhang, C. S. Lu and X. N. Li, *Chin. J. Catal.*, 2009, **30**, 613.
110 X. Han, W. Chu, P. Ni, S. Z. Luo and T. Zhang, *J. Fuel Chem. Technol.*, 2007, **35**, 691.
111 D. A. Hansgen, L. M. Thomanek, J. G. Chen and D. G. Vlachos, *J. Chem. Phys.*, 2011, **134**, 184701.
112 F. R. Garcia-Garcia, Y. H. Ma, I. Rodriguez-Ramos and A. Guerrero-Ruiz, *Catal. Commun.*, 2008, **9**, 482.
113 A. D. Carlo, A. Dell'Era and Z. D. Prete, *Int. J. Hydrogen Energ.*, 2011, **36**, 11815.
114 G. Li, M. Kanezashi and T. Tsuru, *Catal. Commun.*, 2011, **15**, 60.
115 G. Li, M. Kanezashi, H. R. Lee, M. Maeda, T. Yoshioka and T. Tsuru, *Int. J. Hydrogen Energ.*, 2012, **37**, 12105.

Applications of electron beam lithography in surface science and catalysis – model-nano-array catalysts

Uwe Burghaus
DOI: 10.1039/9781849737203-00141

Applications of electron beam lithography (EBL) in surface science and catalysis are detailed. Advantages and disadvantages of EBL in that field are critically discussed. Emphasis is placed on ultra-high vacuum model studies utilizing so-called model nano array catalysts which consist of a simple predetermined perriodic arrangement of clusters on a support. Discussed are surface reactions as well as the kinetics and dynamics of the interactions of gas-phase species with EBL catalysts. In addition, physical properties of these model catalysts are describes including theire cleaning, thermal stability, and composition.

1 Introduction

Historically, in heterogeneous catalysis/surface science, single crystal surfaces of metals have initially been studied followed by an increasing interest in metal oxides (including thin films) and, finally, by the interest in metal-on-metal oxide systems. The later samples, typically referred to as model catalysts, are thought to be more realistic models of technical catalysts that can still be handled experimentally and theoretically. Similarly, and very recently, oxides-on-metals (so-called inverse model catalysts), including supported oxidic clusters, gained attention. Simple techniques, such as physical vapor deposition in an ultra-high vacuum environment, have typically been utilized to fabricate a model catalyst with little control of the morphology. Consequently, a significant number of work hours were spent on just characterizing the morphology of the supported model catalysts. However, one must ask whether this approach is still timely for surface science in the 21st century of nanoscience.

In the past, single crystals have typically been purchased by academic research laboratories, not manufactured in them. Unfortunately, nanofabrication techniques are often too complicated, expensive, and too time consuming to allow for combining nanofabrication and surface science research easily in a single research group. Furthermore, only a few "nanosamples" are commercially available. However, despite these difficulties, which may require an interdisciplinary approach, it appears to be the right and timely strategy to consider nanofabricated model catalysts in surface science and catalysis, at least as a complementary approach to the studies of traditional model catalysts.

One particularly interesting class of nano-model catalysts are the so-called model-nano-array catalysts, which consist of a predetermined array of clusters written on a support using electron beam lithography. Naturally,

North Dakota State University Fargo, North Dakota, USA. E-mail: uwe.burghaus@ndsu.edu

the same old strategy is followed again to reduce the complexity of the system to a controllable level, but in addition with perfect control of the morphology of the catalyst utilizing nanofabrication techniques. Electron beam lithography allows for pre-determination of cluster size, shape, height, rim length, chemical composition, and lattice type (*e.g.*, square *vs.* hexagonal cluster arrays).

Size and shape effects are studied both to satisfy scientific curiosity and because they are pertinent for a variety of applications of general interest as diverse as microelectronics and biology. Studying these effects requires perfect control over the morphology, which can be accomplished using model-nano-array catalysts.

Recent projects that combine modern nanofabrication techniques and traditional surface science methods toward applications in heterogeneous catalysis are reviewed. Since this field is literally exploding, this review is restricted to model-nano-array catalysts fabricated by electron beam lithography. While no acronym exists for model-nano-array catalysts, we suggest using MNAC or even shorter NAC (for nano-array catalyst), although the term nanoassembled model catalyst has been used in a more general context.

The field of nanocatalysis is developing rapidly. The literature search for this review was completed in the summer of 2012. Even as this draft is going into print, I would not be too amazed that new and exciting results are being published which could simply, technically, not be included here. In addition, this literature survey focuses only on EBL (see list of acronyms) fabricated systems with an application to surface science/catalysis, preferentially at UHV conditions. A few high pressure catalysis studies are, however, described, as long as these relate to EBL samples. Only a few liquid phase electrochemistry studies and photo-electro chemistry projects are included.

Earlier reviews or comprehensive articles addressing related aspects of nanocatalysis can be found in refs. 1–14.

1.1 What are model-nano-array catalysts (NAC)?

Heterogeneous catalysts are complex, multi-component systems consisting of dispersed nanoparticles supported on silica, alumina, graphite, or other supports. Catalytic properties are often sensitive to particle size, particle number density, nature of the support, and metal support interactions. The dependence among these simultaneously-present factors is currently poorly understood. Thus, there is still an urgent need to study model systems for supported catalysts by means of surface chemistry techniques with the aim of obtaining a deeper understanding of adsorption dynamics, kinetics, and reaction mechanisms with the final goal to tune catalyst performance.

The combination of nanotechnology and surface science towards heterogeneous catalysis is a promising challenge towards this goal. It is, however, still in its infancy – perhaps even more so than nano-electronics and nano-sensor applications. Rather few surface chemistry studies have been conducted that truly attempt to bridge nanotechnology and heterogeneous catalysis.[1–14] A most promising approach in fundamental research is to apply lithographic techniques to manufacture model-nano-array catalysts. (Apparently, no acronym exists. We suggest using NACs.) Such

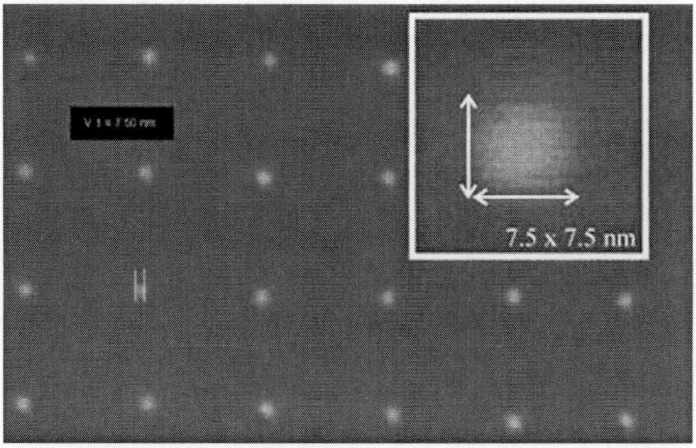

Fig. 1 SEM images of a model-nano-array catalyst consisting of 7.5 nm Au particles in a square array. The sample has been manufactured by Stefano Cabrini (Molecular Foundry) utilizing electron-beam lithography (EBL).[58–60,128]

systems consist of a predetermined geometrical structure of (metal) nanoclusters on supports (Fig. 1). It has been demonstrated that these structures are, indeed, catalytically active and show long-term and high temperature stability for some reactions. Additionally, unique features such as communication effects and bistabilities as well as metal-metal oxide interface effects have been disclosed. Furthermore, catalytically active sites can be identified in a unique fashion. This approach has, however, thus far, been limited by technical difficulties. Most NACs studied so far consist of rather large cluster sizes which are often no more active than single crystal surfaces. Nevertheless, lithographic techniques for manufacturing catalytically highly active structures over macroscopic distances in which cluster sizes are smaller than 10 nm became recently available (see Fig. 1).

1.2 Advantages of model-nano-array catalysts (NAC)

One common criticism of projects on model catalysts (metal clusters on metal oxide supports, see Fig. 2) obtained by conventional UHV techniques (*e.g.*, vapor deposition, single crystals) is that the morphology of the catalyst could not be controlled and characterized precisely enough. At least, such referee reports were obtained from NSF and DoE.

EBL techniques for manufacturing samples provide the ultimate tool for generating nanostructures with perfect control over particle size, shape, and dispersion, at least in principle. The EBL method is a state-of-the-art nanofabrication technique; there is literally no better way to control the morphology of model catalysts. In addition, the particle size, shape, height, rim length, and dispersion as well as the type of materials (cluster/support) can be systematically varied which is basically impossible with conventional techniques such as metal vapor deposition. Template techniques using, for example, porous silica have been developed. However as today, the particle sizes accessible are too large to study the mechanism of chemical surface reactions at the nano-scale. In addition, the type of support cannot be varied easily.

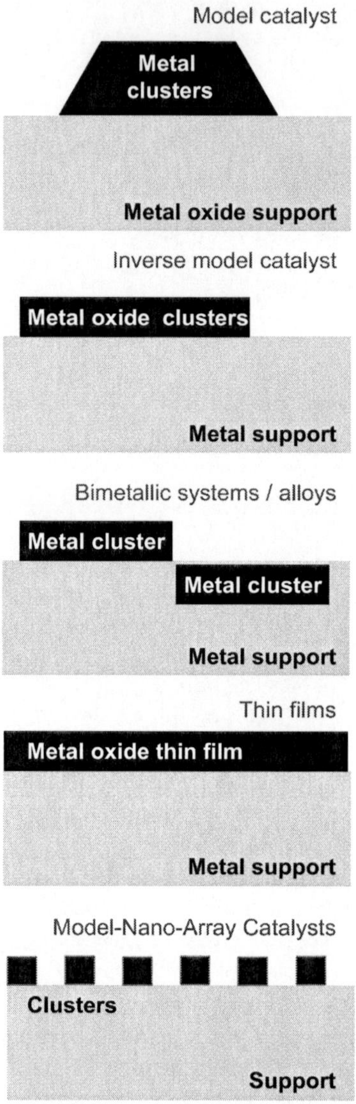

Fig. 2 Concepts of model catalysts.

The second concern is that model catalysts are not of relevance for industrial catalysis. Thus, the obtained results could not be applied to industrial processes. Although high-flux and high-energetic molecular beams can simulate, to some extent, industrial processes,[15] those studies are certainly model studies aimed to obtain mechanistic information that can help improve industrial catalysts. However, in the not-too-distant future, mass production of specifically designed array catalysts (*e.g.*, by nano-imprint techniques), with long-term stability, is not out of the question. In addition, recently the legendary pressure-gap between model studies and real-world catalysts has been objected for a few systems.[16]

It is noted that manufacturing nanostructures is important, but can only be the first step. Thus, a collaboration of groups specialized in nanofabrication and groups devoted to applications of these structures will highly advance both fields.

1.3 Other concepts

Other techniques besides EBL have certainly been developed and are still under development. However, for most of them either the cluster sizes are not variable or the clusters are rather large and/or the pattern cannot easily be extended over macroscopic distances (see also Section 1.5). Recently, solution-based synthetic procedures have been developed which appear to allow for tuning of cluster sizes[17] down to a few nanometers as well as tuning of particles shapes. The surface properties of these nanoparticles are not yet well studied; the procedures have not been tested for a larger variety of materials. The disadvantage of the serial nature of EBL could be overcome using combinations of techniques. For example, manufacturing a master sample by EBL which could be duplicated using imprint techniques (soft lithography). That approach could already be applied to catalysis applications, assuming resources (funding, manpower, instrument time, ...) become available. Most other parallel nanofabrication techniques (*e.g.* LASER interference nanolithography) presently lack sufficient spatial resolution. One may envision, however, to use fast and parallel techniques to prescreen catalytic processes and subsequently EBL samples to determine mechanistic details.

However, it is beyond the scope of this short book chapter to summarize all of the different "bottom up", "top down", and hybrid nanofabrication techniques. Therefore, only a brief glossary, including some references, has been attached to this review. In addition, please, consider the following reviews or more comprehensive papers given in refs. 3–9. Figure 2 illustrates common classes of model systems studied traditionally in surface science and catalysis.

1.4 What fundamental information can be obtained by studying MANCs?

One of the ultimate goals in catalysis is certainly to obtain a microscopic molecular level understanding of heterogeneously catalyst reactions. The perfect control over the cluster morphology using EBL allows contributing to this endeavor by addressing the following questions in a unique way:

- A systematic variation in the cluster size allows for studying, for example, a particle size dependent enhancement of reaction rates systematically. What is the mechanism which leads to the enhanced reactivity seen for nanocluster systems? Especially, this topic is very much in the focus of the surface science community.[18,19] Variations in cluster sizes while keeping a narrow particle size distribution can be accomplished, in principle, with EBL.
- The particle size distribution is extremely narrow using EBL, for example, cluster sizes of (20 ± 1) nm have been manufactured (Table 1). This capability of EBL will allow addressing particle size effects in a unique way. Yes, rather narrow metal-cluster size distributions can also be

Table 1 Studies about model-nano-array catalysts manufactured mostly by electron beam lithography. (Hy. Reveres to hydrogenation/dehydrogenation reactions.) Typically square arrays of dots have been used.

Particle size (nm)	Particle height (nm)	Pitch (nm)	Shape	Material	Supports	Techniques	Applications	Ref.
Physical properties (thermal, mechanical stability, *etc.*)								
25–30	30	100	Dots	Ag	Alumina	TEM, SEM, XPS, AFM, reaction rates	Thermal stability	20
100	20	2 μm	Dots	Ag	Silica	SEM, AFM, SAM	Thermal stability	36
750	36	750 μm	Dots	Ag	Silica	SEM, AFM, SAM	Thermal stability	36
28 ± 2	15	150	Dots	Pt	Alumina	AFM, SEM	Cluster adhesion	73
25–40 ± 1	15	150/200	Dots	Pt	Silica	TEM, AFM	Thermal stability, adhesion	70
28 ± 2	15 ± 2	100 ± 2	Dots	Pt	Silica	AFM, SEM, TDS, AES, TDS, GC	Sample cleaning	32, 95
30–100			"Dots"	Pt	STO(100)		Detailed particle shape analysis	72

Particle size (nm)	Particle height (nm)	Pitch (nm)	Shape	Material	Supports	Techniques	Applications	Ref.
Adsorption kinetics/dynamics								
50	15	200	dots	Pt	silica	High-pressure TDS	H_2, D_2	29
12, 35, 63	5	100	dots	Cu		Beam scattering, XPS, AES	CO, CO2	24, 25, 60, 86
12, 50	5	100	dots	Au	silica	Beam scattering, XPS, AES	CO	58, 62
34 × 150	5	100	cigars	Au	silica	Beam scattering, XPS, AES	butane	68

Particle size (nm)	Particle height (nm)	Pitch (nm)	Shape	Coverage (ML)	Material	Supports	Techniques	Applications	Ref.
Surface reactions, reaction kinetics									
20 × 5	5	250	Wires		Pt	Zirconia, silica alumina, ceria	SEM, AFM, XPS	CO oxidation	92, 93
500	450	1500	Dots	0.11	Pd	silica	SEM, AFM, QMA	CO oxidation Bistabilities etc.	28, 51, 52, 96
750	20	1400	Dots	0.29	Pt	Silica, ceria	SEM, AES, XPS, μ-flow-reac.	CO oxidation	64
20 ± 2	15 ± 1	100 ± 1	Dots	0.04	Pt	Silica Alumina	SEM, AFM, XPS, AES, reaction rates	Ethylene hy.	30
28 ± 2	15 ± 2	100 ± 1	Dots	0.08	Pt	alumina	AFM, SEM, XPS, AES, GC	Ethylene hy.	66, 94
50	15	200	Dots	0.06	Pt	Silica		Ethylene hy.	29
50 ± 0.5	20 ± 0.5	230 ± 2	Dots	0.06	Pt	Silica	TEM, AFM, UPS, SEM	Ethylene hy.	74
50	15	200	Dots	0.06	Pt	Silica	AFM, SEM, XPS, AES	Ethylene hy.	67
30 × 40	7.5	200	Shape tuned	0.03	Pt	Strontium titanate	SEM, x-ray scattering	Oxygen reduction	75

Particle size (nm)	Particle height (nm)	Pitch (nm)	Shape	Material	Supports	Techniques	Applications	Ref.
Optical properties								
44 to 95	23	107–140	Dots	Ag	ITO/glass	SEM, AFM, Absorption spec.	Optics	53
50 to 260	40	350	Tri-angles	Au	ITO/glass	Surface Plasmon spectroscopy	Optics	21
100	50	—	Tri-angles	Ag	Glass	UVvis, AFM, SPR	Biosensors	129
40–520 6–200	—	100–250 100–250	Holes dots	4-MP on Au	Silica ITO	SERS	Size effects	102

obtained by other techniques (see glossary) but typically with less control over particle sizes and only over distance of a few 100 nm.

• A systematic variation of the particle dispersion *via* a variation of the pitch (cluster-cluster distance, lattice constant) can allow for studying the mechanism of diffusion controlled reactions and spill-over effects.[20] In addition, rim effects have been studied using this advantage of EBL samples. The interparticle distance can hardly be controlled precisely with traditional catalyst preparation techniques. Using EBL, lattice constant and cluster size can be adjusted independently.

• For applications in nano-optics also the shape of the nanoclusters can be important.[21] Studying those effects is very difficult with traditional sample preparation techniques although the particle shape dependence of surface kinetics has recently been studied too, however, so far utilizing the Langmuir-Blodgett technique.[22]

• EBL allows for changing the type of metal the clusters consist of systematically while keeping structural parameters constant. Is nano-gold unique, or not?[23]

• Metal oxide clusters can be studied applying appropriate pre/post-processing of the samples.[24,25]

• How does the cluster-support-interface affect the reaction rates? Changing clusters' sizes and pitch (*i.e.* changing the length of the cluster-support interface) and the type of the materials could help to address this question. Recently, we could show that rim effects dominate CO_2 adsorption for Cu-oxide EBL clusters.[24,25]

• The as-prepared clusters are amorphous, but (a partial) crystallization has been observed after annealing the NACs (see Section 5.2). However, it is experimentally quite difficult to determine the crystallinity of small metal clusters. Note that the crystal structure of the model catalyst is often unknown as well. Thus, the same difficulties faced when studying traditional model catalysts is certainly also present when using NAC.

• Communication effects (see Section 6.3.3) are related to locally different reaction rates and adsorption kinetics. Different patches on the catalyst surface "communicate" (*i.e.*, they are coupled) *via* surface diffusion. Although local effects on surfaces have been considered many years ago,[26] correlation and communication effects have only recently been studied experimentally which requires perfect control over catalyst morphology (see *e.g.* refs. 27, 28).

• Variable aspect ratios of NACs, while keeping the cluster sizes constant, could be used to study dynamic effects (angular distributions) including communication effects further (see Section 6.3.3).

• Rim effects, as already mentioned, can be studied in a unique way utilizing EBL samples. Directly decoupling the rim length and cluster area is probably impossible when using simple shapes for the clusters. However, comparing different cluster sizes allows for separating rim and surface effects when using EBL samples (see Section 6.2). This approach does not require a sophisticated morphology characterization.[24,25]

• Other potential applications include: data storage, photonics, (bio)-sensors, single electron devices, nano-electronics, optical-switching devices, nano-optics, and surface enhances Raman scattering. (See Section 7.1).

1.5 Concerns about the concept of NACs – experimental pitfalls, difficulties, and limitations

The main concerns about model-nano-array catalysts may include the following:

- *Can these samples really be made?* Yes, it is, in the meantime, possible to manufacture NACs of macroscopic size (For example, 8 by 5 mm samples, see ref. 29) with particle sizes small enough to be catalytically interesting. For example, in 2006 the fabrication of 10^9 nanoparticles required a writing time of about one day according to ref. 30. In 2008, with a JEOL JBX9300 system (at Argonne National Laboratories) the writing time to cover a 4 by 2 mm area with metal dots amounts to 11 h for 50-by-50 nm particles with a pitch of 60 nm and 35 h for 14-by-10 nm particles with a 60 nm pitch.[31] Larger metal coverages can also be obtained by increasing the thickness of the nanoparticles without increasing the fabrication time, *i.e.*, the aspect ratio is variable. Cluster sizes down to 7 nm have been nanofabricated over macroscopic distance (see Fig. 1).
- *Is the morphology of such small particles controllable?* Yes, this has been shown in a number of cases for MANCs (see Table 1 and Section 5.2) and it is not too astonishing since the morphology of traditional model catalysts could also be "controlled" in the sense that these samples are stable at UHV conditions and sintering can be suppressed by working at low enough temperatures. A nice comparison of SEM figures of the morphology of industrial, model catalysts, and MANCs is given in ref. 32.
- *Can these samples be cleaned without destroying the pattern?* Yes, this has also already been shown; see Section 3. Obviously, traditional sputter-anneal cleaning procedures may not be the first choice although they also have been applied to NACs. However, it is plausible that chemical cleaning procedures such as using oxygen plasma work; see Section 3.
- *Is the sensitivity of kinetics/dynamics techniques good enough?* This is mostly a matter of the writing speed of the EBL system and the advancement of this technology. Most experimental surface science techniques require a total metal coverage of a few percent in order to obtain good enough signal-to-noise ratios. Therefore, dropping the particle sizes down requires also that the pitch (the distance between the metal clusters) be reduced and the e-beam writing time increased dramatically (Fig. 3). However, new EBL systems are already fast enough to tackle this problem successfully. In addition, surface reactions with large reaction rates can be studied and/or high pressure chemical micro-reactors can be utilized (see *e.g.* refs. 33, 34).
- *Can samples which have been transferred through air form one lab to another be studied?* The EBL technique requires "wet-chemistry" procedures and a transfer of samples through air. However, the sample problematic applies to some studies on thin film samples and can be handled experimentally; see *e.g.* ref. 35.
- *Are model-nano-array catalysts heavily affected by sintering effects?* Sintering of metal nanoparticles is a problem in any study of metal cluster systems independent of the fabrication process. Note, however, that Ag clusters in MANCs remain stable up to 700 K under UHV conditions

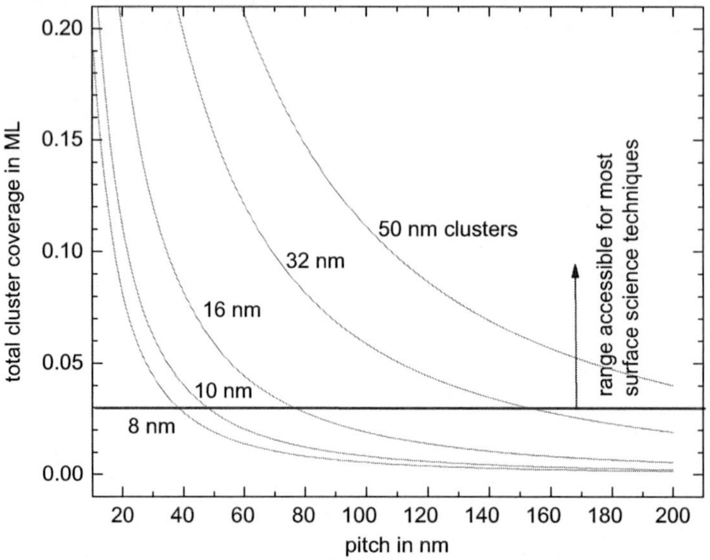

Fig. 3 Total metal coverage *vs.* pitch and parametric in cluster size for square shaped clusters in a square array. What samples can be made using EBL and what samples can still be analyzed by standard surface science techniques?

according to ref. 36 (see also Section 5.1). Furthermore, identical deactivation of Pt foil and Pt nanoparticles has been observed for, for example, the cyclohexene hydrogenation/dehydro-genation reaction.[32] Reactions of choice may be low temperature reactions such as CO oxidation which yields to large CO_2 formation rates at temperatures even below 100 K for Au, Ag, and Cu systems.[37–39] In practice, in our own studies we have not seen any unusual sintering of Cu, Au, and Mo EBL samples, unless temperatures close to the bulk Tammann temperatures were used. On the other hand, studying sintering effects can be desirable since real world catalysts are also affected.

1.6 Expected impact

The following brief list provides a further justification of the approach to study NACs:

- Heterogeneous catalysis is a multibillion dollar industry ($11.2 Bil. in 2006)[40] of pertinent importance for the (U.S.) economy.
- Strategic plan of the NSF[41] states, "...nanotechnology could become a $1 trillion/year industry by 2015".
- Studies devoted to catalyst improvements, even in the longer run, are part of the effort towards "Technology for a Sustainable Environment," which is one of the core research areas of the 21th century probably not limited to the US.
- Nanoscience is highly inter-disciplinary. Therefore, the results obtained in surface chemistry studies will be of great significance outside the surface science and catalysis communities. For example, the gas-surface interaction of O_2 with nanostructured surfaces has been studied as a potential technique for O_2-doping of electronics devices.[42] Thus, we can safely assume that studies on heterogeneous catalysis also will be significant for material science studies.

- It is expected that, in the not-too-distant future, a mass production, *via* nano imprint techniques, of samples tailored towards a given surface reaction (designer catalysts) is achievable. The economic impact will be enormous. In the meantime, NACs are a perfect model for supported catalysts, providing the ultimate control over catalyst morphology.

Some skepticism was encountered in regard to using EBL to nanofabricate samples for catalysis model studies. Therefore, the last two subsections summarize critically potential advantages and concerns of this concept. In summary, it is my believed that NACs are an interesting alternative to traditional model catalysts worthwhile to study.

1.7 Facilities and equipment for fabricating NAC

EBL is currently mostly restricted to large-scale research centers due to the high equipment costs ($\sim 10^6$\$-US). Some of the centers are open-access user facilities (based on peer-reviewed user proposals) at least for non-commercial research applications. I am aware of the following facilities:

- Ernest Orlando Lawrence Berkeley National Laboratory (LBNL) - Nanofabrication Molecular Foundry - (for an info about the center see ref. 43; for projects on NACs see, for example, refs. 30, 44–47)
- IBM research center, Yorktown Heights (project description see ref. 48)
- Center for Nanoscale at Argonne National Laboratory (Chicago)[49]
- Oak Ridge National Laboratory[50]
- Facilities in Europe include Chalmers University of Technology - Microtechnology Laboratory MC2- (Göteborg, Sweden) (see refs. 28, 51, 52 for projects).
- Standard electron microscopes (JEOL6400, CM20-Philips, *etc.*) modified and/or upgraded with a pattern generator have been used, as well (see refs. 53–55).

A spatial resolution down to 3–5 nm (see ref. 54) and 2 nm (= dot size, pitch = 75 nm, see ref. 56), obtained by EBL-based techniques, has been reported although, in these cases, not over macroscopic distances. It appears that nano-model-array catalysts are, meanwhile, also commercially available (see ref. 57), however, currently at a price level that is unacceptable for a project in academia.

2 Nano fabrication of NAC: an example

Many variations of EBL techniques have been developed. (See the reviews given in refs. 3–5 as well as the short glossary attached to this book chapter.) As an example, a typical procedure (used at LBNL Molecular Foundry) to fabricate NACs samples is given below. This version has been used to fabricate the samples studied by our group.[24,25,58–63]

The Vistec VB300 e-beam nanowriter at LBNL allows for control of the pattern with accuracy better than 1 nm and a resolution better than 10 nm. The theoretical spot size is $\sim$4 nm which can be reduced further by post-processing the samples (see below). The high resolution pattern can be written on a large area (5 by 5 mm) within a reasonable time. In addition, the sample size is not too critical applying, for example, molecular beam scattering techniques: 1) The beam spot at the sample position of the molecular beam scattering system can be focused to a diameter below 5 mm; 2) support effects have to be studied anyhow by characterizing blank samples. Another approach is using a micro-flow reactor[64] which allows for studying very small samples ($<1\,mm^2$). Typically less expensive silica/silicon substrates are used to prepare the samples. However, other substrates (silicon nitride membranes, alumina, and graphite) can be used as well. A number of fabrication steps are required which are briefly outlined in the following. This is a generic description. All of these steps were not always used.

- A 50 nm layer of ZEP 520A, an e-beam resist, was deposited on the substrate before the electron beam exposure.
- The pattern was written using a 100 kV e-beam exposure (Vistec system).
- After the exposure, the sample was developed (by ultrasonication in amyl-acetate). In so doing, all e-beam exposed areas were removed.
- With a metal evaporator, a thin layer of typically 3–5 nm of *e.g.* Cu/Au/Mo was deposited on the patterned and 20° tilted substrate. Tilting the metal evaporator reduces the spot size further. That trick has, however, not been used for all of our samples.
- A UV ozone plasma was sometimes used to clean the surface for the following lift-off process. (The same process can be used for sample pre-cleaning before starting UHV experiments.)
- Next is the so-called lift-off process (in dichloromethane). The lift-off removes the residual (none-developed) e-beam resist as well as excess metal.
- The metal deposited on the surface forms initially small dots of variable size. Annealing the sample (at 500 K) collapses all the small dots in a single drop leading to an extremely narrow particle size distribution and a patterned surface (Fig. 1).
- Argon ion plasma etching can be used to further reduce the dot size. Finally, dots smaller than 10 nm can be obtained (see Fig. 1).

In 2012, an area of 5×5 mm of those small dots can be written by the electron beam in a few hours. This time is short enough to guarantee good stability of the nanowriter in order to obtain good uniformity of the sample.

3 UHV sample cleaning and sample stability

Developing reliable cleaning procedures is necessary for any subsequent experimental study.[65] Therefore, cleaning procedures are outlined here in detail. The systems studied so far are summarized in Table 1. The cleaning procedures for NACs are similar to cleaning thin film samples which have been transferred through air.[35] Thus, chemical cleaning is preferred so as not to destroy the sample although mild sputtering has also been used successfully. Examples are explicitly listed in the following.

3.1 Platinum NACs

Main impurities were CO and oxygen which could be removed by annealing at 573 K in 1×10^{-6} torr NO_2 and subsequent removal of oxygen by dosing 10 L of CO and subsequently flashing the sample to 573 K.[30,66] It has been pointed out that AES scans can lead to carbon impurities. In another study, Ne ion sputtering (1×10^{-5} torr, 1 kV, emission current 20 mA, sample current 2.6 μA, 5 min) followed by annealing in oxygen (1×10^{-6} torr at 800 K, 5 min) and flash to 950 K has been used.[32,67] The sputtering process replicated the Pt pattern in the silica support as determined by moving the Pt cluster with an AFM tip. No indications for a contamination of the support by sputtering Pt off the clusters onto the support have been seen. Cleaning procedures using an oxygen plasma (10–100 mTorr at 300 K) have been used which lead, apparently, to an encapsulation of the Pt clusters by Si from the silica support.[67]

3.2 Palladium NACs

Cleaning in a flow reactor (1 h, 770 K, 5%O_2 in Ar; 1 h, 820 K, 2%H_2 in Ar, 500 ml/min) as well as UHV (4×10^{-4} mbar from a gas doser for 2 h, 650 K; CO exposure).[28] For the rather large sized NACs studied here, the formation of a Pd ring around Pd "towers" has been observed as an effect of the annealing procedures.

3.3 Silver NACs

No sintering of Ag/alumina-NACs has been seen for annealing temperatures as high as 773 K based on a TEM/SEM analysis.[20] At 873 K, some mobility of the Ag clusters was present accompanied by Ag sublimation, but still no agglomeration of the clusters has been seen. Certainly, sintering will depend on the pitch of the metal clusters as well as their size.

3.4 Gold NACs

NO_2 annealing turned out to be less efficient and generated a heavy load on the pumps of our vacuum system. The gold clusters sample was cleaned using an atomic hydrogen doser, AES and XPS were used to verify the cleanliness.[61,62] Please note that an unintentional contamination by W from hot filaments is difficult to identify by AES alone.

3.5 Copper-oxide NACs

The 12 and 63 nm Cu/SiO_2 model-nano-array catalysts studied were initially cleaned by annealing in a flux of atomic hydrogen ($p(H_2) = 7 \times 10^{-8}$ mbar, $T_s = 480$ K) using a commercial thermal hydrogen doser. This removes carbon containing impurities as judged by AES. In some cases, sulfur impurities were present which could be removed by annealing in 2×10^{-6} mbar of molecular oxygen at $T_s = 500$ K.[24,25,60]

3.6 Molybdenum NACs

Due to the large initial C impurity level, atomic hydrogen treatments were too inefficient. Therefore, the Mo EBL clusters were cleaned by mild sputtering down to C impurity levels of 10% judged by AES. The final cleaning included cycles of oxygen annealing at 400 K.[68] Figure 4 depicts the Mo cluster samples made at Molecular Foundry (Berkeley). The strategy here was keeping the total metal coverage about constant, but changing the total rim length in order to identify rim site and cluster-support effects.

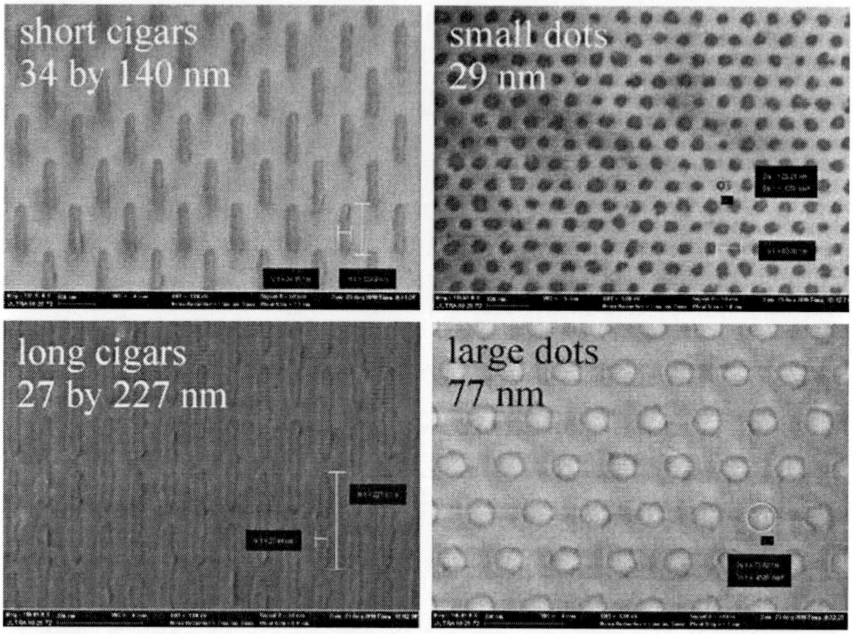

Fig. 4 Scanning electron microscopy image of (silica supported) Mo model-nano-array catalysts fabricated at Molecular Foundry (Cabrini, Dhuey, Schmidt) using electron beam lithography.

4 Characterization of morphology and chemical composition

AFM and STM, although widely used for model catalyst, are perhaps not the techniques of choice for NAC since it is difficult to obtain the true size of the nanoclusters due to convolution effects with the tips of these microscopes.[32] Furthermore, tip-induced morphology changes have been observed.[32] On the other hand, electron microscopy can be used easily to obtain this information. Charging effects of the support can be handled even with less expensive supports such as silica wafer.[69] For the smallest structures, silicon nitride membranes can be used as a support which allows a TEM characterization. TEM systems have a better spatial resolution than SEM. TEM studies have also been conducted on thinned silica supports.[70]

AFM appears to be useful in determining the height of NACs, see refs. 30, 53. In some cases, it has been reported that the height of the particles cannot be varied independently from their lateral dimensions.[53]

AES, which is available in most surface science laboratories, can be used to determine the cleanliness and total metal coverage on the surface. In addition, AES also allows for determination of the oxidation state of metal clusters. XPS has been applied to NACs in several studies (see Table 1) and may allow better to distinguish metal contaminations (from hot filaments) from the actual clusters.

5 Physical properties

Physical properties, as far as they may affect catalytic behavior, are summarized in this section. Note that the optical properties of NACs have been studied quite extensively. However, this brief review focusses on catalysis and, therefore, includes only a very brief summary of other related topics, see Section 7.

5.1 Thermal stability

Pt/alumina-NACs were reported to be thermally stable up to annealing temperatures of 773 K in UHV[20] and up to 680 K in oxygen, hydrogen, or ethylene (1×10^{-6} torr).[29] Detailed experiments about the thermal stability also have been conducted for Ag/silica-NACs.[36] The Ag/NACS were stable in H_2 up to 980 K, but evaporated at 1030 K. In one atmosphere air, Ag oxidized below 500 K; the oxide remained stable up to 650 K and the sample was destroyed at annealing temperatures of 700 K. The Ag clusters have a greater stability in oxygen-free environments, as expected. Thus, the Pt/Ag-NAC demonstrates excellent thermal stability at reducing and oxidizing conditions. In our own studies we collected SEM images before and after the UHV experiments.[58] Sintering was absent for the Au and Cu-oxide clusters which were annealed up to 500 K in UHV[58] similarly to Mo clusters. The samples were stable for months even when transferred through air from one lab to another. However, sintering studies on very small EBL clusters (perhaps available in the future) could be another application of NAC since no macroscopic samples would be required for such projects, using TEM and STM as analytic tools.

5.2 Crystallinity and particle shapes

One criticism of studies on NACs concerned a possibly bad crystallinity of samples obtained by EBL. However, in the meantime, several examples in the literature indicate that simple annealing of the samples in vacuum can lead to crystalline metal nano clusters also when fabricated by EBL. The annealing does not necessarily lead to particle sintering when NACS of rather large pitch are used.

5.2.1 Predicted particle shapes and crystal structures. What crystal structure small aggregates form and how their gas-phase structure (unsupported cluster) changes due to the deposition on supports have been studied extensively by theoretical techniques, see *e.g.* ref. 71. The Wulff theorem, which dates back to a publication in 1901 (Z. Krislallographie), describes the equilibrium shape of unsupported (free) clusters. This has been extended to include the effect of a support (Wulff-Kaischew theorem, 1952). Examples, of modern DFT calculations can be found in *e.g.* ref. 72.

5.2.2 Platinum-NACs. Several studies address the crystallinity of metal clusters obtained by EBL techniques. For example, the crystallinity of Pt/alumina-NACs has been studied with TEM.[73] The as-prepared 28 nm Pt clusters were polycrystalline with crystallite domain sizes of 5 nm. Annealing the sample (for 3 h at 800 K in UHV) leads to the formation of a mixed array of amorphous and polycrystalline Pt clusters. According to ref. 73, some Pt clusters appear to be crystalline and the averaged crystallinity improved while annealing the samples. Bright field (BF) TEM and dark field (DF) TEM images of individual particles revealing the temperature dependent crystallinity of Pt/silica-NACS are given in refs. 70, 74 together with a discussion of possible crystallization mechanism (melting *vs.* diffusion). Interestingly, a detailed theoretical and experimental study (SEM) revealed the effect of the support on the crystallinity of Pt nanoclusters fabricated by EBL.[72] TiO_2-terminated and SrO-terminated STO ($SrTiO_3$) has been used to write 30–100 nm Pt dots by EBL. The Pt clusters were crystalline. Two different particle shapes were observed, one of them characterized by a reconstruction of the Pt clusters.[72] It appears that using a single crystal support may facilitate the crystallization of the metal clusters. Similarly, by using single crystalline supports of different orientation, crystalline Pt clusters with a support dependent shape were nanofabricated using EBL.[75] X-ray scattering revealed their crystalline nature: cuboctahedron and hexagonal platelets formed which were dominated by (111) planes.[75]

5.2.3 Nickel-NACs. Instead of analyzing the supported metal clusters directly by TEM, in ref. 76. Ni clusters have been chemically removed from the support in order to characterize the crystallinity of the nanoclusters by BF-TEM. The Ni needles (20 by 400 nm) were polycrystalline.

5.2.4 Shape effects. Utilizing EBL fabrication techniques, the effect of the particles shapes on catalytic properties has, to the best of my knowledge, not been directly studied at UHV yet. Experiments of this kind are in progress in our group, see Fig. 4. Perhaps related to that is the effect of rim sites *vs.* terrace sites which could indeed be identified at UHV in a unique way using EBL fabricated samples, see Section 6.2.

However, the fact that the shape of nanoclusters affects surface catalytic processes has been reported before using other sample fabrication techniques. For example, using traditional "wet-chemistry" procedures the shape of nanoclusters could be tuned in some cases. In a study about butadiene hydrogenation on Al_2O_3 powders impregnated with shape selected Pd particles, the selectivity and activity increased from nanocubes (dominated by (100) facets) to elongated nanoprisms (with (111) facets).[77] These effects are certainly expected since structure sensitive surface reactions are known since the beginning of surface science.

In ref. 75, crystalline supports of different orientation allowed for shape tuning of supported Pt clusters using EBL. Oxygen reduction electrocatalytic activity probed in the liquid phase did indeed depend on the shape of the Pt clusters. The authors propose a, what they call "division of labor", mechanism. The different facets of the Pt clusters have different adsorption and reduction properties for oxygen. The close proximity of these active sites result in a synergy promoting the oxygen reduction activity of the electrocatalyst.[75] That mechanism may be refereed to inverse-spillover effects in heterogeneous catalysis.

5.2.5 Other techniques. The formation of single crystalline metal clusters has been reported utilizing gas-phase techniques. For an example, see ref. 78 and MECS/MCCR in the glossary.

5.3 Adhesion

AFM experiments in contact mode have been used to estimate the adhesion of Pt clusters on alumina.[73] As prepared Pt clusters could be displaced with the AFM tip applying a normal force of 25–31 nN depending on the scanning speed. Interestingly, the morphology of annealed Pt-NAC remained unaffected applying much larger forces. An order of magnitude estimate of the interfacial energy was provided. An increase from 5 mJ/m^2 to 0.65 J/m^2 was estimated by annealing the samples; two possible mechanisms were proposed. A hydrocarbon layer, resulting from the wet-chemistry procedures as part of the EBL process, can affecting the adhesion, or, the improved crystallinity of the metal clusters may affect their adhesion.

5.4 Sputtering

The sputtering of a Pt-NAC with Ne ions replicated the Pt pattern in the silica support as determined by moving the Pt cluster with the AFM tip, *i.e.*, the Pt-NAC were acting as a mask.[32] Ion sputtering yields are given in ref. 36. Mild sputtering can be used to clean EBL cluster samples while monitoring with AES/XPS the carbon impurity level.[68] Similar procedures were used to clean thin film samples.[35]

6 Chemical properties and applications in surface chemistry

6.1 Adsorption kinetics of small molecules on NAC samples

On EBL model catalysts, the adsorption kinetics of small molecules has only been studied for a few systems, as described in this section, see also Table 1.

6.1.1 Hydrogen/deuterium adsorption. TDS of hydrogen and CO (at very large exposures) has been studied on Pt-NACs in order to investigate spill-over effects.[29] Accordingly spill-over of hydrogen on the silica support has been concluded.

On Mo EBL samples, H_2 TDS data were collected, but indicated perhaps amazingly very small adsorption probabilities ($S_0 < 0.05$).[68] Mo clusters are traditionally employed as hydrodesulphurization catalysts.

6.1.2 CO adsorption kinetics. The adsorption kinetics of CO was studied on Au EBL cluster samples. The continuing interest in Au systems is motivated *e.g.* by the importance of the low temperature CO oxidation reaction.[18,19,79] The adsorption of CO is the first elementary step in this bimolecular surface reaction. To obtain a mechanistic understanding of the CO oxidation and related surface reactions, a characterization of the kinetics and dynamics of CO adsorption on Au clusters are pertinent. In the majority of studies the Au clusters are, however, formed by PVD with less control of the cluster's morphology. In most experimental works (utilizing TDS, IR, XPS, and STM), universal kinetics of CO adsorption has been reported which is fairly independent of the support.[23,80] Two features are present in TDS and one IR peak has been measured. These results have in most cases been related to CO adsorption on low-coordinated sites of Au clusters.[23,80–82] DFT calculations indicate CO adsorption on 7-fold and 6-fold coordinated sites which may be associated with on top adsorption of CO on Au step edges and kink sites of Au clusters.[80] The DFT calculations also reproduce the presence of only one IR structure.[80]

In refs. 59, 62, 83 silica supported Au cluster samples fabricated by PVD and EBL were directly compared. CO was dosed on the catalyst surface using a molecular beam system avoiding sample holder effects. The uptake of CO on the silica support is very small. TDS data for the Au PVD samples were consistent with prior work by other groups, as described in the preceding paragraph. Interestingly, for the Au EBL samples, the TDS curves are dominated by a single feature, which appears to coincide approximately with the position of the low temperature TDS peak obtained for the PVD samples. The cluster size and most likely the shape of the Au clusters of the PVD and EBL samples will differ significantly. In addition, it is known from calculations (using the so-called Wulf construction, Section 5.2.1) that the relative number of defect sites depends on cluster size.[84]

6.1.3 CO_2 adsorption kinetics – rim effects. Silica supported 12, 35, and 63 nm Cu EBL clusters were analyzed by TDS and molecular beam scattering. The Cu NAC clusters were subsequently oxidized in UHV. CO_2 was used as the probe molecule since it is part of the syngas for the industrial methanol synthesis which utilizes copper based catalysts. Unexpectedly, the CO_2 saturation coverage does not simply scale with the clusters' area, but rather with the rim length of the clusters, as described in more detail in Section 6.2 since molecular beam scattering techniques were used. That result was supported by kinetics experiments: in CO_2 TDS only one low temperature desorption feature is seen, whereas for CO two TDS peaks were detectable. Thus, kinetically distinct adsorption sites were present for CO_2 and CO adsorption on Cu-oxide NAC.

6.1.4 Adsorption kinetics of small organic molecules. Motivated by HDS catalysis, the adsorption kinetics/dynamics of small organic molecules was studied on molybdenum model-nano-array catalysts.[68] Perhaps unexpectedly, alkanes such as butane, just adsorb molecularly on metallic 77 nm Mo clusters, with initial adsorption probabilities in the range of $S_0 = 0.6–0.3$, depending on impact energy. Furthermore, when oxidizing the Mo clusters, they appear to lose any catalytic activity ($S_0 < 0.05$) for the adsorption of butane. These conclusions were based on multi-mass TDS (kinetics) experiments. Mo cluster are known to interact strongly with thiophene (see *e.g.* ref. 85).

6.2 Adsorption dynamics of CO and CO_2: identifying active sites

A major advantage of NAC is that catalytically active sites can be identified in a unique way. The number and density of these sites depend on the shape of supported clusters. A recent example concerns copper oxide EBL clusters supported on silica.[24,25,60,86] Several EBL samples with different cluster size were analyzed which revealed that CO_2 coverages on these samples scales with their rim length rather than the cluster surface area (Fig. 5). That was in contrast to results obtained for CO on the same samples. That conclusion would be much harder to derive when studying PVD samples or applying "wet-chemistry" sample preparation techniques. Molecular beam scattering techniques were utilized, *i.e.*, not only the adsorption kinetics, but also energy-transfer processes (adsorption dynamics) were studied.

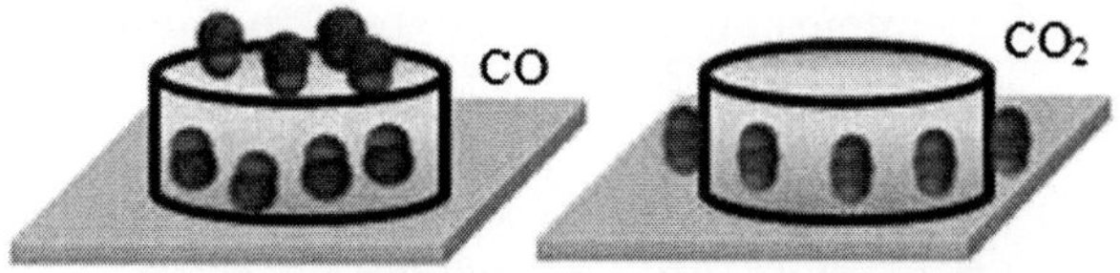

Fig. 5 Identifying active sites: rim sites *vs*. terrace sites.

Briefly, since the morphology of EBL samples is known a priori, simple geometrical considerations allow for separating the effect of edge sites from terrace sites. In so doing, pairs of clusters with different size were compared (*i.e.*, 12 nm *vs*. 63 nm clusters or 35 nm *vs*. 12 nm clusters, *etc*.).[24,25,60,86] The ratio of their rim length to surface area differs significantly. For a given pair of clusters, for example, the cluster area increased by a factor of ~13, whereas the rim length increased only by a factor of ~2. In that project 12 nm and 63 nm copper clusters were studied. (Similarly, combinations -rim plus surface sites- or one dimensional *vs*. two-dimensional rim sites can be considered.) Although the geometry of these samples is know in advance, SEM images were used to independent verify these geometrical predictions.

Now, the number of adsorption sites can be measured using molecular beam scattering techniques. In particular, it is rather simple to determine the ratio of adsorption sites for a pair of EBL samples. One simply measures the saturation coverage of probe molecules such as CO_2 or CO for two different cluster sizes. Only the flux of the beam system (nozzle pressure) needs to be kept reasonably constant. No calibrations are required. In so doing, the experimental ratio of the saturation coverages for CO did amount to 5.5 and 4.5 for metallic and oxidic Cu clusters, respectively. Interestingly, for CO_2, the ratio for oxidic clusters was determined as 2.0. From that it was concluded that CO adsorbs on rim and terrace sites, but CO_2 populates only the rim sites along the Cu clusters. (Statistical analysis of several samples was applied.) This procedure works "so easily" only when studying EBL clusters. Thus, catalytically actives sites could be determined for a model system, which are related, in the widest sense, to the shape of these clusters. DFT calculations from other groups indeed show that CO_2 does not adsorb on terrace sites of copper, whereas CO adsorption is structure insensitive on copper. On a macroscopic scale, the 12 nm clusters are the smallest clusters ever made by EBL.

A similar (although conventional PVD) approach has been used successfully before in an electrochemistry study.[87] Here, however, a sophisticated sample characterization by STM was required since the morphology of PVD samples is unknown *a priori*.

The adsorption dynamics of CO and CO_2 was also studied in that project utilizing molecular beam scattering techniques. The initial adsorption probability of CO, S_0, decreases with increasing impact energy and XPS/AES did not indicate carbon residuals after the experiments. Thus, the adsorption is non-activated and molecular. The S_0 of Cu clusters did depend on cluster size, which can be described in the framework of the capture zone model (CZM).[88–90] Interestingly, differences in the adsorption dynamics depending on the oxidation state of the clusters were evident. S_0 of the metallic clusters was larger than for the oxidic clusters. In addition, cluster size effects were distinct only for the metallic clusters. Simple mass matching models would indeed predict the first observation. The second result would be consistent with a cluster size expansion while oxidizing the copper clusters.

CO_2 adsorption turned out to be molecular and non-activated. Adsorption transients (King and Wells technique[91]) were recorded as a function of surface temperature and CO_2 impact energy. The transients are consistent with precursor models, as expected from the capture zone model (CZM).[88,89] Cluster size effects are evident, in agreement with the CZM.

The EBL concept illustrated in this example concerned a single system, *i.e.*, a methanol synthesis model catalyst. However, the same technique could be applied to a variety of different model catalysts. The practical conclusion may be to optimize the catalyst synthesis such that the rim length of the supported clusters is maximized. This would result in larger coverages of reactants on the surface and probably larger reaction rates. Following that idea, Haldor-Topsoe provide, at the present time, what is referred to as 'brim site' (*i.e.* rim site) catalysts for HDS applications. The hypothesis that rim sites along catalysts show special properties is, perhaps, evident. It has long been discussed, but it is quite difficult to prove experimentally.

6.3 Surface reaction kinetics

A number of examples exist, meanwhile, where surface reactions have been studied experimentally taking advantage of "designer" lithography catalysts with pre-determined geometrical structure. Some of the phenomena studied were motivated by theoretical investigations.[27,97]

6.3.1 CO oxidation on Pt-nanowire-catalysts at high pressure. The CO oxidation reaction has been studied on Pt nanowires fabricated by UV lithography combined with nanoimprint techniques.[92] Various supports have been used (see Table 1). However, this study focused on the nanofabrication rather than on a characterization of the reaction kinetics. The reaction has been studied on Pt nanowires supported on zirconia in a combined high-pressure-UHV system. An apparent activation energy of (1.6 ± 0.1) kcal/mol and a pre-exponential of 1.5×10^3/s has been determined at the following reaction conditions: 100 torr $O_2 + 40$ torr $CO + 620$ Torr He within the temperature range of 513–593 K. The kinetics experiments have later been extended to other supports such as alumina, silica, and ceria (thin films on silica wafer) at similar high-pressure measuring conditions.[93] Reaction activation energies have been determined. Pt nanowires on silica had the smallest activation energy and Pt/ceria the largest following the trend: silica < zirconia < alumina < ceria. The activity of Pt/silica is quite similar to the one for Pt(111) single crystals. A reaction induced morphology change and a loss of Pt metal has been reported.

6.3.2 Hydrogenation reactions of small organic molecules on Pt-NACs. A series of papers about hydrogenation/dehydrogenation reactions of small organic molecules at high pressure on model-nano-array catalysts has been published by G. Sormorjai's group following his tradition of studying related problems on single crystal surfaces.

Hydrogenation of ethylene has been studied on Pt-NACs (28 nm square array) using different oxide supports (silica, alumina) in order to address how the metal-oxide interface affects the poisoning of the hydrogenation by coadsorbed CO.[30] In this study a UHV-high pressure combination set up was used to dose ethylene-hydrogen mixtures at pressures as high as 100 torr on the sample at rather high temperatures of 473 K. CO has been added to the reaction gas mixture in order to study the CO tolerance of the NACs (catalyst poisoning). The reaction product (ethane) formation has been studied with a gas chromatograph. It turn out that the reactivity of NACs is very similar than the one for Pt single crystals, but that the NACs have a much better CO tolerance. Attempts to correlate the reactivity (turnover frequency) with the number of metal-oxide interface sites have been undertaken. It appears plausible that the enhanced CO tolerance is related to metal-oxide interface sites as the active sites. Results of related studies have been published in refs. 29, 66, 94.

The reactivity of Pt-NACs (28 nm) towards cyclohexane hydrogenation has been compared with the one of Pt foil: the selectivity of Pt-NACs was three times larger and the reactivity about twice as large as for the Pt foil, whereas the deactivation behavior of the two systems was comparable.[32] Again, metal-support interactions and interface effects have been invoked as an explanation for the enhancement in reactivity. In another study, the formation of platinum silicide on the Pt-SiO_2 interface of Pt-NACs has been detectable by XPS and was related with the increased catalytic activity of Pt-NACs as compared to Pt foil.[95] Control experiments with silane dosed on Pt foil have been conducted.

6.3.3 Communication effects in the CO oxidation reaction on Pd-NACs. The interplay of local reaction rates and diffusion controlled reactions (*vice versa*) can be studied by tailoring the right model system by means of EBL. These studies are truly exciting examples that illustrate the concept of using NAC in basic science. A series of publications by the groups of B. Kasemo and H.J. Freund investigated those effects experimentally and theoretically, mostly on Pd-NAC.[28,52,64,96] (See also the review cited as ref. 27).

Locally different reaction rates have been generated by a variation of the reaction condition for the CO oxidation reaction on Pd clusters. Using a combination of effusive and molecular beams (of O_2 and CO) the CO flux could be varied such that the CO_2 formation took place at oxygen-rich or CO-rich conditions (large/small oxygen/CO coverage). At oxygen-rich condition the effective diffusion length of oxygen on the catalyst surface is larger than at CO-rich conditions. (The diffusion of CO is always fast and the diffusion length long for the system considered here.) In addition, Pd particles with an aspect ratio of about one (500 nm Pd particles, 450 nm high), *i.e.*, Pd "towers" have been fabricated using EBL. This leads to a shadow zone behind the Pd towers when the oxygen beam strikes the surface at glancing incidence angle. Thus, at CO-rich condition the CO_2 formation rate is larger on the Pd-tower side which is directly exposed to the oxygen beam. Oxygen cannot efficiently diffuse "around the corners of the Pd towers". Thus, to continue in simple terms, the CO_2 formation is greater in the light (in the oxygen beam) as compared with the reaction rate in the shade of the Pd towers. Both effects: the oxygen diffusion length and the shadow effect, lead to local variations in the CO_2 formation rates and non-equilibrium effects. Amazingly this simple concept works and results in significantly different angular distributions of the formed CO_2 molecules (which desorb from the catalyst) at oxygen rich and CO rich measuring conditions. (For flat nanoclusters – 5.5 nm diameter and 1.7 nm height- no (or much smaller) asymmetries have been observed.[96,97]) A microkinetic (diffusion-reaction) model has been developed to simulate the measured angular distributions. Diffusion rates and local reaction rates can be determined.

6.3.4 Bistabilities in the CO oxidation reactions on Pd and Pt-NACs. Related with communication effects described in the last section are bistabilities and chaos effects in kinetics experiments which have been investigated extensively on single crystal surfaces by G. Ertl's group.[98] (Bistabilities do not require "tower" NACs and shadow effects to occur; they are mostly pure kinetic effects.) Two stable kinetic states can exist for the same system and at identical measuring conditions depending on the direction along the reaction coordinate chosen to approach this measuring condition. The steady-state can be approached starting from oxygen or CO rich measuring conditions in the case of the CO oxidation reaction. Within the "bistability" window these two different experiments lead to different steady-state CO_2 formation rates. Thus, a hysteresis is seen in the CO_2 formation rate *vs.* the reactant concentrations. These effects have been seen a long time ago in studies about single crystal surfaces. However, using NACs allows setting up precise kinetic models for a rather complex model catalysts -which maps more closely technical catalysts than single crystals- since the morphology of the Pd-NACs catalyst is know precisely (as precisely as for a "simple" single crystal surface).[28,51,52,64]

Similar experiments have been conducted for Pt-NAC/silica and Pt-NAC/ceria by means of a micro-flow reactor[64] which allows for using very small samples ($<1\,mm^2$). Bistability diagrams have been determined and possible spillover effects of oxygen; ceria catalysts are used as oxygen storage components in the three-way-catalyst.

6.3.5 Coverage fluctuations and quenching of bistabilities in the CO oxidation on Pd-NACs. Interestingly, the kinetic bistabilities (see above) are only observed for single crystals or NACs consisting of rather large cluster sizes; they disappear with decreasing cluster size.[51,96] By a combination of traditional techniques (vapor deposition) and EBL, the cluster size has been varied within the range of 1.8 to 500 nm. Pd-NACs have been utilized to reveal the mechanism which leads to the quenching of the bistabilities: coverage fluctuations *vs.* defects.[51] The effect of coverage fluctuation-induced kinetic transitions between the two bisatabilities increases with decreasing cluster size which can lead to a quenching of the bistabilities for small clusters. However, also the defect density of clusters increases with decreasing size. Therefore, the coverage dependence of binding energies is more distinct for small clusters which can lead to a quenching of bistabilities. Detailed modeling of the results led the authors to conclude that a combination of both effects, fluctuation and defects, are required to quantitatively understand the kinetics.

7 Non-catalytic applications

As a supplement, a few examples of further applications and research areas are mentioned which take advantage of EBL samples.

7.1 Nano-optics

Numerous projects on NACs are motivated by applications such as the design of plasmonics waveguids and optical sensors as well as a further development and of experimental techniques such as surface enhanced Raman spectroscopy, fluorescence spectroscopy, and second harmonics generation.[99] For these applications, also, the shape of the nanoparticles is very important. Thus, NACs are an excellent tool. In a number of these studies, localized surface plasmon resonances are investigated. In addition, a geometric resonance can occur when the scattered light is commensurate with the periodicity of the NACs.[100] Some of these studies using EBL have been added to Table 1 (see also refs. 6, 101, 102); other techniques have been applied, too, such as ion-beam milling.[103] However, this review focuses on applications in heterogeneous catalysis.

7.2 Nano-electronics (nanowiring)

Direct electron-beam writing in a film of Au nanoparticles has been demonstrated in order to connect electrically gold electrodes; 30 nm width wiring has been written.[104] With a modified electron microscope Ni/Au gate and gratings structures have been written on GaAs supports.[55] Structures as small as 10 nm have been obtained, but they were restricted to a total size of 150 by 250 μm.

7.3 Nano-magnetism

The miniaturization of magnetic storage media is apparently hampered by the thermal stability of magnetic metal clusters and the sensitivity of the heads (which are reading the data bits). One approach to circumvent this difficulty is to fabricate large aspect ratio magnetic metal clusters using, for example, EBL. In ref. 76 high-aspect-ratio Ni clusters (20 nm, pitch 100 nm, 65 Giga columns per inch^2) have been synthesized and characterized by means of a magnetic force microscope.

8 Summary

Because the concept of NACs introduced for catalysis applications primarily by G.A. Somorjai (Berkeley)[67,105] and B. Kasemo[106] is still rather novel, the opportunity to outline the advantages and disadvantages of this approach (from my perspective) was taken. In the following, the cleaning procedures used and the physical properties of MANCs have been summarized as well as the applications demonstrated in surface science and catalysis studies have been compiled. A brief glossary lists a number of related nanofabrication techniques. Many other applications exist in such areas as nano-electronics and nano-optics. However, they have only been mentioned since this review is restricted to surface chemistry applications related to heterogeneous catalysis. Efforts to overcome the sequential nature of techniques such as EBL by developing high-throughput patterning methods are under way in many laboratories[3,4] which promises that high-quality NAC will become available in larger quantities soon.

Abbreviations

Measuring techniques

BF TEM	bright field TEM
DF TEM	dark field TEM
EDX	energy-dispersive X-ray spectroscopy
FTIR	Fourier transform infrared spectroscopy
HREELS	high resolution electron energy loss spectroscopy
HDS	hydrodesulphurization
IR	infrared spectroscopy
LEED	low energy electron diffraction
NEXAFS	near edge X-ray absorption fine structure
NMR	nuclear magnetic resonance
RR	retarded reflector (King and Wells) technique – adsorption transients
RAIRS	reflection absorption infrared spectroscopy
STM	scanning tunneling microscopy
TDS/TPR	thermal desorption spectroscopy/temperature programmed reactions
UVvis-NIR	ultra violet– visible-near infrared spectrophotometer
XPS	X-ray photoelectron spectroscopy
TEM	transmission electron microscopy
SEM	scanning electron microscopy
AFM	atomic force microscopy
SAM	scanning Auger microscopy//self-assembled monolayer
GC	gas chromatograph
2PP	two photon photoemission
PEEM	photoemission microscopy
(L)SPR	(localized) surface plasmon resonance
SERS	surface-enhances Raman spectroscopy
MFM	magnetic force microscope

Further symbols and acronyms

S_0	initial adsorption probability
L	Langmuir (one sec gas exposure at 1×10^{-6} torr)
EBL	electron beam lithography
SAR	structure activity relationship
STO	$SrTiO_3$
UHV	ultra-high vacuum (typically pressure $<5 \times 10^{-10}$ mbar)
NSF	US national science foundation
DoE	US department of energy

Short glossary of related nanofabrication techniques

Brief glossary of techniques used to fabricate model catalysts. This summary is mostly restricted to procedures which allow for manufacturing ordered catalysts. In addition, some nano-vocabulary words are included as used in this draft. For completeness and considering the diversity of readers this brief glossary appears appropriate.

Top-down: Physically/chemically manipulate molecules into nanostructures. Carve out or add aggregates to form a nanostructure, for example by EBL. Typically an expensive and slow fabrication procedure.

Bottom-up: Molecules assembling themselves into nanostructures (Langmuir-Blodget films, self-assembled monolayer (**SAMs**), *etc.*). Less expensive and fast fabrication technique, but cannot produce complicated predetermined patterns.

Chemical vapor deposition - CVD: Somewhat similar to the hydrothermal method used to impregnate nanocatalyst such as carbon nanotubes. In contrast to physical vapor deposition, metal containing precursor (such as organometallic compounds) are deposited on a support and chemically/thermally decomposed. Problems are related with expensive specialty chemicals, impurities, sometimes harsh conditions necessary to decompose the precursor, and difficulties to control the cluster size.

Colloidal lithography: Similar to nanosphere lithography, except that colloids are used to generate a mask. It is a fast and less-expensive technique which is, however, so far limited to sub 100 nm structures.[130,131]

Dip-pen lithography: (AFM - atomic force microscopy - based technique)

Electron beam lithography - EBL: Basically direct electron beam writing of a pattern in a photoresist.

Hydrothermal method: Typically used for powders. The powders (for example carbon nanotubes) will be impregnated with a metal containing precursor (by sonication of the precursor-powder solution), dried, and annealed in *e.g.* an hydrogen ambient (or water, or acid) in a sealed autoclave container. Conceptually similar to chemical vapor deposition.

LASER interference nanolithography: The interference pattern of a LASER is used to pattern a photoresist layer. Structures of 300 nm have been made over macroscopic distances as a parallel and fast process.[5]

Low energy cluster beam deposition - LECBD: Beams consisting of metal clusters can be formed by techniques such as metal-vaporization,[107] LASER[108] or magnetron sputtering of metal targets in an inert gas beam. A supersonic expansion[109,110] of the rare gas – metal vapor leads to clusters with a narrow kinetic energy distribution. Mass selection of the clusters can be accomplished by a variety of techniques including conventional quadrupole mass spectrometers or time of flight systems. In order to conserve the original cluster sizes "soft landing" of the clusters (utilizing low impact energy clusters) on cold surfaces has been used. A related technique is **Ionized cluster beam deposition (ICBD):** Here, high impact energy ionic cluster beams can generate defects on impact (on HOPG) which pin down the clusters on the support and prevent sintering. (See, for example, the reviews given in refs. 8, 9, 111).

Model catalysts: Refers historically[112] to systems consisting of metal clusters on metal oxide supports, *i.e.*, model catalyst can be fabricated basically by all techniques included in this glossary. However, historically model catalysts in surface science are obtained by physical vacuum vapor deposition of metals on metal oxide supports with no, (or little) control of the cluster morphology. For reviews mostly about traditional model catalysts see refs. 12, 113–116. Narrow size distributions and small cluster sizes have been obtained by "patterning" supports such as HOPG (which interact only weakly with metals) by, for example, ion sputtering. However, only randomly distributed clusters pinned at the defect can be formed on the support.[117] Oxygen etching/patterning of HOPG also has been used.[118]

Multi expansion cluster source/metal cluster colloid reactor - MECS/MCCR: The formation of single crystal Au clusters has been reported using a multi expansion cluster source. This is a gas-phase technique somewhat similar to an effusive beam set up but including heating and cooling zones. In addition, surfactants are used to encapsulate the Au clusters in order to prevent cluster agglomeration as well as to self-assemble them on a support. In this way, 4 nm Au clusters have been deposited on silica, forming more or less ordered structures with a size of a few 100 nm. Size tuning of the clusters appears possible, but no perfectly ordered arrays are formed (see ref. 78 and references therein).

Natural lithography: Is sometimes used as a synonym for nanosphere lithography.[119]

Nanocrystalline: In the meanwhile a common term (nano-/-micro-crystalline), it refers to small grains of crystallites within an amorphous phase.

Nano imprint technique: See soft lithography and ref. 3.

Nanosphere lithography - NSL: (Polystyrene) nanospheres can be used as a mask for metal vapor deposition. Subsequent annealing of the deposited metals can result in different and ordered morphologies after chemically removal of the spheres. Size and shape tuning over macroscopic dimensions are possible, however, very small structures (<10 nm) have apparently not been made, yet.[6]

Pulsed LASER deposition/LASER ablation: A variety of variations have been developed, the main idea is to evaporate a metal by means of a pulsed LASER in inert gas. The gas pressure allows for cluster size tuning. Particle sizes below 10 nm and rather narrow size distributions have been reported, see ref. 120 and references therein.

Physical vapor deposition: Vaporizing (typically) metals in a vacuum chamber and condensing the vapor on a support which leads to the formation of metal clusters with a morphology given alone by the metal-support interactions. Low melting point metals can easily be vaporized using a metal filament which is resistively heated.[121,122] For high-meting point metals electron beam evaporators are typically used in order to obtain chemically clean metal clusters.[123] This techniques has so far mostly been used to fabricate model catalysts. Unfortunately, the cluster size and dispersion is difficult to control. Variations of this technique such as "**buffer layer assisted growth**" of metal clusters has been used where an inert gas film is condensed on the support prior to vapor deposition of the metal clusters.

This technique may allow to change the cluster dispersion while keeping the cluster size distribution constant.[124]

"wet" chemistry nanofabrication: A large variety of ("wet chemisty") techniques has been developed to synthesize small metal clusters to functionalize, for example, carbon nanotubes which are mostly based on the decomposition of metal containing precursors. (Such as the hydrothermal method, ethylene glycol solid reductions, solid-state reactions, electro deposition, fast evaporation, galvanic displacement reactions, *etc.*) Some techniques allow for obtaining quite narrow cluster size distribution. In addition, these nanoparticles can simply be deposited on supports by the drop-and-dry technique or utilizing the Langmuir-Blodgett technique. However, the formation of ordered (self-assembled) layers on a support is often based on the effect of the organic surfactance which raised issues such as the cleaning of those samples (high temperature annealing, oxygen/hydrogen plasma etching) in order to obtain clean and still well ordered 2D metal cluster systems. (See *e.g.* ref. 125).

Reactive ion etching (or plasma reactive ion etching – RIE): Many different experimental set ups are used. For example, rf glow discharge can be used to generate ions which are accelerated towards the target. This technique has been used to reduce the size of nanoparticle arrays, *etc*.

Soft lithography: The main idea is to fabricate a master die by photolithography/e-beam lithography and replicate the pattern by stamping it into "soft matter" (PDMS – polydimethylsiloxane "rubber polymer") which generates a stamp. This stamp is then used to mass produce the original pattern by nano-imprint lithography. Many variations of this technique have been developed, see the review given in ref. 3.

Template techniques: A large variety of nanofabrication techniques take advantage of templates often in a sophisticated way. For example, the reconstructed surface of AgBr(111) films has been used as a template to grow 5.6 nm and 8.3 nm Au clusters with a coherence length of a few thousand nm (see ref. 126 and references therein). Or, porous silica gel has been used to obtain ordered copper-oxide particles.[127] See also nanosphere lithography.

Acknowledgments

Discussions and joint projects with J.A. Liddle, Stefano Cabrini, and Scott Dhuey at Molecular Foundry (Berkeley) as well as Leonidas Ocola and Dan Rosenmann at Argonne National laboratories (Chicago) are acknowledged. Over the years, numerous samples were fabricated at these two locations for joint projects. SEM imaging was mostly provided by Lu Ming at Brookhaven National Labs. (Upton). Many thanks to him. Most of our works which are also included in this draft were part of the thesis projects of S. Funk, J. Goering, M. Komarneni, and A. Chakradhar at North Dakota State University (NDSU). In addition, many thanks to the postdocs E. Kadossov and J. Shan as well as numerous undergraduates, namely Jordan Schmidt who supported the nanofabrication of some of the EBL samples.

Financial support of the EBL projects by an NSF CAREER grant is acknowledged.

References

1 U. Heinz and U. Landman (Eds.), Springer, 2007, ISBN 978-3-540-74551-8.
2 C. P. Vinod, *Catal. Today*, 2010, **154**, 113.
3 L. J. Guo, *J. of Physics D: Appl. Phys.*, 2004, **37**, R123.
4 B. D. Gates, Q. Xu, J. C. Love, D. B. Wolfe and G. M. Whitesides, *Ann. Rev. Mater. Res.*, 2004, **34**, 339.
5 M. Schildenberger, Y. Bonetti, M. Aeschlimann, L. Scandella, J. Gobrecht and R. Prins, *Catalysis Lett.*, 1998, **56**, 1.
6 C. L. Haynes and R. P. Duyne, *J. of Chem. B*, 2001, **105**, 5599.
7 G. M. Wallraff and W. D. Hinsberg, *Chem. Rev.*, 1999, **99**, 1801.
8 R. E. Palmer, S. Pratontep and H. G. Boyen, *Nature Materials*, 2003, **2**, 443.
9 U. Heiz and W. D. Schneider, *J. Phys. D: Appl. Phys.*, 2000, **33**, R85.
10 I. Lee, M. A. Albiter, Q. Zhang, J. Ge, Y. Yin and F. Zaera, *Phys. Chem. Chem. Phys.*, 2011, **13**, 2449.
11 F. Zaera, *Surf. Sci.*, 2002, **500**, 947.
12 L. J. Gunter, J. W. Niemantsverdriet, F. H. Ribeiro and G. A. Somorjai, *Catal. Reviews - Science and Engineering*, 1997, **39**, 77.
13 P. C. Thune and J. W. Niemantsverdriet, *Surf. Sci.*, 2009, **603**, 1756.
14 M. Bowker, *Surf. Sci.*, 2009, **603**, 2359.
15 S. T. Ceyer, *Science*, 1990, **249**, 133.
16 S. M. McClure and D. W. Goodman, *Chem. Phys. Lett.*, 2009, **469**, 1.
17 S. Peng, Y. Lee, C. Wang, H. Yin, S. Dai and S. Sun, *Nano Res.*, 2008, **1**, 229.
18 M. Haruta, *Catal.Today*, 1997, **36**, 153.
19 T. V. Choudhary and D. W. Goodman, *Topics in Catal.*, 2002, **21**, 25.
20 A. Avoyan, G. Rupprechter, A. S. Eppler and G. A. Somorjai, *Topics in Catal.*, 2000, **10**, 107.
21 N. Felidj, J. Grand, G. Laurent, J. Aubard, G. Levi, A. Hohenau, N. Galler, F. R. Aussenegg and J. R. Krenn, *J. of Chem. Phys.*, 2008, **128**, 094702.
22 K. M. Bratlie, H. Lee, K. Komvopoulos, P. Yang and G. A. Somorjai, *Nano Lett.*, 2007, **7**, 3097.
23 C. Lemire, R. Meyer, S. Shaikhutdinov and H. J. Freund, *Angewandte Chemie Int. Ed.*, 2004, **43**, 118.
24 J. Shan, A. Chakradhar, M. Komarneni and U. Burghaus, *J. Chem. Phys. C*, 2012, **116**, 18930.
25 J. Shan, A. Chakradhar, M. R. Komarneni, M. Lu and U. Burghaus, *Chem. Phys. Lett.*, 2012, **544**, 70.
26 T. Matsushima, *Surf. Sci. Rep.*, 2003, **1**, 1.
27 V. P. Zhdanov and B. Kasemo, *Surf. Sci. Rep.*, 2000, **39**, 25.
28 M. Laurin, V. Johanek, A. W. Grant, B. Kasemo, J. Libuda and H. J. Freund, *J. of Chem. Phys.*, 2005, **122**, 084713.
29 P. W. Jacobs, S. J. Wind, F. H. Ribeiro and G. A. Somorjai, *Surf. Sci.*, 1997, **372**, L249.
30 A. M. Contreras, J. Grunes, X. M. Yan, L. Liddle and G. A. Somorjai, *Topics in Catal.*, 2006, **39**, 123.
31 Leonidas Ocola, Argonne National Laboratory, Center for Nanoscale Materials, personal information (March 2008).
32 A. S. Eppler, J. Zhu, E. A. Anderson and G. A. Somorjai, *Topics in Catal.*, 2000, **13**, 33.
33 K. McCrea, J. S. Parker, P. Chen and G. Somorjai, *Surf. Sci.*, 2001, **494**, 238.
34 S. Johansson, E. Fridell and B. Kasemo, *J. Vac. Sci. Technol., A: Vacuum, Surfaces, and Films*, 2000, **18**, 1514.

35 J. Goering, E. Kadossov, U. Burghaus, Z. Q. Yu, S. Thevuthasan and L. V. Saraf, *Catal. Lett.*, 2007, **116**, 9.
36 M. X. Yang, D. H. Garcias, P. W. Jacobs and G. A. Somorjai, *Langmuir*, 1998, **14**, 1458.
37 T. Sueyoshi, T. Sasaki and Y. Iwasawa, *Surf. Sci.*, 1995, **343**, 1.
38 T. Sueyoshi, T. Sasaki and Y. Iwasawa, *Surf. Sci.*, 1996, **365**, 310.
39 W. W. Crew and R. J. Madix, *Surf. Sci.*, 1996, **356**, 1.
40 http://www.hyperioncatalysis.com/emerging.htm
41 http://www.nsf.gov/pubs/2006/nsf0648/nsf0648.jsp
42 H. Ulbricht, G. Moos and T. Hertel, *Surf. Sci.*, 2003, **532–535**, 852.
43 http://foundry.lbl.gov/user_program/nanofab_capabilities.html.
44 S. Kwon, X. Yan, A. M. Contreras, J. A. Liddle, G. A. Somorjai and J. Boker, *Nano Lett.*, 2005, **5**, 2557.
45 J. A. Liddle, P. Naulleau and G. M. Schmid, *J. Vac. Sci. Technol. Part B*, 2004, **22**, 2897.
46 G. M. Schmid, L. E. Carpenter and J. A. Liddle, *J. Vac. Sci. Technol.*, 2004, **22**, 3497.
47 X. M. Yan, A. M. Contreras, M. M. Koebel, J. A. Liddle and G. A. Somorjai, *Nano Lett.*, 2005, **5**, 1129.
48 G. A. Somorjai and M. X. Yang, *J. of Molecular Catal. A: Chem.*, 1997, **115**, 389.
49 http://www.anl.gov/welcome.html
50 http://www.cnms.ornl.gov/ or http://www.ornl.gov/.
51 V. Johanek, M. Laurin, A. W. Grant, B. Kasemo, C. R. Henry and J. Libuda, *Science*, 2004, **304**, 1639.
52 M. Laurin, V. Johanek, A. W. Grant, B. Kasemo, J. Libuda and H. J. Freund, *J. of Chem. Phys.*, 2005, **123**, 054701.
53 W. Gotschy, K. Vonmetz, A. Leitner and F. R. Aussenegg, *Appl. Phys. B*, 1996, **63**, 381.
54 C. Vieu, F. Carcenac, A. Pepin, Y. Chen, M. Mejias, A. Lebib, L. Manin-Ferlazzo, L. Couraud and H. Launois, *Appl. Surf. Sci.*, 2000, **164**, 111.
55 P. B. Fischer and S. Y. Chou, *Appl. Phys. Lett.*, 1993, **62**, 2989.
56 A. Javey and H. Dai, *J. Am. Chem. Soc.*, 2005, **127**, 11942.
57 http://www.mateck.com/ http:/www.ntt-at.com/products_e/nanopattern/index.html
58 E. Kadossov, J. Justin, M. Lu, D. Rosenmann, L. E. Ocola, S. Cabrini and U. Burghaus, *Chem.Phys.Lett.*, 2009, **483**, 250.
59 E. Kadossov, S. Cabrini and U. Burghaus, *J. of Mol. Catal. A: Chemical*, 2010, **321**, 101.
60 M. Komarneni, J. Shan, A. Chakradhar, E. Kadossov, S. Cabrini and U. Burghaus, *J. Phys. Chem. C*, 2011, **116**, 5792.
61 J. Shan, M. Komarneni and U. Burghaus, *Chem. Phys. Lett.*, 2011, **517**, 59.
62 E. Kadossov and U. Burghaus, *Catalysis Lett.*, 2010, **134**, 228.
63 J. Shan, A. Chakradhar, M. Komarneni, E. Kadossov, J. Schmidt, U. Burghaus, D. Rosenmann, L. E. Ocola, S. Dhuey and S. Cabrini, *Am. Chem. Soc., Div. Fuel Chem.*, 2012, in press.
64 S. Johansson, L. Osterlund and B. Kasemo, *J. of Catal.*, 2001, **201**, 275.
65 R. G. Musket, W. McLean, C. A. Colmenares, D. M. Makowiecki and W. J. Siekhaus, *Appl. Surf. Sci.*, 1982, **10**, 143.
66 J. Grunes, J. Zhu, E. A. Anderson and G. A. Somorjai, *J. Phys. Chem. B*, 2002, **106**, 11463.
67 P. W. Jacobs, F. H. Riberio, G. A. Somorjai and S. J. Wind, *Catal. Lett.*, 1996, **37**, 131.

68 J. Shan, A. Chakradhar, K. Anderson and U. Burghaus, *Nanotechnology for Sustainable Energy*, vol. 2, (Eds.), Y. Hu, S. Qiao, U. Burghaus, ACS book series, 2013, in press.
69 S. Funk, B. Hokkanen, T. Nurkig, U. Burghaus, B. White, S. O'Brien and N. Turro, *J. of Phys. Chem. C*, 2007, **111**, 8043.
70 A. S. Eppler, G. Rupprechter, E. A. Anderson and G. A. Somorjai, *J. of Phys. Chem. B*, 2000, **104**, 7286.
71 B. Nacer, C. Massobrio and C. Felix, *Phys. Rev. B*, 1997, **56**, 10590.
72 H. Iddir, V. Komanicky, S. Ogut, H. You and P. Zapol, *J. Phys. Chem. C*, 2007, **111**, 14782.
73 T. Tsirlin, J. Zhu, J. Grunes and G. A. Somorjai, *Topics in Catal.*, 2002, **19**, 165.
74 A. S. Eppler, G. Rupprechter, L. Guczi and G. A. Somorjai, *J. Phys,. Chem. B*, 1997, **101**, 9973.
75 V. Komanicky, H. Iddir, K. C. Chang, A. Menzel, G. Karapetrov, D. Hennessy, P. Zapol and H. You, *J. Am. Chem. Soc.*, 2009, **131**, 5732.
76 R. O'Barr, S. Y. Yamamoto, S. Schultz, W. Xu and A. Scherer, *J. of Appl. Phys.*, 1997, **81**, 4730.
77 L. Piccolo, A. Valcarcel, M. Bausach, C. Thomazeau, D. Uzio and G. Berhault, *PCCP*, 2008, **10**, 5504.
78 R. G. Osifchin, R. P. Andres, J. I. Henderson, C. P. Kubiak and R. N. Dominey, *Nanotechnology*, 1996, **7**, 412.
79 M. Haruta, B. S. Uphade, S. Tsubota and A. Miyamoto, *Res.Chem.Intermed.*, 1998, **24**, 329.
80 W. L. Yim, T. Nowitzki, M. Necke, H. Schnars, P. Nickut, J. Biener, M. M. Biener, V. Zielasek, K. Al-Shamery, T. Kluener and M. Baeumer, *J. of Phys. Chem. C*, 2007, **111**, 445.
81 M. Mavrikakis, P. Stoltze and J. K. Norskov, *Catal. Lett.*, 2000, **64**, 101.
82 L. Vattuone, L. Savio and M. Rocca, *Surf.Sci.Rep.*, 2008, **63**, 101.
83 E. Kadossov and U. Burghaus, *Chem. Phys. Lett.*, 2009, **483**, 250.
84 S. Schimpf, M. Lucas, C. Mohr, U. Rodemerck, A. Bruchner, J. Radnik, H. Hofmeister and P. Claus, *Catal. Today*, 2002, **72**, 63.
85 M. Komarneni, E. Kadossov, J. Justin, M. Ming and U. Burghaus, *Surf. Sci.*, 2010, **604**, 1221.
86 M. Komarneni, J. Shan and U. Burghaus, *J. Phys. Chem. C*, 2011, **115**, 16590.
87 T. F. Jaramillo, K. P. Jorgensen, J. Bonde, J. H. Nielsen, S. Horch and I. Charkendorff, *Science*, 2007, **317**, 100.
88 F. Rumpf, H. Poppa and M. Boudard, *Langmuir*, 1988, **4**, 722.
89 M. Bowker, *Surf. Rev. Lett.*, 1994, **4**, 549.
90 U. Burghaus, *Trends in Physical Chemistry*, 2001, **8**, 21.
91 D. A. King and M. G. Wells, *Surf. Sci.*, 1972, **29**, 454.
92 X. M. Yan, S. Kwon, A. M. Contreras, M. M. Koebel, J. Bokor and G. A. Somorjai, *Catalysis Lett.*, 2005, **105**, 127.
93 A. M. Contreras, X. M. Yan, S. Kwon, J. Bokor and G. A. Somorjai, *Catal. Lett.*, 2006, **111**, 5.
94 J. Grunes, J. Zhu, M. Yang and G. A. Somorjai, *Catal. Lett.*, 2003, **4**, 157.
95 J. Zhu and G. A. Somorjai, *Nano Lett.*, 2001, **1**, 8.
96 V. Johanek, M. Laurin, J. Hoffmann, S. Schauermann, A. W. Grant, B. Kasemo, J. Libuda and H. J. Freund, *Surf. Sci.*, 2004, **561**, L218.
97 J. Hoffmann, S. Schauermann, J. Hartmann, V. P. Zhdanov, B. Kasemo, J. Libuda and H. J. Freund, *Chem. Phys. Lett.*, 2002, **354**, 403.
98 R. Imbihl and G. Ertl, *Chem.Rev.*, 1995, **95**, 697.
99 F. J. G. de Abajo, *Rev. Mod. Phys.*, 2007, **79**, 1267.

100 B. Auguie and W. L. Barnes, *Phys. Rev. Lett.*, 2008, **101**, 143902.
101 E. M. Hicks, S. Zou, G. C. Schatz, K. G. Spears, R. P. v. Duyne, L. Gunnarsson, T. Rindzevicius, B. Kasemo and M. Kall, *Nano Lett.*, 2005, **5**, 1065.
102 Q. Yu, P. Guan, D. Qin, G. Golden and P. M. Wallace, *Nano Lett.*, 2008, **8**, 1923.
103 R. Gordon, A. G. Brolo, A. McKinnon, A. Rajora, B. Leathem and K. L. Kavanagh, *Phys. Rev. Lett.*, 2004, **92**, 037401.
104 J. L. Plaza, Y. Chen, S. Jacke and R. E. Palmer, *Langmuir*, 2005, **21**, 1556.
105 F. H. Ribeiro and G. A. Somorjai, *Recl. Trav. Chim. Pays-Bas.*, 1994, **113**, 419.
106 K. Wong, S. Johansson and B. Kasemo, *Faraday Discussions - 114*, 1996, **105**, 237.
107 K. Sattler, J. Muehlbach and E. Recknagel, *Phys. Rev. Lett.*, 1980, **45**, 821.
108 T. G. Dietz, M. A. Duncan, D. E. Powers and R. E. Smalley, *J. of Chem. Phys.*, 1981, **74**, 6511.
109 S. Abbet, A. Sanchez, U. Heiz, W. D. Schneider, A. M. Ferrari, G. Pacchioni and N. Rosch, *Surf. Sci.*, 2000, **454456**, 984.
110 U. Heiz, A. Sanchez, S. Abbet and W. D. Schneider, *J. Am. Chem. Soc.*, 1999, **121**, 3214.
111 S. Lee, C. Fan, T. Wu and S. L. Anderson, *J. Am. Chem. Soc.*, 2004, **126**, 5682.
112 M. Thomas, J. T. Dickinson, H. Poppa and G. M. Pound, *J. Vac. Sci. Technol.*, 1978, **15**, 568.
113 D. W. Goodman, *J. Chem. Phys.*, 1996, **100**, 13090.
114 C. R. Henry, *Surf. Sci. Rep.*, 1998, **31**, 235.
115 G. Ertl and H. J. Freund, *Phys. Today*, 1999, **1**, 32.
116 J. Libuda and H. J. Freund, *Surf. Sci. Rep.*, 2005, **57**, 157.
117 I. Lopez-Salido, D. C. Lim and Y. D. Kim, *Surf. Sci.*, 2005, **588**, 6.
118 K. T. Nicholsen, T. K. Minton and S. J. Sibener, *J. Phy.Chem.B*, 2005, **109**, 8476.
119 H. W. Deckman and J. H. Dunsmuir, *Appl. Phys. Lett.*, 1982, **42**, 377.
120 Z. Paszti, Z. E. Horcath, G. Petra, A. Karacs and L. Guczi, *Appl. Surf. Sci.*, 1997, **109/110**, 67.
121 J. Wang and U. Burghaus, *J. of Chem. Phys.*, 2005, **123**, 184716.
122 S. Funk, B. Hokkanen, J. Wang, U. Burghaus, G. H. Bozzolo and J. E. Garces, *Surf. Sci.*, 2005, **600**, 1870.
123 J. V. Lauritzen, M. Nyberg, R. T. Vang, M. V. Bollinger, B. S. Clausen, H. Topsoe, K. W. Jacobsen, E. Laegsgaard, J. K. Norskov and F. Besenbacher, *Nanotechnology*, 2003, **14**, 385.
124 E. Gross, M. Asscher, M. Lundwall and D. W. Goodman, *J. Phys. Chem. C*, 2007, **111**, 16197.
125 J. R. Heath, C. M. Knobler and D. V. Leff, *J. of Phys. Chem. B*, 1997, **101**, 189.
126 H. Hofmeister, S. Grosse, G. gerth and H. Haefke, *NanoStructures Materials*, 1995, **6**, 529.
127 B. White, M. Yin, A. Hall, D. Le, S. Stolbov, T. Rahman, N. Turro and S. O'Brien, *Nano Lett.*, 2006, **6**, 2095.
128 S. Funk, E. Kadossov, S. Cabrini and U. Burghaus, 2007, unpublished.
129 J. A. Haes and P. R. V. Duyne, *J. Am. Chem. Soc.*, 2002, **124**, 10596.
130 M. A. Wood, *J R Soc Interface*, 2007, **22**, 17.
131 S. M. Yang, S. G. Jang, D. G. Choi, S. Kim and H. K. Yu, *Small*, 2006, **2**, 458.

High throughput technology: approaches of research in homogeneous and heterogeneous catalysis

Stephan A. Schunk,* Natalia Böhmer, Cornelia Futter, Andreas Kuschel, Eko Prasetyo and Thomas Roussière
DOI: 10.1039/9781849737203-00172

High throughput experimentation (HTE) approaches and the choice of the design of experiment (DoE) tools are discussed with regard to their convenience and applicability in homogeneous and heterogeneous catalysis as a concerted workflow. Much attention is given to diverse methodologies and strategies, which are fundamental for the experimental planning. For two target reactions in two case studies presented in this chapter, HTE methods were applied to create and evaluate catalyst libraries. A homogeneous catalyst case study is illustrated first, which deals with parallel synthesis and screening of organometallic catalysts in the polymerisation of ethylene. The second case study (heterogeneous catalysis) focuses on coherent synthesis and testing of dopant effects on the performance of oxidation catalysts in a reaction of transformation of n-butane to maleic anhydride. Supporting examples from the literature described here show that careful planning of libraries and test conditions is vital in high throughput experimentation in order to deliver meaningful results leading to performance improvements or disruptive new findings.

1 Introduction

Nowadays, almost all chemical industrial processes involve the use of catalysts. We can therefore regard industrial catalysis by far the most important means of producing chemicals, oil products, fertilisers, plastics, drugs and pharmaceuticals.[1] Moreover, continually increasing demand for more environmentally friendly, cleaner and energy efficient technologies require prompt identification of highly active and selective catalysts and the elaboration of suitable reaction corridors to tackle new approaches for improved or alternative value chains. If we take an extreme viewpoint one could say that traditional methods of catalyst synthesis and testing usually lack sufficient speed and therefore leave room for improvements with regard to their efficiency and the resulting velocity rendering impulse to process development.[2,3] If we follow the train of developments based on the roots of combinatorial chemistry and high throughput experimentation we can say the demand for shorter realisation time for chemical processes favoured the implementation for new research strategies that were developed in order to produce a larger number of breakthroughs and also would in a revitalisation in the field of industrial and academic chemical research.[1]

The up-to-date screening methods used in high throughput experimentation (HTE) give an opportunity to perform at once several, in some cases even more than hundreds of experiments involving perturbation of a

hte Aktiengesellschaft, Kurpfalzring 104, 69123 Heidelberg, Germany.
E-mail: Stephan.Schunk@hte-company.de

large number of variables. It is for certain that the adequate use of suitable equipment and integrated data management systems are the underlying aspects of successful workflows. Still it is evident that the development and implementation of appropriate screening tools and parallel testing equipment requires enhanced engineering but can only be brought to its full value when elaborate software architectures are available to propel the entire effort.[4–6]

With the technology for both, synthesis and screening and the resulting workflows available up to date which offer a broad coverage of all aspects of modern catalytic processes for chemical, petrochemical and environmental catalytic applications the majority of development work going on in academia and industry can be supported with HTE.[1,7–10] From our point of view an established technology basis is readily available today and has been reviewed at high qualitative level by other authors.[1,9,11] The question that has not been addressed to a broader extent in literature is what the right underlying approach is for the right choice of catalyst candidate libraries and suitable reaction conditions with which we can tackle challenging questions in combinatorial chemistry and high throughput experimentation. The question of taking the appropriate approach is vitally important especially in the initial or planning phase as the approach sets the base on how fundamental questions are addressed and finally may become decisive of success or failure of an experimental campaign. The answer to the challenges raised here will in many cases not be monolithic but have more than one appropriate answer. But it is evident that the core aspects with regard to successful solutions for challenging problems in catalysis are vitally connected to appropriate experimental planning and the right choice of the design of experiment tools in conjunction with the workflows applied. Not only the choice of the right tools of DoE, the strategic setting of the experimental points defining the multidimensional experimental space impacts the solution functions that can be identified. In general confinement strategies will be followed therefore the experimental points and the corresponding variations of the variables connecting them can always be interpreted also as the boundaries of an experimental plan. The experimenter always has to be aware of the fact that the creation of the experimental boundaries using DoE tools is a vital step of the same importance as the number and the inter-distance in between experimental points that cover the entire search space. It is of course debatable whether equivalent distance between points, as depicted in Fig. 1(a) may in all cases be desirable or not. In many cases we will try to divide our search space evenly by choice of suitable distance between experimental points and for sure with regard to independent variables like temperature or pressure this is an easy task. With regard to inherent physicochemical properties of catalyst candidates like for example acidity or basicity also depending on the tools applied for the DoE it may be that finally the search space may look different from the "ideal" case (Fig. 1(b)).

The size of the experimental space has also a decisive impact on the effort that has to be undertaken in exploring it: the number of points required to cover the entire search space increases with the number of dimensions as exponent.[12] The choice of DoE tools and their appropriate use is therefore key element in any workflow solution.

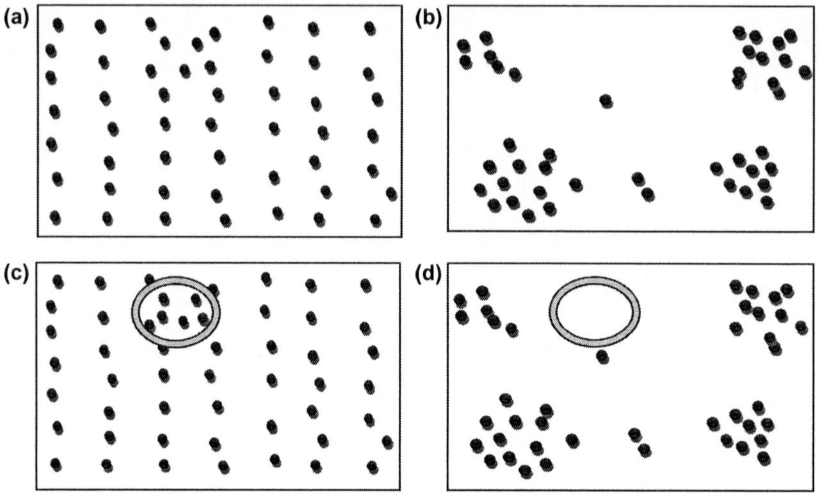

Fig. 1 Illustrative example of (a, c) continuous systems (ideal systems) and (b, d) discontinuous systems of experiment design.

Several groups have devoted minor or major efforts to the above mentioned challenges; what we would like to do within the framework of this article is to critically review selected approaches, for both homogeneous and heterogeneous catalytic examples. In some cases we will also discuss examples of interest from other fields like pharmaceutical research or materials research outside of catalysis. It is clear that the tendency to publish successful approaches over approaches that have not paid off may overrate the suitability of certain tools within the field and falsify the picture. We may also not forget that principally a direct comparison of DoE approaches applied to the very same problem will in many cases be impossible and can hardly be found. This review is therefore not considered to serve as benchmarking study to different solution approaches, but should rather be seen in the light as summarising and critically commenting selected examples that have proven to pay off with regard to their efficiency for selected topics.

In this chapter we would like to make a division for our discussion of the general high throughput experimentation approaches of research and specific design of experiment tools based on the fields of homogeneous and heterogeneous catalysis.

2 Homogeneous catalysis

2.1 High throughput experimentation in homogeneous catalysis

Originated in the pharmaceutical industry, where the combination of highly efficient assays and parallel synthetic methods allowed a major increase in throughput, high throughput experimentation (HTE) has also been successfully employed to improve organometallic catalysts for metal-catalysed transformations.[13–15] With HTE, new catalyst systems can be efficiently identified and optimised with regard to optimum reaction conditions.

Depending on the structure and functionalities of the ligand systems; monodentate or polydentate, the nature of bridging groups and the

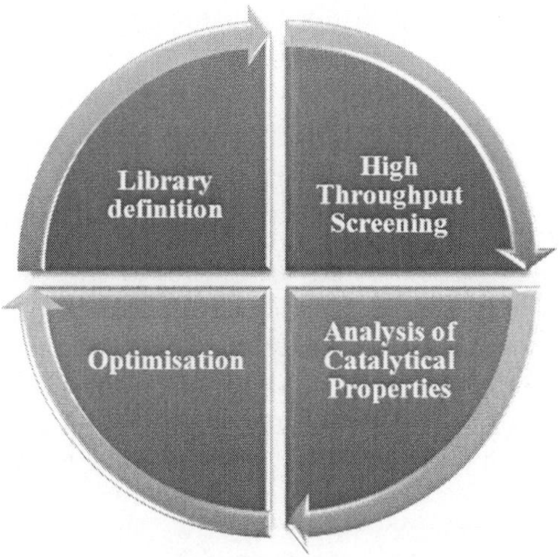

Fig. 2 Iterative approach in designing novel homogeneous catalyst.

complexity of the residual or functional groups, billions of possible structures exist. Nevertheless, the final properties of a molecular entity as catalyst are also governed by other parameters: Such as the nature of the transition metal, solvent, counter ions of the complexes as well as other reaction parameters like temperature and pressure also the performance of a molecular entity which serves as catalyst in any homogeneously catalysed reactions, resulting in widely different and often even opposing effects. From the point of view of the screening effort in many cases the synthesis of a requisite number of candidate molecules may become the decisive bottleneck, hindered by the time-consuming and cost-intensive ligand preparation as well as the purification steps and vitally necessary characterisation of organometallic intermediates and complexes. Finding an optimal high performance homogeneous catalyst for a given reaction is thus often difficult and challenging.

Common approaches for discovery and optimisation of homogeneous catalysts involve combinations of parallel synthesis, screening and modelling (Fig. 2).[16] Generally, both combinatorial and computational methods as well as a combination of both serve as powerful HTE tools to advance catalysis research.

Combinatorial approaches may be part of HTE programs in homogeneous catalysis, but classical parallel approaches using parallel synthetic workstations are much more prominent.[11,17–19] The main reason for the popularity of classical parallel versus combinatorial approaches relates to the fact that small focused libraries are nowadays more favoured than extensive combinatorial ones.[20–23]

One of the evident drawbacks is that combinatorial chemistry for sure allows the rapid synthesis of a large number of different but always structurally related compounds. One can see the routes of combichem being based on the solid-phase synthesis like the efforts propelled for peptides by Merrifield in the 1960s,[24] combinatorial chemistry is now routinely used in the field of natural product synthesis,[25] chemical biology,[26] catalyst discovery,[27] materials science,[28] and drug discovery.[29] Combinatorial synthesis may involve synthetic techniques like solid- and solution-phase synthetic methods, including parallel synthesis,[11,17,23,30–33] split-and-pool techniques,[34–37] encoding/deconvolution techniques,[38–40] polymer-supported reagents and scavenger resins.[41]

The library generation within any HTE campaign in homogeneous catalyst development is usually led with a rapid and efficient synthesis of a catalyst library followed by the characterisation and quality control of synthesised compounds using in many cases high throughput NMR techniques.[7,42–44] The synthetic effort is followed by high throughput screening of the different catalyst candidates in order to obtain detailed information on the catalytic properties, *e.g.* their activity (which may be expressed as Turn Over Frequency (TOF)), selectivity (in many cases also enantioselectivity) and lifetime (which may be expressed as Turn Over Number (TON)). In a final step, the results from the HTE screening need to be related with the chemical properties of the catalysts in order to gain insight on structure and properties, deduct hypothesis on a potential reaction mechanism and to potentially predict catalytic properties of hypothetical structures. In an ideal world the knowledge obtained can be used to prove or develop a hypothesis which may lead to new libraries with promising candidate structures.

In the last decades, parallel synthesis has shared substantial benefits from automation technologies, which enable the synthesis of a huge number of unique compounds within a short time.[44–47] In general, HTE synthetic approaches towards homogeneous catalysts have been reviewed in detail for a wide range of reactions,[11,32] especially within the field of asymmetric transformation.[48–63]

High throughput screening of catalyst libraries can be performed with a broad range of testing equipment and technologies. For a comprehensive list of screening methods, the reader is referred elsewhere.[16,58,61,62,64–106]

Even with increasing HTE screening capacity, the numbers of potential experiments that can be performed may nearly be infinite due to the huge variety of possible homogeneous catalyst structures and the potentially large variation reaction conditions. Therefore measures have to be taken to limit the number of experiments in order to increase the experimental efficiency. Statistical design of experiment approaches may help limit experimental points in the diversity space of continuous experimental variables, but for chemical structures or discontinuous variables other approaches may be more successful. Again here the pioneering work has been done in pharmaceutical research; it is evident that molecular modelling has had an immense effect in the way research is now done concerning the optimisation of structural entities.[107,108] One still has to bear in mind that the demands for modelling in pharmaceutical research and catalysis are different. In pharmaceutical research in many cases binding sites are modelled and these sites also represent thermodynamic minima. In homogeneous catalysis usually the transition states will make the difference between catalyst candidates and are of greatest help for understanding critical components within the catalytic cycle.[109,110] In some cases though design and development of novel homogeneous catalysts is impeded by a limited understanding of the relationships between molecular structure and catalytic activity similar to the field of drug discovery, especially for "hypothetical" catalytic cycles.

A systematic design of experiment (DoE) and a complementary structure activity relationship based on modelling assumptions are thus crucial to a) create meaningful libraries of catalyst candidates that enable the verification of essential hypothesis and b) increase the HTE efficiency with regard to effective choice of experimental points in the field of homogeneous catalysis.

In this review, we focus on DoE approaches and emerging computational approaches in detail (Section 2.2) and we will finally present a case study of a high throughput approach in the development of novel polyolefin catalysts (Section 2.3).

2.2 Computational approaches

The intensive scientific effort to improve computational catalyst design approaches making them available for a larger community over the last years is reflected by the number of publications in the field. Computer power has immensely increased and helps in carrying out calculations faster, but the choice of the computational model remains a core challenge for computational approaches applied in the field of modelling for homogeneous catalysis.

The following section summarises some trends in computational approaches in conjunction with catalyst library design in homogeneous catalysis research that we consider important. We also describe approaches that have until now not found their way into application in homogeneous catalysis, but seem so powerful in the field of pharmaceutical research that we included them. The determination of the relationship between the structure of the catalysts and the performance of this entity in a target reaction is discussed as a general and holistic approach to identify and to understand a promising lead catalyst structure and thus a highly performing catalyst.

2.2.1 Molecular descriptors. Molecular descriptors are variables that aim at the representation of the physicochemical features of a class of compounds. They are commonly classified into four descriptor types: (i) constitutional (molecular formula) descriptors, (ii) topological (molecule connectivity) descriptors, (iii) geometrical (3D molecular models) descriptors and (iv) quantum-chemical (3D semi empirical or *ab initio* quantum mechanics) descriptors.[111] Molecular descriptors of an organometallic compound represent uniquely each catalyst structure, *e.g.* the effect of bite and cone angles of the ligating groups, backbone flexibility, lipophilicity, hydrophobicity and partial charge of the metal atom. Generally, 3D descriptors contain the highest information density and offer the perspective of highest computational accuracy. However, 3D descriptors are often accompanied with high computational cost necessary for geometry optimisation. Therefore, more simple 1D and 2D descriptors are preferred especially when huge catalyst diversity spaces are involved.[112,113] These advantages of 1D and 2D descriptors are restricted by their information density as well as by their inadequate capability to address parameters such as conformational information or interactions including chirality.

Rothenberg *et al.*[113] developed a set of topological descriptors for a series of bidentate phosphine and phosphite ligands. The study revealed that the 2D models were capable of identifying the similar trends (bite angle) as the 3D models albeit with more noise.

Generally, other reaction parameters such as interactions with solvents or dependence on reaction temperature will also influence the overall structure of active compounds in organometallic catalysed reactions. Because solvent properties such as polarisability, pH, solubility or boiling point play an important role in the catalytic cycle, it is reasonable that these are represented by their own descriptors. The solvent dependence of a reaction can be expressed by the solvent cohesive energy density ε_c, while the ability to dissolve the solute is represented by the Hildebrand solubility δ_H.[114,115] In some structure-activity relationship studies the authors have been successful in defining these solvent properties by simpler 2D descriptors.[16]

A catalyst can be represented with a set of descriptors, some of which may add value due to their relevance and deliver unique information about the catalyst-substrate interaction. Both the inclusion of irrelevant parameters and calculations of descriptors, which are similar, are contra productive processes that hinder efficient catalyst development. Finding suitable, non-redundant descriptors is challenging and often remains a major challenge in constructing an appropriate Quantitative Structure-Activity Relationship (QSAR) model. Appropriate descriptors must therefore fulfil several requirements: They (i) should be easy to calculate, (ii) can be related to figures of merit, (iii) represent structural variations within the data set and in addition (iv) should be chemically meaningful.[116]

The construction of descriptor vectors based on chemical intuition is often useful for chemical challenges which have already been explored to a certain extent. Ranking approaches such as principal component analysis (PCA) are used to avoid redundancy of variables in case of very large number of possible descriptors or when multiple inter-correlation effects predominate.[16] Alternatively, molecular descriptors can be pre-processed using the supervised calculation of the information content,[117] the unsupervised Forward Selection (UFS) algorithm or other computer-aided algorithms.[118]

In an elaborate computational study, Bajorath *et al.*[119,120] calculated the information content of 143 descriptors with the Shannon entropy (SE) and differential Shannon entropy (DSE) concepts and detected the intrinsic differences between them. A formalism to calculate and compare the information content has been introduced in digital communication theory but not in conventional chemistry. The most relevant molecular descriptors in databases consisting of over 300.000 synthetic compounds, drug-like, and natural molecules could be identified and thus redundant information can be avoided.

Whitley *et al.*[121] employed UFS method to generate a subset of relevant, non-redundant descriptors from three data sets: (i) 21 steroid compounds; (ii) 19 pyrethroid insecticides; (iii) 31 antifilarial antimicyn analogues. UFS selects two least-correlated descriptors and extends them with additional variables on the basis of their multiple correlations. This approach is shown to produce simple, robust and easily interpreted QSAR models for all three data sets.

Various customised computer-aided algorithms may simplify the determination of the relevant descriptors out of a huge number of possibilities in a decent time frame. There are a few methods employing the computer's calculating power such as the simulated annealing,[122] generalised simulated annealing,[123] supervised neural networks approach,[124] evolutionary algorithms,[125] and genetic algorithms. Especially pharmaceutical research is a prime example of the successful implementation of computer-assisted calculations to solve specific challenges of drug development. One prime example is the successful determination of the most important 150 descriptors out of 398 molecular descriptors for three diverse sets of bulky compounds (463 oestrogen receptor ligands, 337 carbonic anhydrase II inhibitors and 1608 monoamine oxidase inhibitors). An improved prediction accuracy of the binary QSAR model of up to 23% was reported.[126] Currently no such extensive studies could be found for homogeneous catalysis although the close connection is evident.

Well-chosen molecular descriptors may enable the absolute or relative quantification of physicochemical characteristics of the catalysts, redundancy and thus the spread of experiments can be optimised. Moreover, molecular descriptors can be used to draw relationships between the catalyst structures and their performance, so that a promising structure can be traced back and new candidate structures can be generated from the collected experimental data.

Rothenberg and co-workers successfully identified important descriptors for metal-ligand-complexes and applied their findings towards the design of new nickel-based hydrocyanation catalysts.[127] To achieve this, representative descriptors were chosen from the following three key groups:

a) Chelating effect descriptors, which cover the size, flexibility, denticity and possible conformational changes of the ligands.
b) Steric descriptors, which provide information on the ligands' shape, backbone group size and metal centre accessibility.
c) Electronic descriptors, which measure the charge distribution and the orbital energies of the complexes.

A total of 39 chelating, steric and electronic descriptors were investigated and calculated in this study. Using Partial Least Squares (PLS) regression analysis, the structure-activity correlation was determined and the following descriptors were found to be most important: (i) the charge at the metal ligating atoms, (ii) the rigidity of the backbone groups, (iii) the steric surrounding groups around the metal atom and (iv) the bite angles. The derived model provided a rational and quantitative means to screen optimised virtual libraries of potential *in silico* hydrocyanation catalysts.

In a subsequent study, Rothenberg *et al.*[128] calculated a total of 168 topological, constitutional, geometric and electrostatic descriptors for 96 Ni-based hydrocyanation catalysts. Consistently, they observed that the topological and electrostatic descriptors including the bite angle, the ligand backbone flexibility and the charge density are the most important descriptors for the considered chemistry.

Furthermore, the group of Rothenberg collected open literature data representing 253 Heck-type reactions performed with 58 different catalysts under different reaction conditions.[129] With the calculated 31 2D descriptors, the researchers used a PLS approach to obtain a QSAR model, correlating the descriptors and the reaction conditions with the Figure of Merit (FOM), yield, TON and TOF. The most important descriptors were the number of H and C atoms and electronic charge density of the ligands.

The above examples show that PLS regression analysis is a powerful tool that can be used to investigate the structure activity relationship based on experimental results *e.g.* provided by literature data and to determine the respective importance of each calculated descriptor. The insight into structurally favoured properties can subsequently be used to minimise further experimental approaches with respect to the often time-consuming and expensive synthesis of huge catalyst libraries.

2.2.2 Quantitative structure activity relationship (QSAR) modelling. One possible approach to minimise the number of samples to be tested while maximising the knowledge and the chance of finding a highly active catalyst is represented by QSAR methods. With these models, the relationship between a combination of different descriptors and the FOM, which expresses the degree of suitability of the catalysts, such as selectivity, TON, TOF, can be quantified and uncovered. The model may enhance a better understanding and more importantly determination of high performance catalysts stemming from extensive catalyst libraries.

There are several methods to generate a QSAR model. The QSAR model estimation can be subdivided into continuous regression methods, which try to create functional dependencies and classification or discrete methods, which predict the affiliation to a certain group.[118] Most of the methods discussed in this review are available in many commercial QSAR software solutions.

In order to arrive at a suitable model, a QSAR approach requires consistent data sets of catalytic screening results from HTE using DoE methods. The amount of data sets as well as huge diversity and spread of the data define the quality of the models.[16]

The validity of the model depends in particular on the information accuracy and a suitable descriptor choice. The more descriptors and the less redundant information are used, the more accurate the model will be. The determination of the accuracy of the generated model is performed *via* cross-validation or rotation estimation.[16] Cross-validation involves dividing the dataset into two parts: 1) a training set to develop the QSAR model and 2) a test set for the verification purpose. A special type of cross-validation - the leave-one-out cross-validation - uses one dataset for testing and the remaining dataset for training. This is repeated until all dataset is used as the test set once.

By combining molecular connectivity properties with Partial Least Squares (PLS) and an Artificial Neural Network (ANN) statistical analysis and utilising 26 unique descriptors, Rothenberg introduced a concept to predict bite angle and backbone flexibility of 80 bisphosphine and bisphosphite ligands, which can be determined from the metal complexes' crystal structures.[113] He managed to improve his structure-property models with 93% prediction accuracy for bite angles and 90% for flexibilities.

Landrum *et al.*[130] used data in publications of olefin co-polymerisation study from Symyx Technologies,[131] to build a model to predict the molecular weight of the resulting polymer. The novel group IV polymerisation catalysts used in the reaction were formed by combining arrays of 96 amide ether ligands with reactive metal complex $Hf(CH_2Ph)_4$ and activators. By applying ten instead of only five descriptors, they could increase the model accuracy from 70% to 80%.

Based on the above examples, QSAR modelling again proves to be a powerful tool to predict the importance of catalyst properties such as bite angle or backbone flexibility and therefore catalyst activity. Using structure-property models derived from data sets generated within high throughput experimentation, high performing catalysts can be determined from extensive libraries successfully minimising the number of samples to be screened.

2.2.2.1 Continuous regression analysis. Depending on the problem complexity and the number of parameters in the dataset (solvent, catalyst, ligands, *etc.*), continuous regression analysis can be accomplished with linear or non-linear methods.

The most simple linear regression analysis aims to fit a straight line through several points so that the squared deviations of the points from the regression line are minimised:

$$\mathbf{y} = a_1 \cdot \mathbf{x_1} + a_2 \cdot \mathbf{x_2} + \cdots + a_n \cdot \mathbf{x_n} \quad (1)$$

A sophisticated linear regression analysis dealing with multi variable data is the PLS analysis. This analysis method is characterised by finding a score vector X and Y that provides maximal covariance.[132] This method is especially suitable when the number of descriptors is very high or when the reaction mechanism or transition states are not yet known. PLS has the ability to incorporate several figures of merit into the model as well as reduce the interco-linearity between descriptors while determining the most dominant one.

Van der Linden *et al.*[133] employed PLS to derive a relationship between substrate structure and product enantiomeric excess (*ee*) in asymmetric benzophenone hydrogenation reactions with Noyori's catalyst. Despite the limited size of the training set of only 13 ketones, the model was able to point towards the conformation of a transition state as well as to predict valid *ee* values for 21 selected benzophenones. The trends in the model predictions agreed with the chemists' intuition: ortho-substituted compounds yield high *ee* values while compounds with para-substituents yield low *ee* values.

Carlson and Gautun developed a PLS model that proved to be capable of identifying critical properties of a reaction system so that both favourable and non-favourable conditions can be identified. The substrates, Lewis acid catalysts and the solvents were varied to achieve combinations of 600 unique experiments for a regio-selective indole synthesis. The researchers selected 256 experiments with maximum spread in the descriptor space. A PLS analysis was performed with the score values, their squares and the interaction terms including those which account for nucleophilic/electrophilic properties of the ketones and the size of one side chain. It was concluded that Lewis acid hardness and the solvent polarisability defined the reaction success.[134]

Following the study of ligand descriptor analysis in Ni-catalysed hydrocyanation reaction, Rothenberg and co-workers investigated 96 Ni-based complexes and derived appropriate QSAR model using the PLS method. The researchers concluded the bite angle and the backbone flexibility of the ligands as well as the charge densities as the most important descriptors among 168 calculated descriptors. The derived model predicted the selectivity and the yield within $R^2 = 0.6937$ and $R^2 = 0.5878$ respectively.[128]

Continuous regression analysis can therefore be used to minimise the number of parameters and thus the complexity of the investigated structure-activity model. Moreover, favourable as well as non-favourable conditions are identified and subsequently verified based on the success of the investigated reaction.

2.2.2.2 Discrete classification methods. Classification trees,[135] supported vector machines (SVM),[136] and Artificial Neural Networks (ANN),[137] are some of the few methods that can be well applied to discrete questions. These methods are particularly useful when the chemical system is complex and many factors affect the overall reaction. ANN is the most flexible and remarkable method, capable of deriving relationship from complicated and difficult data, which are barely accessible with other computational techniques.[16]

Farrusseng and Rothenberg analysed the data of 412 Heck-type cross-coupling reaction experiments collected from worldwide published literatures. The relationship between calculated descriptors (ligands, substrates, catalysts and reaction conditions) and the TON and TOF were analysed using ANN, classification trees and linear regression models, yielding a 93% prediction accuracy. The linear method was only able to identify the role of Pd loading in the reactions while the non-linear methods were able to detect higher order effects concerning the steric and electronic properties of the ligands.[129,138]

Cundari *et al.* applied ANN and classification methods in data analysis of transition metal imido complexes, which are associated with nitrogen fixation and C-H activation reactions. The study confirmed the inversely proportional relationship between the metal-nitrogen bond length and the metal-nitrogen-substituent angle. Metals, ligands and substituents which favour a double bond between the nitrogen and the metal atom would give an active well performing catalyst.[139–141]

A continuous development of robust and fast computational methods along with accurate determination of experimental errors and precise but less complex models are crucial and major issues are still yet to be addressed.

Nevertheless, discrete classification methods are a well understood alternative to investigate complex structure activity relationships where linear methods are unsatisfactory.

2.2.3 Homogeneous catalysis virtual libraries. To be able to fully exploit HTE advantages, appropriate methods are required to handle and extract useful information from the flow of experimental data. Planning of libraries is essential as well as careful design and planning of numerous parallel experiments. DoEs, which interpolate the results for the different experiments by the help of appropriate descriptors quantitatively, limit the number of experiments that must be optimised in order to avoid redundancy of the experiments.

Another way to develop efficient libraries and test campaigns in order to pinpoint the range of superior catalysts can be achieved by screening of catalyst libraries virtual *in silico*. *In silico* prediction utilise QSAR/QSPR model generated from HTE to scan vast virtual catalyst libraries. The group of Rothenberg has developed a useful approach: To simplify *in silico* screening, libraries of virtual catalyst candidates can be developed by assuming that an asymmetric catalyst complex may contain one metal atom M and one mono- or polydentate ligand, which includes ligating groups LN, a backbone (bridge) group B and residue groups RM.[142] Grey indicates ligating groups, black indicates residue groups, and light grey curved line indicates backbone groups (Fig. 3). This assumption enables a straight forward computation of a variety of structure combinations. The library-generated sets of $L_1(R_1) - B_1(R_2) - L_2 - B_2(R_3) - L_3(R_4)$ strings can be assembled: Two connection points are selected for each ligating group, which are for the backbone and the residual groups. Two or three points are selected for the bridging groups in order to connect the two ligating groups and the backbone-bonded residue group. The geometries of these structure combinations are optimised and can be linked with the catalyst descriptors and reaction conditions derived from the HTE. The selection of the set of compounds is crucial. Optimally, diverse and representative choice of compounds affords non-redundant libraries for virtual screening purposes.

The QSAR model validated with the real experiments is used to predict the performance of virtual catalysts and reaction conditions. Coupling *in silico* screening with high throughput experimentation enables the study of a large catalyst diversity space in a short time. It may serve as a tool which helps to preselect candidate regions in the catalyst space, which will become the foundation for the next generation catalyst library.

In silico screening has been widely used for rational drug design approaches. The most recent application of *in silico* research used the protein docking algorithm EADock to find potential inhibitors to an enzyme associated with cancer activity.[143,144] This approach presented a

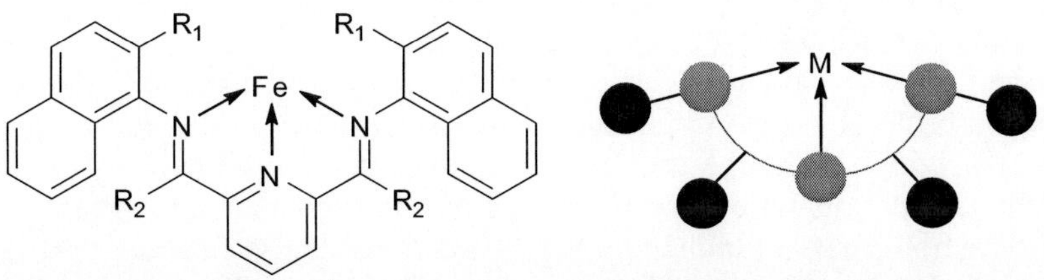

Fig. 3 Iron catalyst with tridentate ligand and its derived model for library. Grey indicates ligating groups, black indicates residue groups, and light grey curved line indicates backbone groups.

significant advantage to the expensive high throughput screening robotic labs which physically test thousands of diverse compounds a day with less than 1% hit rate. Scientists found that fifty percent of the molecules designed *in silico* were active indoleamine 2,3-dioxygenase (IDO) inhibitors *in vitro*.

Rothenberg *et al.* introduced the concept of virtual library to his study of *in silico* Rh-catalysed hydroformylation, which successfully predicted the bite angle in a library of 600 ligand Rh complexes as well as the linear:branched aldehyde product ratio in the hydroformylation of 1-octene.[142]

The same group exploited the best calculated artificial neural network model derived from 412 Heck cross-coupling reactions data to predict the performance of 60.000 Heck reaction combinations representing 61 new phosphines, four olefins, four arylhalides, five catalyst precursors, four solvents and three Pd loadings *in silico*. Four percent of the combinations were predicted to have TON > 10,000 (see also Section 2.2.2.2).[138]

In another study, Burello and Rothenberg screened a very large ligand library *in silico* to characterise the bite angles and backbone flexibilities for a set of 80 bisphosphine and bisphosphite complexes (see also Section 2.2.2). Non-linear methods combined with fundamental chemical topology principles yielded a satisfactory model with a prediction accuracy of over 90%.[113]

The above examples show the successful application of *in silico* screening methods. To validate those systems, input from high throughput experimentation is necessary, but guides to further and even more precise and promising catalyst systems to be synthesised and screened thereof.

2.3 Case study

Another approach in finding suitable and highly active catalysts on the basis of data provided by literature is the efficient design of a simple synthesis workflow, leading to a huge number of different, but compositionally related catalyst systems. A multitude of structurally diverse compounds is obtained using different building blocks in a defined synthetic route. The following case study describes the successful application of high throughput experimentation regarding the synthesis and testing of late transition metal complexes with bidentate diazabutadiene or tridentate diiminopyridine ligands in the polymerisation of ethylene.[7] Variations of the ligand, the metal center but also the type and quantity of activator based on the original catalyst entity reported by Brookhart and Gibson resulted in 302 different complexes with different influence towards the catalytic activity and selectivity.

2.3.1 Catalyst library design and synthesis. To simplify the synthetic concept and subsequent analysis, any given complex within the study can be fully described by three different fragments: (i) the central metal atom, (ii) the dicarbonyl backbone group and (iii) the suitable amine providing the donor atoms. Ligand synthesis was realised in a condensation reaction of the dicarbonyl compound (Fig. 4) with two equivalents of an amine (Fig. 5) to yield diazabutadiene (DAB) or diiminopyridine (DIP) ligands for the subsequent complexation with metal halides to give the respective late-transition metal catalysts thereof. The combination of readily available and

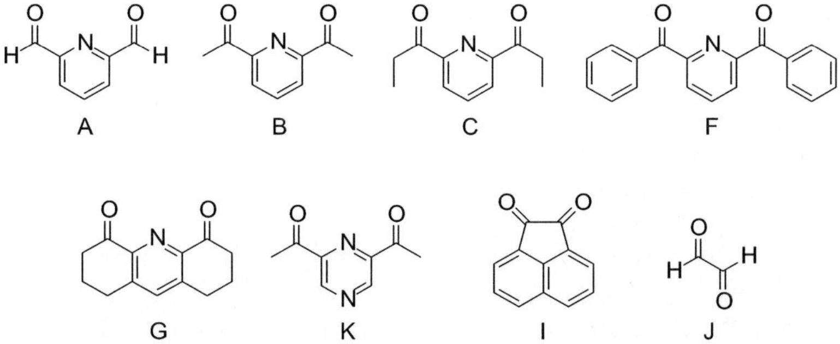

Fig. 4 Dicarbonyl compounds used for ligand synthesis.

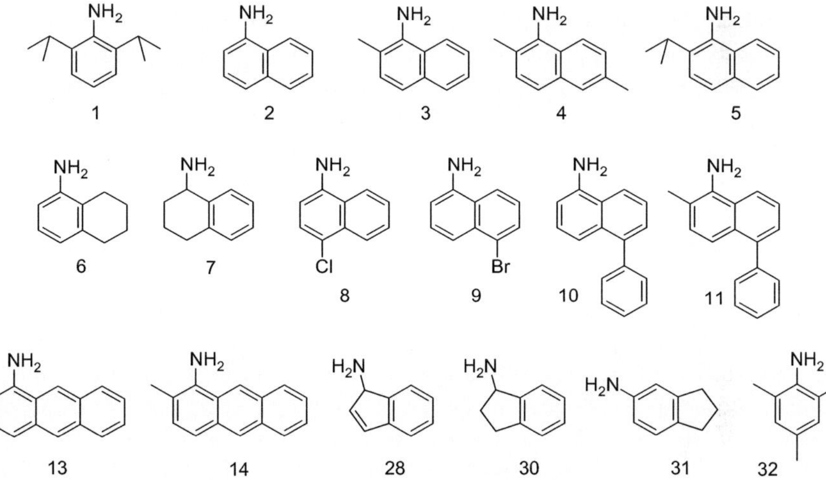

Fig. 5 Amines used for ligand synthesis.

accessible but versatile building blocks was performed using a Chemspeed ASW 2000 parallel synthesiser to obtain a manifold library with minimum redundancy to investigate the effect of rigidity of the ligand system and the steric congestion around the metal centre towards the catalytic activity.

Interestingly, the aldehyde backbones proved to be most versatile as condensation with all amines from the library design led to the desired ligands, meanwhile the ketone backbones were only compatible with aromatic amines to just yield diimines. Nevertheless, the applicability and yield of the described ligands is mainly determined by the solubility of the precursors limiting their conversion and the solubility of the products, as isolation was performed by filtration of the respective precipitates.

Screening was performed using an 8-fold parallel autoclave reactor system developed by hte AG at 50 °C with a monomer pressure of 10 bar and 30 ml reaction volume under moisture and oxygen free conditions.

In total, 420 polymerisation experiments involving 302 different organometallic complexes were carried out during the study. The catalyst library consisted of 86 Fe(II)-, 87 Fe(III)-, 48 and 16 Cr(II)- and Cr(III)-,

31 Co(II)- and 31 Ni(II)- complexes.[7] Methylaluminoxane (MAO) was used as co-catalyst to the above library but also for additional conditioning of the autoclaves prior to the reaction run.

2.3.2 The structure-property relationship analysis. Focusing on iron as metal centre, the catalytic activity strongly depends on the structure of the ligand backbone. As complexes with 2,6-diformylpyridine ligand backbones A proved to be mostly unsuitable for the polymerisation of ethylene, using the respective 2,6-diacetylpyridine ligand led to the most active catalyst (Fe(II)B1) from the entire screening campaign bearing almost 140 kg $PE \cdot mmol^{-1}$ $Fe \cdot h^{-1}$ besides other promising candidates. It is noteworthy that the majority of Fe(II) complexes using the motives B and C turned out to be suitable catalysts approaching 100 kg $PE \cdot mmol^{-1}$ $Fe \cdot h^{-1}$. Interestingly, the pyrazine analogon K did not produce any active catalyst. Remarkably, amines with substituents in β position to the nitrogen were identified as promising building blocks with one exception. Unexpectedly, 5-bromo-1-aminonaphthalene led to Fe(III)C9 as one of the major competitors to lead finding Fe(II)B1 regarding the catalytic activity. Both complexes gave polymers with molecular weights above 20 kDa the latter even 74 kDa. The highest molecular weight was obtained for Fe(III)B5 with 110 kDa. This is in good agreement with the findings from the activity screening proving that 2,6-diacetylpyridine forms suitable ligands with β substituted amines to obtain iron catalysts highly active in the batch polymerisation of ethylene.

According to previous studies by Gibson *et al.*[145] and Brookhart *et al.*,[146] the molecular weight of the polymers (FOM) can be approximately correlated with the degree of steric congestion around the imino nitrogen atom (molecular descriptor) implying sterically less demanding substituents to result in a decrease of the molecular weight. Moreover, Gibson *et al.*[145] added that the polymerisation conditions (reaction descriptor) may also affect the polymer weight. Increasing amount of co-catalyst methylaluminoxane (MAO) as well as high polymerisation temperature favour the formation of low-molecular weight polymer while increasing polymerisation time will facilitate the formation of high-molecular weight polyethylene. This is well aligned to HTE's findings and the use of novel naphthylamine-based building blocks, short polimerisation runtimes and high amounts of applied co-catalyst.

The co-catalyst MAO activates the organometallic catalyst by substituting both chlorine atoms bonded to the metal centre with methyl moieties and drawing one methyl anion, forming a cationic metal complex with methyl moiety which allows ethylene insertion.[147] Beside the catalyst-activating function of the MAO, it also acts as water scavenger in the reactor through its coordinated trimethylaluminium.[148–150] The hydrolysis of this coordinated trimethylaluminium with water affords huge oligomeric MAOs which allow further stabilisation of the cationic active site.[7,147,151,152]

Nevertheless, based on the lead findings regarding molecular weights above 20 kDa, all candidates have ligands with at least one substituent in β positions to the nitrogen atom but the above mentioned bromo-substituted exception. This is in total agreement with Brookhart *et al.* and the fact that complexes prepared from ortho-substituted anilines give highly active oligomerisation catalysts to obtain α-olefins.

Furthermore, we managed to elucidate the relationship between the catalyst activity and the steric effect of the organic ligand based on Fe(II) as central metal. In general, complexes in which the ligand is rotationally hindered were observed to be more active than complexes with free rotating residual groups. For example, in a highly active complex with C1 ligand (Fig. 6 left and Fig. 7), the rotation of the aromatic residual groups is obstructed, thus the aromatic rings of the aniline units 1 are forced out of plane relative to the diketone backbone C, giving better substrate accessibility to the metal centre. On the other hand, a complex with A1 ligand (Fig. 6 right and Fig. 8) is less active due to its free rotating aromatic rings, resulting in less accessible metal centre.

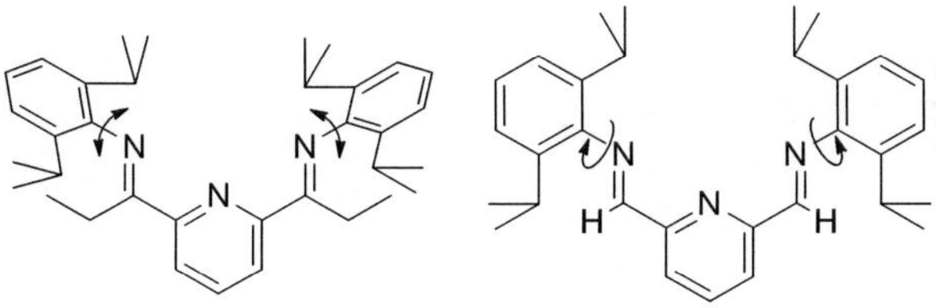

Fig. 6 Highly active ligand C1 with backbone group C and ligating groups 1 (left); Low active ligand A1 with backbone group A and ligating groups 1 (right).

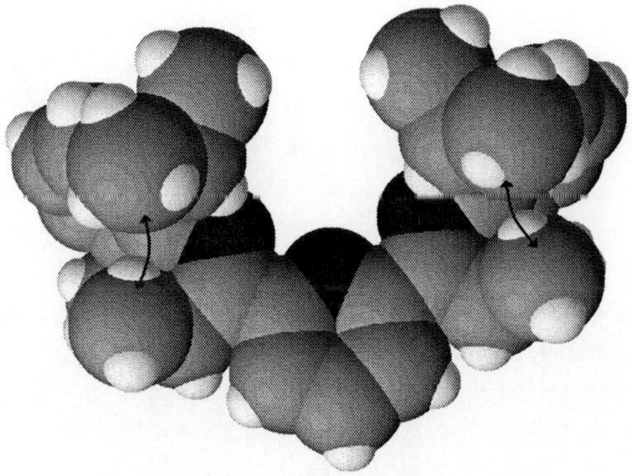

Fig. 7 3D balls and sticks model of highly active ligand C1.

Fig. 8 3D balls and sticks model of low active ligand A1.

The combination of less bulky, symmetrical amines with additional alkyl substituents on the aromatic ring and high alkyl-substituted diketones leads to a rigid tridentate ligand with rotation-hindered aromatic rings. The structure-property relationships analysis elucidates the criteria of an active catalyst and gives an insight in how to design a suitable catalyst for the ethylene polymerisation purpose. The systematic library allows an easy handling of the experimental data so that a clear relationship between the ligands and the experimental results can be derived.

Finally, the present case study proves the successful application of high throughput experimentation towards the synthesis and evaluation of complex catalyst libraries regarding the results well aligned to the literature. Using a simple synthetic but efficient workflow to combine structurally different building blocks leads to a large number of compositionally equivalent catalysts. The diverse libraries systematically generated thereof are almost predestinated for high throughput investigations to elucidate structure relationship properties.

3 Heterogeneous catalysis

3.1 High throughput screening in heterogeneous catalysis

The incentive use of HTE methods in the field of heterogeneous catalysis aims at optimising and/or improving the chances of discovery of new or improved catalysts and to reduce the time-to-market of catalysis technologies and process improvements. Thanks to modern high performance processors and advanced robotic technology, HTE has nowadays a wide application in the field of heterogeneous catalysis.[153–156] The aim of any screening effort is to manage and rank the materials within the respective libraries and track them down from a very large number of possible combinations to a few promising lead candidate systems in a reasonable time with as few as possible experiments. Not only suitable synthetic methods, but also advanced testing environments and characterisation technologies that allow for quality checks and the establishment of structure-property relationships are especially important in the scope of this process.

In general, the scheme for the development of novel catalysts and processes is similar in both heterogeneous and homogeneous catalysis (see Fig. 2). Firstly, promising potential candidates need to be identified based on given hypothesis. The candidates are then synthesised and potentially also analysed in high throughput characterisation and appropriate analytical methods. The 1st generation of candidate materials are prepared and exposed to the catalytic test in the test environment. As scalability for catalyst recipes plays a major role in heterogeneous catalyst synthesis and production it is advisable to have a focus on scalable unit operations for candidate materials synthesis. Fortunately, today the choice of proper equipment for synthetic and characterisation efforts is easy due to commercial availability; the same is true for screening equipment where a number of established providers make validated equipment available.[1,111,157–160] The experimental loop is terminated by the generation of the experimental data, preferably being processed in suitable data warehouses and workflows. It is evident that the feedback loop can only be put into action if the original hypothesis is validated against the achieved results. With the combined input from the actual test results and improved models the next step is to generate the next generation of candidates with optimised composition and experimental parameters. For each loop different features can be varied and optimised such as composition of the material, synthetic route, changes in secondary parameters like porosity or crystallinity and also not to forget the testing parameters. The statistical treatment of the results should be carefully handled due to the fact that the identification of hits requires both positive and negative results: the achievement of negative results cannot be undervalued! For critical pathway analysis it is vital that the greater number of approaches can be excluded in order to avoid so called "hit-explosion".

The differences with regard to homogeneous catalysis are evident: solid state materials represent complex components which are in many cases much harder characterised than molecular catalyst entities. Characterisation of the catalyst surface is even harder, but finally this is where the catalytic reaction takes place. This of course reduces the validation of hypothesis in many cases to a phenomenological point of view, which nevertheless may help in identifying appropriate technical solutions. In our review on approaches for DoE in heterogeneous HTE we will especially focus on approaches which try to go beyond the phenomenological "boundaries". It is debatable to whether or not the models employed in these approaches are to a minor or larger extent predictive. What is in our view important is that they allow a differentiation of every library also with regard to the diversity of the library created and therefore may be an important quality criterion for any approach as such. We will later come back to that point and try to elaborate a bit further.

3.2 Experimental design

There are a variety of methods and different mathematical tools for the Design of Experiments (DoE) used in HTE, examples in literature are random elements selection, statistical methods, genetic algorithms, descriptors and the quantitative structure-activity relationship (QSAR), which can be used for library design and data evaluation.[1,161] As we have already illustrated above "intelligent" design of libraries and efficient design methods which allow "knowledge/data extraction" become more important in HTE compared to conventional development work as much larger data sets have to be treated and the essence supporting or corrupting the experimental hypothesis has to be done in an efficient manner (Fig. 9).

The final target of a suitable Design of Experiment is to optimise information gain from experimentation with the minimal number of experiments needed.[162] Creation and development of catalyst libraries for specific target reactions face us with the question: "how should members of a library be selected?" and "which methodology allows the design of libraries in the most efficient manner?". If the boundaries of the search space are given by technical limitations or other restrictions, then well-known "design of experiment" tools can be used and are probably in many cases the most efficient workhorses for optimal mapping of the search space.[163] These tools of design of experiment (DoE) can be applied towards optimisation of

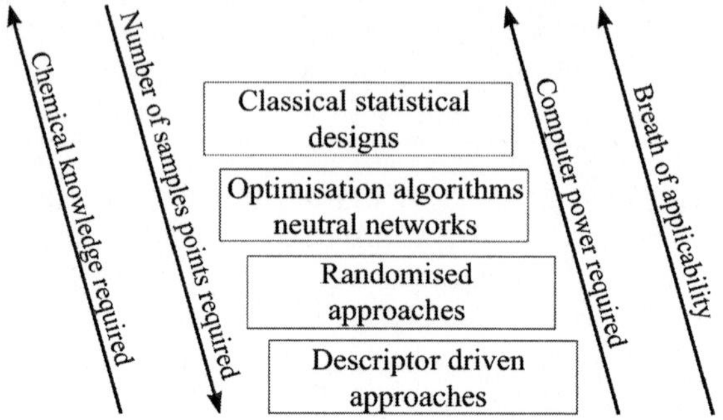

Fig. 9 Success through adequate experimental planning.

different objectives using high throughput experimentation always bearing in mind that the mechanics of these tools is based on algorithms utilising simple linear regression. DoE can be used to efficiently explore even large search spaces defined by discrete variables, while guaranteeing maximal efficiency in terms of the information gained from experiments. The methodology of DoE involves the simultaneous modification of variables (usually called factors) and the avoidance of redundant experiments, has found its way into chemical catalysis, especially for industrial applications, in the domain of process engineering.[1] By reason that, without an efficient informatics environment, it will be practically impossible to plan and design a larger number of experiments as they appear in HTE, tools of DoE are typically computer based.

DoEs are usually applied in order to quantify influences of different variables on an objective such as the catalytic performance represented by values activity, selectivity or life-time of a catalyst. Not many studies go deep on comparing complex methodologies in DoE. As an example using complex DoEs as assisting tool for designing libraries, the exemplary work of Farrusseng is recommended where the first library of diverse catalysts for CO oxidation and selective oxidation (selective CO oxidation in the presence of hydrogen) was tested to evaluate the impact of operating conditions and nature of the catalyst on performance.[162,164,165] An analysis of the effect induced by four selected factors (alkali addition: Li, Cs, Ca; type of support: Al_2O_3, CeO_2, ZrO_2, ZnO, C; nature of the transition metals: Cr, Co, Mn, La, Sm, Mo; and of the noble metals: Pt, Pd, Ru, Rh, Au) was carried out following different models from a dataset with the CO conversion as determining variable. An exhaustive way to present the results was to compute the effects when moving from one factor to another one, other variables being kept constant and assuming a linear relationship. In other words, the intention was to indicate in a quantitative manner which changes in CO conversion are induced by substituting one element by another one and are they significant or not. Summarising results of those studies,[162,164,165] it was established that adding any alkali/alkali earth improves activity, mainly for CO conversion.[164] Doping of Cs had the largest effect on activity with increases of the conversion of CO of about 12% in the absence of H_2. The highest impact compared to all other supports brought CeO_2, especially in the absence of H_2.[164] Using noble metals as catalytically active component had by far the largest effect on the response of the selective oxidation in the presence of hydrogen.[162] It is important to note that apart from the lead finding also general trends could be extracted, for catalyst activity such as the effect of temperature, of supports relation to catalyst activity ($CeO_2 > ZrO_2 > TiO_2$), the effect of the inclusion of metal oxides in the system (V > Nb > Mo) and the general trends for the active metals as such (Pt >> Cu > Au) were determined. It is important to notice that some trends did not always show linearity, especially in the presence of H_2, the conversion did not always increase with temperature. Curiously, for the ternary system Pt–V–Zr, raising the temperature leads first to an increase in activity followed by a decrease of activity for the oxidation reaction at higher temperatures.[162,165] Although not all effects were understood to an exhaustive level the results from the preliminary mapping could be used as guidelines and design criteria for the subsequent libraries.[164] We find that in this case study, it is well illustrated that suitable DoEs can minimise the number of experiments and help to identify governing factors within a given experimental design.

3.2.1 Factorial design of experiments. The reason why factorial designs are often used is that they are much more efficient than systematic studies where dependent or independent parameters are varied one at a time; factorial designs take into account interactions between factors and they are very efficient help to identify the most influencing factors aiding to optimise the studied system, saving time and redundant experiments. Many examples about the efficient use of factorial designs and high throughput experimentation are reported in the scientific literature.[12,166–175]

In order to get a better comprehension of the system to which the functional design should be applied to, a model (Fig. 10) is required.[176]

The output variations are analysed according to the change of the controllable factors. Uncontrollable factors involve for instance the human or other sources of errors. All factors need to be correlated with possible interactions between them and can be represented in the case of two controllable factors in a regression model representation in the form of:

$$\mathbf{y} = a_0 + a_1\mathbf{x_1} + a_2\mathbf{x_2} + a_{12}\mathbf{x_1}\mathbf{x_2} + \varepsilon \quad (2)$$

where $\mathbf{y}$ is the response, a_0 is the average of all the responses, a's are parameters which need to be determined numerically, $\mathbf{x_1}$ and $\mathbf{x_2}$ represent the effect of the factor 1 respectively factor 2 and ε is a random error. $\mathbf{x_1x_2}$

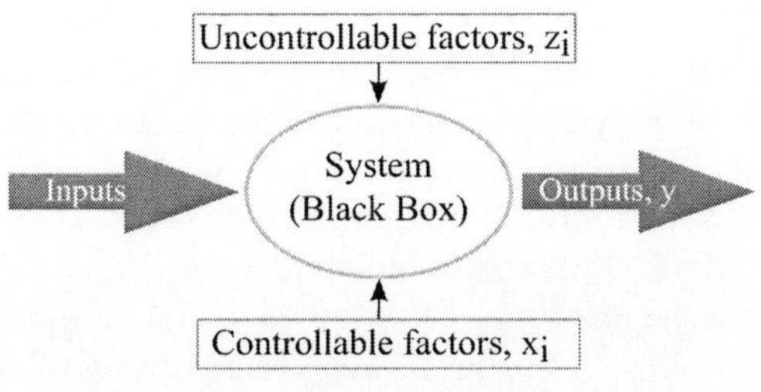

Fig. 10 Model of the process to be analysed.

represents the interaction between $\mathbf{x_1}$ and $\mathbf{x_2}$. A model with 2 factors and 2 levels is called 2^2 factorial design and can be easily generalised to 2^k factorial design of k factors with 2 levels.

When working with factorial designs one has to be aware of the fact that even though the specific design as such helps reducing the number of experimental points necessary to deduct the governing effects for a given system, the more independent variables are included in the design the more experimental points will have to be recorded prior to obtaining a final result. This has decisive impacts on the way HTE experiments should be conducted in a best mode with regard to factorial designs. Both the length of an experiment and the number of samples that can be exposed to experimental conditions within one experiment are usually limited by constraints of capacity with regard to sensible runtimes and the number of samples that can be accommodated in the testing unit. It is important to adapt the experimental design accordingly to these constraints and to optimise it with regard to its efficiency. Even if the length of an experiment and the number of samples have been taken into consideration it remains debatable and needs careful consideration when it comes to the application of factorial designs over a too large set of subsequent experiments. The advantage of the number of complex aspects that can be elucidated in such "hyper-designs" may easily be overruled by inaccuracies or nuisance factors that may arise during extended campaigns and will not be easily extractable at the end of a campaign. The other aspect that is lost is the factor of iterative "learning" and improvement of a system. Small and efficient factorial designs represent one of the most efficient tools in an "iterative learning" process, where findings of prior designs are used as fundamental basis for subsequent design building. We will come back to that aspect of "iterative learning" when we discuss genetic algorithms.

Factorial designs have been adapted to many applications in heterogeneous catalysis and other materials related studies.[1,166–175] Obviously 2^k factorial designs are in many cases not sufficient, as Farrusseng shows in his comprehensive review.[1] 3^k or 4^k factorial designs, mixed factorial designs (different level possibilities for instance 2 levels for x_1 and 4 for x_2) and even more complex designs have been utilised with success in different applications.[166,167] Symbolically, these examples shall be illustrated by designs comprising 2 or 3 factors as shown in Fig. 11. The different coordinates describe the parameter assignments for one dimension.

The great advantage of factorial designs is that they can take into account uncontrollable factors. The idea is to segregate the different uncontrollable factors each one in a block to get rid of so called "nuisance factors". Blocking a factorial plan is also performed for experiments which are usually conducted with different batches or different fractions of a given set of samples.[177–179] This technique is called the randomised complete block design (RCBD) and is one of the most widely used experimental designs.[166] Randomisation is used to avoid influence of a nuisance factor and complete means that each block undergoes the same tests.

A clear and concise example of a full factorial design was used in the study of Corthals *et al.*[169] in order to explore the chemical diversity space of catalysts for dry reforming at elevated pressure. The experimental design

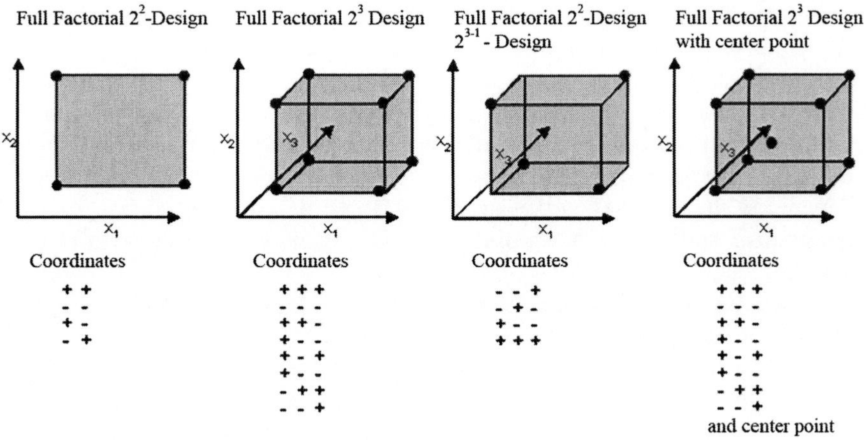

Fig. 11 Representation of common statistical designs.

was set up as a full factorial design and in order to find the best performing catalysts for a given set of reaction conditions. Firstly, Ni, Co and Pt were chosen as active metals on a defined set of supports with different loadings from 0.5 to 5 wt.% that were exposed to test conditions. The analysis of the design brought to bear that Ni/$MgAl_2O_4$ systems proved to be the most active systems at 650 °C, 7 bar, GHSV = 21 000 h^{-1} and CO_2:CH_4 ratio of 1.1. With these catalysts up to 66% methane conversion, 76% carbon dioxide conversion could be obtained and H_2:CO ratio of 0.73 were achieved. The study was further extended to useful promoters (La_2O_3, Y_2O_3, ZrO_2, MnO and BaO) and only La_2O_3, Y_2O_3 and ZrO_2 proved beneficial within the ranges examined in the design of experiment.

The strength of factorial designs lies in the fact that effects and interactions can be identified in a fast and efficient manner. The drawback of the designs though is that in a first approach it can at times be hard to deduct structure property relationships from the results obtained. It may in certain cases be advisable to use full coverage algorithms to explore an extended experimental space and a benefit may be that the response surfaces obtained can be often very elucidating with regard to "chemical" insight.

The simple approach consists in designing full coverage algorithms within the chemical compositional space for materials design, allowing within certain boundaries a large set of all possible compositions. Within the framework of these designs it is important to space experimental points evenly within the total experimental space to arrive at "full coverage". Schunk *et al.*[168] have examined the ternary system of Pd-K-Bi catalysts with full coverage algorithms for the three element system to identify the compositional corridors leading to catalysts with maximised product yields. Based on the study of the ternary Pd-K-Bi system where clear optima could be identified and structure-activity relationships could be deducted, subsequent promoter studies of the base system were also conducted with full coverage algorithms at constant K-loading within Pd-Bi-X chemical space. Here again it could be found for optimum composition of the materials for the target reaction leading to Pd-Bi-K-Pt catalysts that can deliver high yields of

1-acetoxy-2-butene and 1,4-diacetoxy-butadiene were identified within the compositional search domains investigated. From the data and simulation experiments, it can be concluded that Platinum is a vital element in enhancing catalyst reoxidation properties. We have to keep in mind although that the full coverage algorithm appears to be a powerful tool for the cases discussed here, the power of this methodology will also largely depend on the material classes under investigation and the proper choice of parameters, especially the point density, within the given experimental space.

The main reason for cautious planning is that discontinuities in the experimental space may become unexplored if the grid of experimental points is unsuitably chosen. In other words, it remains challenging how a continuous design of libraries can be adapted in a fashion so that discontinuous responses are not missed. Of course one has always the option to increase the density of experimental points but still a full design of experiment can obviously only be planned with regard to reasonable constraints set by timing and/or economic limitations. The open question remains on how to minimise the experimental effort without losing response match.

In another example, we tried a different approach to capture discontinuities in a compositional space caused mainly by structural variability. In our study on 190 complex mixed oxides for the gas phase ammoxidation catalysis, we achieved with the retrospective deconvolution of a complex multi-element lead system into simple binary and ternaries and could track the different components and the structural variances down to structure performance relationships.[5,12] The goal was to find new catalytical systems able to exhibit high activity for the ammoxidation of cyclohexane and benzene to muconitrile. In the first phase of the study the main focus was on element variation, from the lead structures discovered for the challenging conversion structure property relationships could be established and gave insight into the roles of different components in the complex mixed oxides. This led to a series of lead catalysts and structures useful in the ammoxidation of cyclohexane and benzene to muconitrile which could been categorised as based on the elements Mo-Sb-O, Mo-Sb-Ti-O and Mo-Sb-V-Ti-O types of oxides and their structural characteristics. The results of the study show that also retrospectively staying with simple "full coverage" algorithms a hit deconvolution of a multicomponent mixture can be worked up. The retrospective deconvolution can therefore be regarded as a viable way to generate hits and establish structure property relationships with sizable libraries and within more reasonable timeframes as global studies aiming at full structural deconvolution in the first case.

Another approach of dealing with the problem of structural variability is to reduce the number of structures under inspection drastically and only allow for elemental variation within a given structural diversity space. In a so called "focussed" study with mission to elaborate new approaches for the oxidation of acrolein and methacrolein to the corresponding unsaturated acids we implemented the aforesaid design successfully.[170] The focus was to find alternative catalyst materials already known and implemented in industry. Both families the heteropolyacids and the bronzes bear at least from a compositional point of view large similarities, with the bronzes being generally more useful as phosphorous poor components. It was decided to focus the derivatisation on the family of $Cs_{0.5}[Nb_{2.5}W_{2.5}O_{14}]$, Caesium containing bronzes that allow for a broad variation of the elemental composition. Even doping with phosphorous does not lead to a structural modification, if phosphorous contents are kept at a moderate level. The family of compounds therefore offered the unique opportunity to study solids of a formal composition of hetero-polyacids which displayed bronze-type structures. Within the given search space optima could be identified with regard to elemental composition of the bronze and phosphorous doping levels. We will later return to the question of fixing structural variability in our example of a case study.

We have illustrated that factorial designs have the major advantages to make HTE a more powerful and efficient tool, however their application may sometimes not be straightforward in case of highly complex systems depending on the number of parameters that need to be investigated. It has been pointed out that especially with regard to materials screening structural variability has to be taken into account and experimental grids have to be adjusted in order not to miss singularities of the system response. Nevertheless, the future looks bright for new algorithms and unconventional designs of experiments with advanced strategy and experimental tactics.[171] Nevertheless, the challenge of developing suitable experimental designs can only partly be realised by the use of experimental design software tools and statistical methods, *e.g.* organising experiments into blocks to reduce the influence of uncontrolled factors, using reproductions to understand statistical variations in the results and to reduce systematic errors.[180]

We do want to mention the approaches of Holeňa *et al.* and Végváry *et al.*[181–183] as well in the context of statistical methods of DoE. Both have proven that with their statistical design methods parameter spaces can also be screened in a very efficient manner. Although the group claims that the holographic design base represents a special case of experimental design methods, one has to be clear that also these design methodologies follow the lines of DoE methodologies for the "mapping" of parameter spaces. The algorithms used to modulate the specific experimental points within the design are of course unique, a feature with which the authors claim to increase efficiency compared to standard methodologies. It is of course questionable to whether or not the algorithms used in holographic screening are truly more efficient than more "traditional" approaches.

3.2.2 Genetic algorithms. The application of a combinatorial or HTE strategy with an inherent optimisation procedure based on genetic algorithms (GA), which imitate the process of "natural evolution", with the extension of artificial neural networks (ANN), trying to mimic the properties of biological neurons, is described as well suited for the field of discovery of optimal catalyst compositions within a limited number of experiments.[162–164] Many of the pioneering studies of genetic algorithms to catalyst development have their source in Manfred Baerns' group.[184–192] Potentially active catalyst components were used for defining a start library of catalyst formulations which were then further optimised by applying the evolutionary strategy using genetic algorithms for creating new "generations" of catalytic materials.[188]

It has to be made clear that the so called genetic algorithms are in principle optimisation algorithms that help to find maxima or minima within boundaries that are defined by the experimenter. The choice of boundary variables is as critical as for any DoE and may corrupt the success of the experiment, if not correctly arranged. If one considers use of the "evolutionary" technique, one should always outweigh pros and cons compared to conventional DoE methods, which for sure will require more interaction and interpretation with the experimenter in order to take the next step. The origin of these optimisation algorithms lies in fields were multicomponent variable systems have to be optimised and responses of the test system are obtained promptly or at least with very short delay times, a good example are production chains in large scale industries like the automotive industry.[1] It has to be made clear that the large difference compared to the fields where typically genetic algorithms have shown to bear value differ largely from the examples in literature in the field of heterogeneous catalysis: the number of data points and the number of generations of catalyst candidates on which the successful optimisations were achieved were usually much lower for the published examples in heterogeneous catalysis. Again this may illustrate that the choice of starting variables has a major impact on the success of the design as we will discuss later. In one of the first and most frequently cited in the literature of Baerns' papers, the oxidative dehydrogenation of propane was used as a model reaction and as a result of the preliminary optimisation procedure an increase in the propene yield (from 7.5% to 9.0%) was achieved with increasing number of generations (from 1st to 3rd generation).[184] The highest yield of propene of about 9.0% at 57.4% selectivity and a reaction mixture ratio of C_3H_8:O_2 = 3, in the 3rd generation on $V_{0.22}Mg_{0.47}Mo_{0.11}Ga_{0.20}O_x$ was reported.[184]

In the first step, the 1st generation of catalysts are prepared based on input variables which are randomly mixed as individual compounds in such way that each catalytic material consists, for example, of up to three of eight primary components.[184] After parallel testing of the 1st generation of the generated catalytic materials by applying high throughput testing the most active catalysts are chosen to create a 2nd and after their testing a 3rd generation, respectively. The genetic algorithm is used to set the compositions of the catalytic compounds of the 2nd and 3rd generation, again here depending on the input variables cross-overs between "winners" with good performance and "losers" with inferior performance can be made to increase diversity in the mix.[184,186,187]

General principle of genetic algorithms applicable in heterogeneous catalysis is illustrated in the Fig. 12. Genetic algorithms, like in biological systems, can be based on the operators "mutation" and "crossover" as well as selection of the fittest individuals (see Fig. 12(a) and (b) respectively). Applied to catalyst composition new generations are created from the best catalysts of the former generation by qualitative mutation (one or more chemical elements or compounds are replaced by others), quantitative mutation (the concentration of one or more compounds in the inorganic material is changed) and crossover (components are exchanged between two

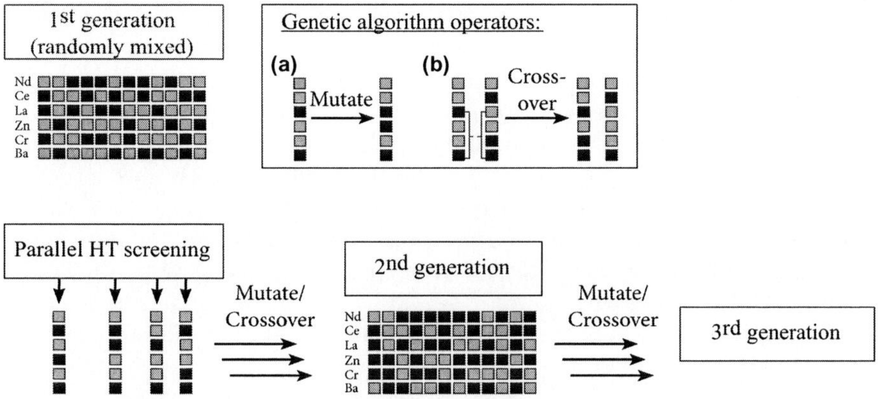

Fig. 12 General principle of genetic algorithm in heterogeneous catalysis.

materials of the former generation generating two new materials).[184] The "survival of the fittest" is reached by using a ranking selection, which gives a higher probability of selection to catalysts with better rank after high throughput testing of a generation of catalysts.[188] The algorithm also may be useful not only in catalyst composition studies but also in the optimisation of reaction condition as illustrated for the case study of propene dehydrogenation. In further studies, the GAs have been successfully applied for discovering new catalytic materials for the oxidative dehydrogenation of ethane to ethylene.[189,190]

As a result a continuous improvement of the catalytic performance from the first to the seventh generation was found. An improvement of the ethylene yield from 9 to 18% under standard conditions (500 °C; C_2H_6:O_2:Ar = 20:10:70) could be achieved.[189,190] By optimising the reaction conditions the yield can be further increased up to 21%.[190] In spite of the evident advantages of genetic algorithm for catalyst and reaction condition optimisation, just a small number of scientific groups have used this approach. The actual reason for the lack of popularity could be related to the absence of available or user-friendly software packages for materials screening.[1]

As shown on example of the water-gas shift (WGS) reaction,[164] from the 1st to the 7th generation an improvement of the best CO conversion from 40% to nearly 70% was observed, though the improvement tended to level out for the last generations. The maximum of thermodynamic equilibrium of CO conversion under used operating conditions (200 °C, 3% CO, 37% H_2, 14% CO_2, 23% H_2O, Ar balance, GSHV = 3000 h^{-1}) should be around 94% (calculated from free Gibbs energy minimisation). So, the goal to reach the thermodynamic equilibrium within the limits of the optimisation procedure, was not achieved.[164]

Several efforts have been made to extend the "chemical insight" of GAs by including the neural networks in the workup of the obtained results.[162,188,193,194] Artificial neural networks are used in this context for modelling the whole parameter space and of identification of compositional areas for materials where high product yields can be expected and to make "virtual" simulation of catalyst performance on the computer possible.[188] Integrating both approaches can have a synergistic effect, allowing one to find a catalyst with high performance more quickly than their independent application.[181] The group of Corma *et al.*[194,195] have proposed the use of an artificial hybrid algorithm consisting of a GA coupled with a neural network.[188,193,194] In this proposal, the soft computing algorithm can design the next catalyst generation in a more efficient way, employing knowledge of the parameter space, which was extracted from the Neural Network modelling of all previous experimental data (previous catalyst generations and the resulting performance data).

This hybrid system seems to be a useful tool for the "intelligent" discovery of new catalytic materials, since it has appropriate tools for high-dimensional optimisation while it can keep in memory the whole "history" of the search, reducing the probability of screening of materials with poor performance due to appropriate materials suggestions by the Neural Network (NN). New generation is still created by the application of crossover and mutation operators of GA but challenged by the NN prior to synthesis and test. GA-NN-hybrid optimisation performance is of course again largely dependent on the quality of the initial generation. Thus, to improve the behaviour of this soft-computing technique, the initial set of materials should be obtained with appropriate procedures that guarantee sufficient population diversity. This complex process consists of creating several random generations and carrying out a statistical population study in order to select the most diverse population (not only the survival of the best-performing catalysts, but also the survival of those that show both relative fitness and are sufficiently different from other good performers are chosen in order to maintain the population diversity). The diversity in the initial generation ensures that the optimisation process gains information for the whole search space.[195] Such GA-NN-hybrid model was applied for the library design of Ti catalysts for epoxidation of different linear olefins. The prediction performance achieved by the employed neural network was trained with data from three successive generations and it was determined that the ANN-model makes it possible to predict qualitatively the epoxidation performance of the different catalysts with a certain probability. It is feasible with this model to distinguish a priori between unfavourable and favourable and beneficial catalysts.[195] Application of this optimisation framework has allowed to identify a highly active and selective Ti-silicate catalyst for the epoxidation of cyclohexene, which can also be applied to the epoxidation of other olefins, especially propylene. Moreover, only three optimisation generations (3×37 samples) were necessary to obtain this new catalyst.[196]

Similar studies were successfully conducted applying of hybrid genetic algorithms in other fields like CO based chemistries (COox, Selox and WGS),[162,164,165,193] the oxidative dehydrogenation of propane by using non-linear benchmarks,[188] or the search for an optimal composition of a catalyst for the ammoxidation of propane,[197] the algorithms also found application for identifying optimal Cu:Zn:Al ratio in mixed oxide catalysts for methanol synthesis from Syngas,[198] and for the optimisation of the process variables of the TS-1-catalysed hydroxylation of benzene.[199]

3.2.3 *Ab initio* quantum chemical methods and calculations.

Ab initio quantum chemical methods are gaining more and more attention due to their ability to describe mechanisms in a molecular scale and to predict reactions. These *in silico* methods can be used prior to an experiment for design purposes in order to predict efficient catalysts or commonly afterwards to describe the mechanisms on the catalytic surface.[170,200–210] From the point of view of high throughput experimentation these approaches are very interesting and important tools that can aid in "designing" libraries of materials that may or may not show prospect in certain reactions. As we have illustrated in the prior chapters these methods are already heavily used in pharmaceutical research and in approaches to HTE in molecular catalysis. One may argue that of course the final goal of any molecular model is to get as close to "chemical reality" as possible. But this highly exact approach may in many cases be very difficult to achieve for the interactions of reactants of the surface of solid state materials and take a lot of computer power, sometimes even experimental data that have to be gained in advance in order to generate insight into reactivity of a chemical system are needed in order to validate such models. A number of groups have taken these highly exact approaches and their results show that for systems with reduced complexity like metal surfaces and small reactant molecules, brilliant insights can be gained. Two factors are currently limiting: first the complexity of certain types of surfaces like oxides where in a large number of cases the question of surface structures is hard to solve and secondly the size and type of reactants that react with the virtual cut-out of the catalyst surface. The examples of stringent studies where molecular modelling has been used successfully to design HTE libraries are sparse. With regard to highly efficient library planning it would be highly desirable to bring the world of molecular modelling closer to the needs of HTE.

Comprehensive models of surface reactions and models of catalytically active surfaces require treatment of smaller or bigger entities on a quantum mechanical level, the aim is to include as much of the electronic degrees of freedom as possible, however in general it will be useful either to limit the size of the clusters or surfaces modelled or to exclude certain aspects of the calculations in order to gain speed.[200]

Nowadays, two very efficient calculation approaches are used to model electronic structures based on density functional theory (DFT) or Hartree-Fock theory, both called first-principles or *ab initio* theories. On one hand, electronic structure calculations based on Hartree-Fock theory allow to solve directly the Schrödinger equation thanks to previous determination of the many-body wave function,[211] on the other hand, these DFT calculations are based on the electron density quantity,[212] and according to the Hohenberg-Kohn theorem,[213] the many-body quantum wave function is replaced by a function which depends only on three coordinates and has to be varied.[200] Local density (LDA) or generalised gradient approximations (GGA) are necessary to perform meaningful DFT calculations. This approximation helps to determine the last unknown, the exact functions for exchange and correlation.

The Hartree-Fock theory is typically used for small systems whereas DFT can be applied to larger systems. Recent advances in processor power and programs and codes based on *ab initio* electronic structure calculation allow now to perform more complex and time consuming calculations.[211] To understand the chemical reactions, the electronic problem for particular cases should be solved and in addition the potential energy regarding the position of the reactants, called potential energy surface (PES), should to be determined.[200]

With increasing efficiency these tools have found their entry into certain aspects of HTE as well. In our work we have used certain algorithms of *ab initio* methods with the aim of interpreting reactivity trends obtained for data sets in order to understand the governing reaction mechanisms,[168] and to compare reactivity between compounds.[170]

Other groups have gone further in the use of such algorithms: Curtarolo *et al.* predicted the structure of binary alloys with DFT calculations. The group has reported results of *ab initio* LDA/GGA computations for 80 binary alloys. The comparison with experimental data shows that only two simulations pinpoint significant discrepancy from the experiments.[201] Certainly the work of Rostrup-Nielsen, Sehested and Nørskov in the field of DFT calculations can be called ground-breaking.[202] Starting ground for the group are reactions of small molecules on metallic surfaces in order to better understand the mechanism of reforming and the activation of CH_4, CO_2, CO, H_2 and H_2O. Certainly, metals like nickel modelled as surfaces from parts of slices of single crystals are the most important approaches to be considered. In addition, coking mechanisms have been studied and it was found that graphite nucleation takes place at the step sites leading to graphene layers.[202,203] The promotion with sulfur, gold and potassium helps to block the step sites because these atoms bind preferentially to them. Jones *et al.*[204] examined with DFT calculations the scaling relationships for adsorption energy of C_2 hydrocarbons on transition metal surfaces. Here *in silico* methods were used to predict the experimental data. It was reported that the adsorption energies for C_2H_x type adsorbates on transition metal surfaces scale with each other according to a simple bond order conservation. The model was applied to the problem of dehydrogenation of ethane over transition metal catalysts and the comparison with full density functional theory calculations proved excellent agreement. The mean average error was only 0.13 eV which was interesting given that the accuracy of DFT is often only expected to be $\pm$0.2 eV. Such a prediction is a step further to the virtual chemical laboratory.

The first principles have proven their potential in understanding the mechanisms of reactions in most of the cases, allowing designing appropriate library for high throughput experimentation however the computing time and also sometimes the calculation are consequent and often experimentations are required to check the accuracy of the calculations.

3.2.4 Descriptors. The method of library and experimental design aided by descriptors can be regarded as an approach that works on the order of a principle component analysis of critical parameters that govern properties of solid materials. These critical parameters may be physical constants reported in literature or values calculated on given data sets, alternatively data may be extracted from experiments and included into the parameter sets. The approach is such that as many data as available for given sets of materials are collected and by comparing these extended data sets with the performance in one or more target reactions trying to extract structure property relationships. The approach therefore circumvents the effort of building a priori models in order to design libraries of materials but allows on a certain level to control the important factor of library diversity based on comparative studies within the libraries chosen. For such comparisons either all or only selected parameters such as atomic radius, electron affinity, normalised formation free enthalpy for the most stable metal oxide, the number of elements in the catalyst *etc.* can be selected as part of the descriptor vectors for comparison. The large difference with regard to other approaches is that descriptors allow comparison of library candidates on a quantitative basis based on the parameter sets recorded and arriving at quantitative descriptions of library diversity. It has to be mentioned here that the methodology as such at a certain point enters a "grey area" of overlap with regard to the QSAR methodology discussed next. For the sake of clarity, we want to discuss both approaches separately though. Schüth and Klanner can clearly claim to be the pioneers of the methodology to develop useful descriptor sets for solid catalysts in HTE.[163,214–217] Descriptors can contain information on elemental composition of the catalysts, on the synthesis procedure, chemicals used in the synthesis, and physicochemical parameters of the materials. On the order of 3000 possible parameters have been used in the studies mentioned. The descriptor vector utilised in the process of testing of various "libraries" of materials are exemplified in the studies of the oxidation of propene.[214–217] From the data sets obtained the catalytic behaviour of the materials tested could be classified in five classes: low, medium or high activity in total oxidation, partial oxidation, and oligomerisation reactions. The catalytic performances obtained for the candidate materials were correlated with the descriptor parameters by different techniques (*i.e.* neural networks or classification trees). For the correlations that could be established in the study the nature of the element ions, respective ionic radius, the expected coordination number, and ionic covalent parameter seem to be the most relevant variables. Inspecting these parameters, chemical intuition would suggest that they could indeed in some complex way correlate in the performance of a catalyst.

Many parameters describing features of the synthesis enter the descriptor, because experience tells, that the synthesis is crucial for the performance of a catalyst. It is evident that any correlation method in heterogeneous catalysis, which does not take into account the synthetic pathways used, will have a high probability to fail in obtaining reasonable prediction.[215,216] Of course it would be more desirable to extract physicochemical properties of the materials obtained *via* various synthetic routes, but the strength of the descriptor method is for sure that it also allows correlation of data on a lower entry level of chemical "understanding". A system like this can also be used for virtual screening of solid catalysts, *i.e.* catalysts can be randomly generated by use of software programs, then the descriptor vectors are calculated and performance of the materials is predicted based on prior experimental data. The prediction of the performance in propene oxidation based on such descriptors gave in various cases correct classifications which could later be proven experimentally. The hope is that this concept can be a first step of virtual screening not based on molecular modelling studies useful for catalyst-discovery programmes and based on the descriptors with a sufficiently broad database, expanded from the initially performed propene oxidation that could be generalised to alkene and alkane oxidation reactions on a more general basis.[215–217] The challenge remains to expand the software and database so that this task can be fulfilled and that the final goal of chemical knowledge extraction can be achieved.[214]

3.2.5 The QSAR method in heterogeneous catalysis. Originally Quantitative Structure-Activity Relationship (QSAR) (sometimes QSPR: Quantitative Structure-Property Relationship) is a methodology of modelling even on molecular basis to quantify relationships between chemical structure and biological activity. The approach is based on the assumption that similar molecules show similar reactivity and display similar activities in target assays – or target reactions in catalysis. The QSAR is up to date, a very modernised and systematised modelling approach. It was firstly successfully utilised in target screening against assays in agro chemistry, drug design, and toxicology, clearly the prime development push was molecular chemistry.[218,219] As mentioned in section 2.2, QSAR finds application in homogeneous catalysis, but is unfortunately at least up to date not a fully appropriate framework for discovery of materials, especially heterogeneous catalysts.[217] There are two reasons for this based on obstacles related to the complexity of heterogeneous catalysts: (i) the surface of complex solids (partially amorphous, microcrystalline domains, metal-oxide interfaces) can hardly be characterised and described at the atomic scale, presenting a serious obstacle to fingerprint encoding as in the case for molecules so severe that estimations have to be made, and (ii) key descriptors are therefore usually not known or can hardly be measured in the case of diverse libraries of materials.[217,220]

The development of a QSAR model requires the next three components: (i) a data set that provides experimental measure of an activity for a group of chemicals; (ii) molecular structure and/or property data (*i.e.* the descriptors, variables, or predictors) for this group of chemicals; and (iii) statistical methods, to find the relationship between these two data sets.[218] The limiting factor in the development of QSARs is often the availability of high quality experimental data. In the QSAR analysis, it is imperative that the input data have to be both accurate and precise to develop a meaningful model or sets of meaningful models. Often data used in QSAR evaluations are obtained either from literature or generated specifically for QSAR-type analyses.[221]

In continuation of the research based on the calculation of descriptors (see section 3.2.4) by Procelewska *et al.* an attempt to apply a QSAR model for solids was done.[220] The property-activity relationship (QPAR) approach consisted in the construction of a correlation between characteristics of heterogeneous catalysts tested in the study, encoded in a descriptor vector, and their experimentally measured performances in the propene oxidation reaction. Selection of adequate input variables was the key issue in the modelling process. The concept of using descriptors utilised like encodes of available and useful native physicochemical properties or calculated features of the solid catalysts as a numerical values (input variables) was performed. The descriptor vectors were represented as a set of all descriptors for a given catalyst. For each of the 467 catalysts submitted to the test, 3296 attributes were systematically computed. Numerous date-driven feature selection strategies were applied (mainly different types of neural networks) to establish multivariant relationships between the descriptor based on chemical codes of the catalysts and their experimentally determined catalytic behaviour. This allowed mapping the dependence of materials properties on their catalytic response. Thereby, a robust classification model, assigning selected attributes of solid compounds as input to an appropriate performance class in the model reaction was obtained. It has been evident that an enhanced mathematical support for the primary attributes set is greatly desirable.[220]

3.3 Case study

Within the case study of heterogeneous catalysis a systematic research comprising catalyst synthesis and testing in high throughput mode in order to classify dopant candidates useful for the performance improvement of the vanadium pyrophosphate catalyst system (VPO-catalyst) in the transformation of n-butane to maleic anhydride oxidation reaction was conducted. The study was carried out using 48-fold parallel reaction system built by hte and is based on our proprietary technology.[157]

3.3.1 Library design. In this study we tried to keep as many as possible parameters constant concerning synthesis, conditioning and testing of the catalyst candidates. The approach chosen from the point of view of the DoE was to utilise a maximum of dopant systems from the whole periodic table of elements in the sense of a "full coverage" algorithm. The solvothermal synthetic route was chosen as the main pathway to the base material, iso-butanol as solvent, the V:P-ratio was adjusted at 1:1. The general batch sizes for the base material were kept in between 50 and 250 grams, the same materials quality could be obtained in the catalytic test and the analytical parameters observed (mainly XRD (X-Ray Diffraction), N_2-sorption, DSC (Differential Scanning Calorimetry) and LOI (Limiting Oxygen Index)). After the solvothermal step, following product separation and a drying step was performed under reduced pressure at 50 °C in which the alcoholic solvent was evaporated. Generally a reduction in oxygen partial pressure and addition of steam is proven to be beneficial for the final performance of the undoped catalyst sample. Generally each a sample of each "masterbatch" of undoped VPO was tested in a catalytic test to evaluate the performance and ensure good quality control. After the thermal treatment step again a series of analytical tests as for the hemihydrate precursors was performed and samples not within the expected threshold for the analytical parameters were discarded. If the qualitative go-criteria for the samples were achieved, the powders were compacted and a particle fraction of 500 to 1000 μm was prepared for the next step. The application of the dopants was done *via* an impregnation technique. The dopant concentrations were adjusted to the specific surface area of the VPO-sample in order to achieve the following theoretical coverage of the dopant on the catalyst surface a) a full theoretical monolayer coverage, b) 1:10 of a theoretical monolayer coverage and c) 1:100 of a theoretical monolayer coverage. After an additional mild drying step at 80 °C for 12 hours the fines of the samples were again removed prior to the catalytic test.

3.3.2 High throughput screening. After the formation procedures the samples were tested in a 48-fold test reactor system at a sample volume of 1 ml. The typical test conditions involved programs around "standard" conditions of 2% n-butane in air at a GHSV of 1000 to 2000 h^{-1}, at temperature levels in between 380 and 420 °C. Test conditions with higher severity were also applied using conditions of 2% n-butane in air at a GHSV of 2000 to 4000 h^{-1}, at temperature levels in between 390 and 430 °C. Typically, during these programs conditions were repeated. As the "sample history" may have a large impact on the performance of a certain sample, here reproducibility could only be checked for the undoped standard samples and in between repeats of doped catalyst candidates. The data were evaluated fully automatic using the commercial software tool "myhte".

In order to check reproducibility a 2-fold strategy was followed: a) within a given set of catalyst candidates, which were tested in the test unit, a number of standard catalysts (typically 3–5) was randomly distributed. These standard catalysts were used as internal comparisons to check deviations within one run and as comparisons in between different runs, b) within every library of catalysts candidates 5% of the candidates were prepared and tested as duplicates. These duplicates were tested within one run to ensure maximum reproducibility of the synthetic method and the test procedure. Typical deviations observed within one run and as comparisons in between runs in activity were in the range of ±2% for the achievement of 80% conversion and ±1% for selectivity values obtained for given conversions (comparisons were made under standard test conditions). To furthermore evaluate and check for scalability of the effects observed for promising catalyst candidates, additional campaigns were run in typical laboratory screening reactor systems operating with catalyst volumes of 10 to 20 ml. From an overall perspective of the prospective candidates that the screening campaign brought to bear, it can be stated that the beneficial effects observed for the "hits" of the study could be scaled-up to the next level.

The effect of the dopants can be classified in a 4-fold matrix: a) dopants which decrease the activity and selectivity of the base catalyst, b) dopants which increase the selectivity but decrease the activity of the catalyst, c) dopants which increase the activity but decrease the selectivity of the catalyst, d) dopants which increase both activity and selectivity of the catalyst (Fig. 13).

From a general point of view it can be said that the number of dopants identified that had a positive effect (matrix b) and d)) under conditions of low severity (low GHSV and low n-butane contents) was generally higher than those which still showed a positive effect at high severity conditions namely high n-butane contents and high GHSV. The dopants summarised under a) and c) were considered to be of minor interest for further studies. The dopants under b) were pursued with low priority, dopants with beneficial effect falling under d) were considered of highest priority.

As a first important fact of the high relevance of the preparation of the samples and the test results obtained in high throughput mode were a

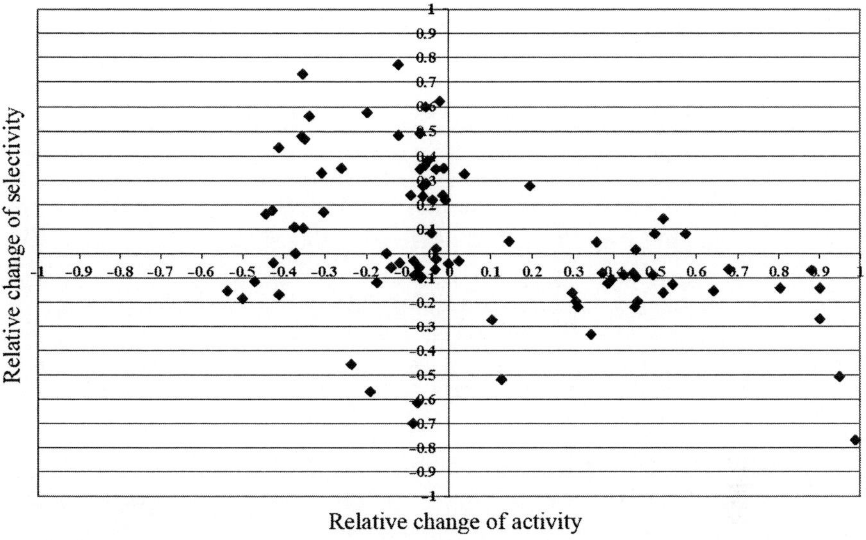

Fig. 13 4-fold matrix of the dopants effect at "standard" conditions (2% n-butane in air at a GHSV of $1000\,h^{-1}$, at 400 °C).

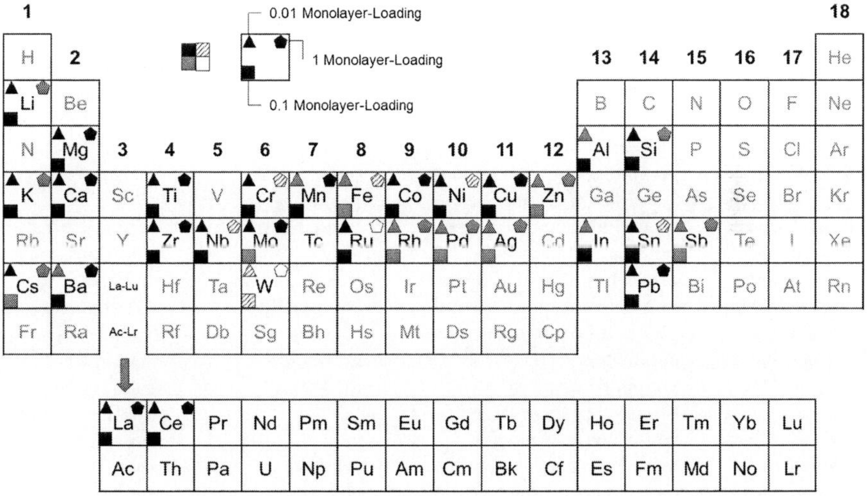

Fig. 14 Projection of the 4-fold matrix of the dopants effect at "standard" conditions on the periodic table of elements (2% n-butane in air at a GHSV of $1000\,h^{-1}$, at 400 °C). Monolayer-loadings of 0.01, 0.1 and 1 are marked as triangle, rectangle and pentagon, respectively. The different quadrants I, II, III and IV (Cartesian coordinates) are colour-coded in grey-striped, black, grey and white, respectively.

number of findings that were reported previously by other groups. HTE over a broad range of conditions with regard to variations in n-butane content, GHSV and temperature dopant elements like lithium, cobalt and cerium proved beneficial, especially at the relevant conversion values around 80%. For these elements all three dopant levels showed positive impact above the noise level of the test (Fig. 14). Selectivity improvements, but only for lower degrees of conversion could be obtained for samples doped with niobium and tungsten. Here dopant levels had to be on the

Table 1 Methodological overview.

	Problems for that the approach is most useful	Advantages of the approach	Potential challenges connected to the approach
Factorial Design of Experiments	Typical examples are: -Optimisation of reaction conditions for homogeneously and heterogeneously catalysed reactions. -Planning and evaluation of libraries where the definition of numeric "levels" is useful. -Optimisation of complex recipes for heterogeneous catalyst synthesis.	Allows easy discrimination of response parameters if boundary conditions are well chosen.	Large experimental campaigns can pose challenges to the methodology and should be divided into experimental designs that allow feedback with sensible response time.
Genetic Algorithms	Library design and candidate discrimination of large candidate libraries or optimisation of reaction conditions can efficiently be evaluated. The method allows easy changes of parameters even during campaigns.	Can efficiently handle large experimental spaces.	As the algorithm focuses on the identification of specific points of interest, additional effort has to be taken to allow for further insight.
(Molecular) Descriptors	Targeted library design and candidate discrimination with correlation of experimental results regarding structural properties or physical parameters of the candidates. Simplified diversity check.	Allows easier setup of campaign compared to *ab initio* molecular modelling.	Choice of descriptors is critical and may have to be adapted in hindsight.
Quantitative Structure Activity Relationship (QSAR) Modelling	Targeted library design and candidate discrimination with correlation of experimental results regarding structural properties or physical parameters of the candidates. Simplified diversity check. Modelling aspect strengthens the recursive character.	Allows easier setup of campaign compared to *ab initio* molecular modelling.	Choice of boundary conditions and structural variables to be evaluated is mission critical. Quantitative aspect must not be overrated.
***Ab initio* Quantum Chemical Methods and Calculations**	Allows high level of insight into the molecular aspects of the catalyst "in action". If correctly applied models can be efficiently verified and predictions for new potential candidates can be made.	High level of insight on a molecular level. Effects of different catalyst candidate structures can be rationalised.	Large computational effort. Needs significant computational head time prior to verification in the laboratory.
Virtual Libraries	Allows large savings in screening and laboratory time. Very efficient way of predicting the diversity space covered by a catalyst library. High level of insight on a molecular level already achieved "*in silico*".	Potentially cost saving. Facile prediction if experimental space is well covered.	Large computational effort. Needs significant computational head time prior to verification in the laboratory.

order of 1:10 of a monolayer at maximum. For both higher degrees of conversion and for samples treated with higher dopant levels the positive effect did not materialise any more.

Two dopants which were not reported in the literature and also showed promising improvements of the catalyst behaviour under high severity conditions were tin and lead. Both dopants proved to be beneficial at all three dopant levels studied and did not show performance degradation even in test runs over 2 weeks. Both dopants, tin and lead, offer an interesting perspective for further developments as they are easy to handle from a chemical point of view and have a rich chemistry for variations in further preparative studies and both have little tendency to form mobile species that can evaporate over the gas phase.

4 Conclusion

In this critical overview we intended to illustrate that design of experiment is critical for HTE in general and may be decisive for the success of any experimental campaign. The choice of tools is immense. Some have matured to a state where they show a large scope of applicability, others still need some fundamental improvements before a broader user group can share their benefits (Table 1).

With our case studies for HTE-workflows adapted for heterogeneous catalysis and homogeneous catalysis we intended to illustrate that even for complex workflows simple designs may propel the effort. Our heterogeneous catalysis study has proven to be a tool of high value in the screening of both 1) potential new dopants for the VPO catalyst system and 2) the optimisation of the complex catalyst synthesis procedure. Moreover, from an overall perspective the workflow developed for the high throughput study, including synthesis, modification and testing of the candidate materials, proved to be highly reproducible. With regard to the screening of dopants, new and known dopant systems could be confirmed and serve as a reference for the meaningfulness of the developed workflow. A classification of the dopant systems was conducted and two new prospective dopant candidates, namely tin and lead could be identified.

This work also illustrates that in case of homogeneous catalysis the synthesis and evaluation of libraries of complex molecular catalysts can be accomplished using high throughput experimentation. The outcomes obtained relate well to results known from literature and prove that combinatorial chemistry is a valuable additional tool for catalyst development.

Acknowledgement

The authors wish to thank the following co-workers for their continued and enthusiastic efforts; T. Gabriel, O. Laus, S. Pautler, H. Kaiser, K. Spielhaupter, A. Strasser, N. Brem, S.K. Weber and P. Kolb all based at hte Aktiengesellschaft Heidelberg.

References

1 D. Farrusseng, *Surf. Sci. Rep.*, 2008, **63**, 487.
2 H. W. Turner, A. F. Volpe Jr. and W. H. Weinberg, *Surf. Sci.*, 2009, **603**, 1763.

3 W. H. Weinberg and H. W. Turner, *High-Throughput Screening in Heterogeneous Catalysis*, ed. A. Hagemeyer, P. Strasser and A. F. Volpe Jr., WILEY-VCH Verlag GmbH & Co. KGaA, Weinheim, 2004, 1.
4 A. Hagemeyer, P. Strasser and A. F. Volpe Jr., *High-Throughput Screening in Heterogeneous Catalysis*, ed. A. Hagemeyer, P. Strasser and A. F. Volpe Jr., WILEY-VCH Verlag GmbH & Co. KGaA, Weinheim, 2004, VII.
5 N. Adams and U. S. Schubert, *Macromol. Rapid Commun.*, 2004, **25**, 48.
6 M. Saupe, R. Födisch, A. Sundermann, S. A. Schunk and K.-E. Finger, *QSAR Comb. Sci.*, 2005, **24**, 66.
7 P. Kolb, D. Demuth, J. M. Newsam, M. A. Smith, A. Sundermann, S. A. Schunk, S. Bettonville, J. Breulet and P. Francois, *Macromol. Rapid Commun.*, 2004, **25**, 280.
8 W. F. Maier, K. Stöwe and S. Sieg, *Angew. Chem., Int. Ed.*, 2007, **46**, 6016.
9 F. Schüth and D. Demuth, *Chem. Ing. Tech.*, 2006, **78**, 851.
10 R. Potyrailo, K. Rajan, K. Stoewe, I. Takeuchi, B. Chisholm and H. Lam, *ACS Comb. Sci.*, 2011, **13**, 579.
11 J. G. de Vries and A. H. M. de Vries, *Eur. J. Org. Chem.*, 2003, **5**, 799.
12 S. A. Schunk, A. Sundermann and H. Hibst, *Comb. Chem. High Throughput Screening*, 2007, **10**, 51.
13 J. N. Cawse, *Experimental Design for Combinatorial and High Throughput Materials Development*, ed. J. N. Cawse, John Wiley & Sons, Inc., Hoboken, 2003.
14 M. Cik and M. R. Jurzak, *Comprehensive Medicinal Chemistry II*, ed. J. B. Taylor and D. J. Triggle, Elsevier, Oxford, 2007, 679.
15 M. T. Reetz, *Comprehensive Coordination Chemistry II*, ed. J. A. McCleverty and T. J. Meyer, Elsevier, Oxford, 2003, 509.
16 E. Burello and G. Rothenberg, *Int. J. Mol. Sci.*, 2006, **7**, 375.
17 S. R. Gilbertson and X. Wang, *Tetrahedron Lett.*, 1996, **37**, 6475.
18 K. D. Shimizu, M. L. Snapper and A. H. Hoveyda, *Chem. Eur. J.*, 1998, **4**, 1885.
19 R. H. Crabtree, *Chem. Commun.*, 1999, 1611.
20 V. Murphy, H. W. Turner and T. Weskamp, *Applied Homogeneous Catalysis with Organometallic Compounds*, ed. B. Cornils and W. A. Herrmann, WILEY-VCH Verlag GmbH & Co. KGaA, Weinheim, 2002, 740.
21 W. Downey, C. Liu and J. Hartigan, *Drug Discovery World*, 2010, retrieved from http://www.ddw-online.com/s/enabling-technologies/p92823/compound-profiling:-size-impact-on-primary-screening-librariesspring-10.html (last accessed June 2012).
22 J. Flarakos, K. L. Morand and P. Vouros, *Anal. Chem.*, 2005, **77**, 1345.
23 M. S. Sigman and E. N. Jacobsen, *J. Am. Chem. Soc.*, 1998, **120**, 4901.
24 R. B. Merrifield, *J. Am. Chem. Soc.*, 1963, **85**, 2149.
25 J. Nielsen, *Curr. Opin. Chem. Biol.*, 2002, **6**, 297.
26 S. A. Poulsen and L. F. Bornaghi, *Bioorg. Med. Chem.*, 2006, **14**, 3275.
27 H. B. Kagan, *J. Organomet. Chem.*, 1998, **567**, 3.
28 R. V. Belosludov, S. Sakahara, K. Yajima, S. Takami, M. Kubo and A. Miyamoto, *Appl. Surf. Sci.*, 2002, **189**, 245.
29 J. C. H. M. Wijkmans and R. P. Beckett, *Drug Discovery Today*, 2002, **7**, 126.
30 T. R. Boussie, C. Coutard, H. Turner, V. Murphy and T. S. Powers, *Angew. Chem., Int. Ed.*, 1998, **37**, 3272.
31 D. J. Jones, V. C. Gibson, S. M. Green and P. J. Maddox, *Chem. Commun.*, 2002, 1038.
32 J. P. Stambuli and J. F. Hartwig, *Curr. Opin. Chem. Biol.*, 2003, **7**, 420.
33 S. R. Gilbertson and X. Wang, *Tetrahedron*, 1999, **55**, 11609.

34 B. C. Goess, R. N. Hannoush, L. K. Chan, T. Kirchhausen and M. D. Shair, *J. Am. Chem. Soc.*, 2006, **128**, 5391.
35 R. Shukla, Y. Sasaki, V. Krchnak and B. D. Smith, *J. Comb. Chem.*, 2004, **6**, 703.
36 J. J. Hwang and J. M. Tour, *Tetrahedron*, 2002, **58**, 10387.
37 K. C. Nicolaou, J. A. Pfefferkorn, H. J. Mitchell, A. J. Roecker, S. Barluenga, G. Q. Cao, R. L. Affleck and J. E. Lillig, *J. Am. Chem. Soc.*, 2000, **122**, 9954.
38 L. L. Martin, B. R. Sculimbrene, M. Nitz and B. Imperiali, *QSAR Comb. Sci.*, 2005, **24**, 1149.
39 R. A. Simon, L. Schuresko, N. Dendukuri, E. Goers, B. Murphy and R. S. Lokey, *J. Comb. Chem.*, 2005, **7**, 697.
40 X. Y. Xiao and M. P. Nova, *Combinatorial Chemistry: Synthesis and Application*, ed. S. R. Wilson and A. W. Czarnik, John Wiley and Sons, Inc., USA, 1997, 135.
41 D. L. Coffen and J. E. A. Luithle, *Handbook of Combinatorial Chemistry. Drugs, Catalysts, Materials*, ed. K. C. Nicolaou, R. Hanko and W. Hartwig, WILEY-VCH Verlag GmbH & Co. KGaA, Weinheim, 2002, 19.
42 M. T. Reetz, *Angew. Chem., Int. Ed.*, 2002, **41**, 1335.
43 C. Schmuck and P. Wich, *New J. Chem.*, 2006, **30**, 1377.
44 V. Murphy, *Comprehensive Organometallic Chemistry III, Volumes 1–13*, ed. R. H. Crabtree, D. Mingos and P. Michael, Elsevier, Oxford, 2007, 341.
45 M. Winter, K. H. Wiesmüller and G. Jung, *Cobinatorial Chemistry: Synthesis, Analysis, Screening*, ed. G. Jung, WILEY-VCH Verlag GmbH & Co. KGaA, Weinheim, 2001, 543.
46 C. Zechel, *Combinatorial Chemistry: From Theory to Application (Method and Principles in Medicinal Chemistry)*, ed. W. Bannwarth, B. Hinzen, R. Mannhold, H. Kubiryi and G. Folkers, WILEY-VCH Verlag GmbH & Co. KGaA, Weinheim, 2006, 519.
47 C. Zechel, *Combinatorial Chemistry: A Practical Approach*, ed. W. Bannwarth and E. Felder, WILEY-VCH Verlag GmbH & Co. KGaA, Weinheim, 2000, 243.
48 A. H. Hoveyda and K. E. Murphy, *Compr. Asymmetric Catal., Suppl.*, 2004, **1**, 171.
49 A. H. Hoveyda, *Handbook of Combinatorial Chemistry*, ed. K. C. Nicolaou, R. Hanko and W. Hartwig, WILEY-VCH Verlag GmbH & Co. KGaA, Weinheim, 2002, 991.
50 M. L. Snapper and A. H. Hoveyda, *Comb. Chem.*, 2000, 433.
51 K. D. Shimizu, M. L. Snapper and A. H. Hoveyda, *Comprehensive Asymmetric Catalysis I-III*, ed. E. N. Jacobsen, A. Pfaltz and H. Yamamoto, Springer-Verlag, Berlin, 1999, 1389.
52 A. H. Hoveyda, *Chem. Biol.*, 1999, **6**, R305.
53 A. H. Hoveyda, *Chem. Biol.*, 1998, **5**, R187.
54 K. Ding, D. Haifeng, Y. Yuan and J. Long, *Chem. Eur. J.*, 2004, **10**, 2872.
55 C. Gennari and U. Piatulli, *Chem. Rev.*, 2003, **103**, 3071.
56 J. F. Traverse and M. L. Snapper, *Drug Discovery Today*, 2002, **7**, 1002.
57 M. T. Reetz, *Angew. Chem., Int. Ed.*, 2008, **47**, 2556.
58 M. T. Reetz, *Angew. Chem., Int. Ed.*, 2001, **40**, 284.
59 A. Merritt, *Rodd's Chemistry of Carbon Compounds Volume V*, ed. M. Sainsbury, Elsevier Science B.V., Amsterdam, 2001, 259.
60 K. Burgess and A. M. Porte, *Adv. Catal. Processes*, 1997, **2**, 69.
61 W. H. Weinberg, B. Jandeleit, K. Self and H. W. Turner, *Curr. Opin. Solid State Mater. Sci.*, 1998, **3**, 104.
62 V. Murphy, X. Bei, T. R. Boussie, O. Brümmer, G. M. Diamond, C. Goh, K. A. Hall, A. M. Lapointe, M. Leclerc, J. M. Longmire, J. A. W. Shoemaker, H. W. Turner and H. Weinberg, *Chem. Rec.*, 2002, **2**, 278.

63 V. Murphy, A. F. Volpe and W. H. Weinberg, *Curr. Opin. Chem. Biol.*, 2003, **7**, 427.
64 B. Kappler, A. Tuchbreiter, D. Faller, P. Liebetraut, W. Horbelt, J. Timmer, J. Honerkamp and R. Mülhaupt, *Polymer*, 2003, **44**, 6179.
65 A. Tuchbreiter, J. Marquardt, J. Zimmermann, P. Walter and R. J. Mülhaupt, *J. Comb. Chem.*, 2001, **3**, 598.
66 S. Garbacia, R. Touzani and O. J. Lavastre, *J. Comb. Chem.*, 2004, **6**, 297.
67 O. Lavastre, R. Touzani and S. Garbacia, *Adv. Synth. Catal.*, 2003, **345**, 974.
68 D. G. Blackmond, T. Rosner and A. Pfaltz, *Org. Process Res. Dev.*, 1999, **3**, 275.
69 D. G. Blackmond, *Angew. Chem., Int. Ed.*, 2005, **44**, 4302.
70 Y. Zhang, X. Gong, H. Zhang, R. C. Larock and E. S. J. Yeung, *J. Comb. Chem.*, 2000, **2**, 450.
71 R. A. van Delden and B. L. Feringa, *Angew. Chem., Int. Ed.*, 2001, **40**, 3198.
72 J. Guo, J. Wu, G. Siuzdak and M. G. Finn, *Angew. Chem., Int. Ed.*, 1999, **38**, 1755.
73 C. M. Stafford, S. Guo, C. Harrison and M. Y. Chiang, *Rev. Sci. Instrum.*, 2005, **76**, 062207.
74 W. Schrader, D. Eipper, J. Pugh and M. T. Reetz, *Can. J. Chem.*, 2002, **80**, 626.
75 A. Horeau and A. Nouaille, *Tetrahedron Lett.*, 1990, **31**, 2707.
76 M. T. Reetz, M. H. Becker, H.-W. Klein and D. Stockigt, *Angew. Chem., Int. Ed.*, 1999, **38**, 1758.
77 M. T. Reetz, M. H. Becker, K. M. Kühling and A. Holzwarth, *Angew. Chem., Int. Ed.*, 1998, **37**, 2647.
78 M. T. Reetz, M. H. Becker, M. Liebl and A. Fürstner, *Angew. Chem., Int. Ed.*, 2000, **39**, 1236.
79 M. T. Reetz, K. M. Kuhling, A. Deege, H. Hinrichs and D. Belder, *Angew. Chem., Int. Ed.*, 2000, **39**, 3891.
80 M. T. Reetz, A. Eipper, P. Tielmann and R. Mynott, *Adv. Synth. Catal.*, 2002, **344**, 1008.
81 M. T. Reetz, P. Tielmann, A. Eipper, A. Ross and G. Schlotterbeck, *Chem. Commun.*, 2004, 1366.
82 M. T. Reetz, K. M. Kühling, S. Wilensek, H. Husmann, U. W. Häusig and M. Hermes, *Catal. Today*, 2001, **67**, 389.
83 N. Millot, P. Borman, M. S. Anson, I. B. Campbell, S. J. F. Macdonald and M. Mahmoudian, *Org. Process Res. Dev.*, 2002, **6**, 463.
84 K. Ding, A. Ishii and K. Mikami, *Angew. Chem., Int. Ed.*, 1999, **38**, 497.
85 P. Abato and C. T. Seto, *J. Am. Chem. Soc.*, 2001, **123**, 9206.
86 S. Dey, K. R. Karukurichi, W. Shen and D. B. Berkowitz, *J. Am. Chem. Soc.*, 2005, **127**, 8610.
87 F. Taran, C. Gauchet, B. Mohar, S. Meunier, A. Valleix, P. Y. Renard, C. Creminon, J. Grassi, A. Wagner and C. Mioskowski, *Angew. Chem., Int. Ed.*, 2002, **41**, 124.
88 L. Zhu and E. V. Anslyn, *J. Am. Chem. Soc.*, 2004, **126**, 3676.
89 D. G. Petra, J. N. H. Reek, P. C. J. Kramer, H. E. Schoemaker and P. W. N. M. van Leeuwn, *Chem. Commun.*, 2000, 683.
90 S. J. Taylor and J. P. Morken, *Science*, 1998, **280**, 267.
91 M. A. Evans and J. P. Morken, *J. Am. Chem. Soc.*, 2002, **124**, 9020.
92 A. Safir, M. Petro, R. B. Nielson, T. S. Lee and J. M. J. Frechet, *US 6,475,391*, November 5, 2002. (See also, U.S. 6406632, 6,416,663 and 6,260,407).
93 T. R. Boussie and M. Devenney, *EP 1,160,262*, May 10, 2001.
94 R. A. Potyrailo, R. J. Wroczynski, J. P. Lemmon, W. P. Flanagan and O. P. Siclovan, *J. Comb. Chem.*, 2003, **5**, 8.

95 A. Leugers, D. R. Neithamer, L. S. Sun, J. E. Hetzner, S. Hilty, S. Hong, M. Krause and K. Beyerlein, *J. Comb. Chem.*, 2003, **5**, 238.
96 M. F. Drenski, E. Mignard, A. M. Alina and W. F. Reed, *J. Comb. Chem.*, 2004, **6**, 710.
97 D. A. Hajduk, E. Carlson and R. Srinivasan, *US 6,484,567*, November 6, 2002.
98 M. B. Kossuth, D. A. Hajduk, C. Freitag and J. Varni, *Macromol. Rapid Commun.*, 2004, **25**, 243.
99 D. A. Hajduk, E. Carlson, J. C. Freitag and O. Kolosov, *US 6,664,067*, December 16, 2003.
100 K. Schneider, N. E. Zafeiropoulos, L. Häußler and M. Stamm, *Macromol. Rapid Commun.*, 2004, **25**, 355.
101 T. Hou and D. Raftery, *Analysis and Purification Methods in Combinatorial Chemistry*, ed. B. Yan, John Wiley and Sons, Inc., Hoboken, 2004, 139.
102 B. Yan, *Analytical Methods in Combinatorial Chemistry,* Technomic Publishing Company, Inc., Lancaster, 2000, 55.
103 A. Hagemeyer, B. Jandeleit, Y. M. Liu, D. M. Poojary, H. W. Turner, A. F. Volpe and W. H. Weinberg, *Appl. Catal. A*, 2001, **221**, 23.
104 S. Dahmen and S. Bräse, *Synthesis*, 2001, **10**, 1431.
105 C. Hinderling and P. Chen, *Angew. Chem., Int. Ed.*, 1999, **38**, 2253.
106 A. Behr, G. Henze, L. Johnen and S. Reyer, *J. Mol. Catal. A: Chem.*, 2008, **287**, 95.
107 G. R. Marshall and C. M. Taylor, *Comprehensive Medicinal Chemistry II*, ed. D. Triggle and J. Taylor, Elsevier, Oxford, 2007, **4**, 13.
108 R. S. Bohacek, C. McMartin and W. C. Guida, *Med. Res. Rev.*, 1996, **16**, 3.
109 M. B. Hall, P. Margl, G. Naray-Szabo, V. L. Schramm, D. G. Truhlar, R. A. van Santen, A. Warshel and J. L. Whitten, *Transition State Modeling for Catalysis*, ed. D. G. Truhlar and K. Morokuma, American Chemical Society, Washington, D.C., 1999, 1, 3 and 6.
110 M. R. A. Blomberg and P. E. M. Siegbahn, *Transition State Modeling for Catalysis*, ed. D. G. Truhlar and K. Morokuma, American Chemical Society, Washington, D.C., 1999, **4**, 49.
111 A. Corma, J. M. Serra, P. Serna and M. Moliner, *J. Catal.*, 2005, **232**, 335.
112 G. Rothenberg, *Catal. Today*, 2008, **137**, 2.
113 E. Burello and G. Rothenberg, *Adv. Synth. Catal.*, 2005, **347**, 1969.
114 http://cool.conservation-us.org/byauth/burke/solpar/solpar2.html (last accessed April 2012)
115 J. L. M. Abboud and R. Notario, *Pure Appl. Chem.*, 1999, **71**, 645.
116 E. Burello and G. Rothenberg, *Adv. Synth. Catal.*, 2003, **345**, 1334.
117 L. Xue and J. Bajorath, *J. Chem. Inf. Comput. Sci.*, 2002, **42**, 757.
118 A. Bocker, G. Schneider and A. Teekentrup, *QSAR Comb. Sci.*, 2004, **23**, 207.
119 J. W. Godden and J. Bajorath, *J. Chem. Inf. Comput. Sci.*, 2002, **42**, 87.
120 L. Xue, J. W. Godden, F. L. Stahura and J. Bajorath, *J. Chem. Inf. Comput. Sci.*, 2003, **43**, 1218.
121 D. C. Whitley, M. G. Ford and D. J. Livingstone, *J. Chem. Inf. Comput. Sci.*, 2000, **40**, 1160.
122 P. Blower, M. Fligner, J. Verducci and J. Bjoraker, *J. Chem. Inf. Comput. Sci.*, 2002, **42**, 393.
123 J. M. Sutter, S. L. Dixon and P. C. Jurs, *J. Chem. Inf. Comput. Sci.*, 1995, **35**, 77.
124 A. Yasri and D. Hartsough, *J. Chem. Inf. Comput. Sci.*, 2001, **41**, 1218.
125 B. T. Luke, *J. Chem. Inf. Comput. Sci.*, 1994, **34**, 1279.
126 H. Gao, M. S. Lajiness and J. Van Drie, *J. Mol. Graphics Modell.*, 2002, **20**, 259.
127 E. Burello, P. Marion, J. C. Galland, A. Chamard and G. Rothenberg, *Adv. Synth. Catal.*, 2005, **347**, 803.

128 A. G. Maldonado, J. A. Hageman, S. Mastroianni and G. Rothenberg, *Adv. Synth. Catal.*, 2009, **351**, 387.
129 G. Rothenberg, J. A. Hageman, F. Clerc, H. W. Frühauf and J. A. Westerhuis, *Catalysis of Organic Reactions*, ed. S. R. Schmidt, CRC Press, Boca Raton, 2006, 261.
130 G. A. Landrum, J. E. Penzotti and S. Putta, *Meas. Sci. Technol.*, 2005, **16**, 270.
131 T. R. Boussie, G. M. Diamond, C. Goh, K. A. Hall, A. M. LaPointe, M. Leclerc, C. Lund, V. Murphy, J. A. W. Shoemaker, U. Tracht, H. Turner, J. Zhang, T. Uno, R. K. Rosen and J. C. Stevens, *J. Am. Chem. Soc.*, 2003, **125**, 4306.
132 http://chemometrics.se/index.php?option = com_content&task = view&id = 35& Itemid = 27 (last accessed March 2012).
133 J. B. van der Linden, E. J. Ras, S. M. Hooijschuur, G. M. Klaus, N. T. Luchters, P. Dani, G. Verspui, A. A. Smith, E. W. P. Damen, B. McKay and M. Hoogenraad, *QSAR Comb. Sci.*, 2005, **24**, 94.
134 R. Carlson and H. Gautun, *Chemom. Intell. Lab. Syst.*, 2005, **78**, 113.
135 W. Buntine, *Stat. Comput.*, 1992, **2**, 63.
136 A. J. Smola and B. Scholkopf, *Stat. Comput.*, 2004, **14**, 199.
137 J. M. Serra, A. Corma, A. Chica, E. Argente and V. Botti, *Catal. Today*, 2003, **81**, 393.
138 E. Burello, D. Farrusseng and G. Rothenberg, *Adv. Synth. Catal.*, 2004, **346**, 1844.
139 T. R. Cundari, J. Deng, H. F. Pop and C. Sarbu, *J. Chem. Inf. Comput. Sci.*, 2000, **40**, 1052.
140 T. R. Cundari and M. Russo, *J. Chem. Inf. Comput. Sci.*, 2001, **41**, 281.
141 T. R. Cundari, C. Sarbu and H. F. Pop, *J. Chem. Inf. Comput. Sci.*, 2002, **42**, 1363.
142 J. A. Hageman, J. A. Westerhuis, H. W. Fruhauf and G. Rothenberg, *Adv. Synth. Catal.*, 2006, **348**, 361.
143 U. F. Röhrig, A. Loay, A. Grosdidier, P. Larrieu, V. Stroobant, D. Colau, V. Cerundolo, A. J. Simpson, P. Vogel, B. Van den Eynde, V. Zoete and O. Michielin., *J. Med. Chem.*, 2010, **53**, 1172.
144 http://www.sciencedaily.com/releases/2010/01/100129151756.htm (last accessed March 2012).
145 G. J. P. Britovsek, M. Bruce, V. C. Gibson, B. S. Kimberley, P. J. Maddox, S. Mastrioanni, S. J. McTavish, C. Redshaw, G. A. Solan, S. Stroember, A. J. P. White and D. J. Williams, *J. Am. Chem. Soc.*, 1999, **121**, 8728.
146 B. L. Small and M. Brookhart, *J. Am. Chem. Soc.*, 1998, **120**, 7143.
147 H. Sinn, W. Kaminsky, H. J. Vollmer and R. Woldt, *Angew. Chem., Int. Ed.*, 1980, **19**, 390.
148 E. Zurek and T. Ziegler, *Prog. Polym. Sci.*, 2004, **29**, 107.
149 H. Sinn, *Macromol. Symp.*, 1995, **97**, 27.
150 M. Watanabi, N. McMahon, C. J. Harlan and A. R. Baron, *Organometallics*, 2001, **20**, 460.
151 A. Andresen, H. G. Cordes, J. Herwig, W. Kaminsky, A. Merck, R. Mottweiler, J. Pein, H. Sinn and H. J. Vollmer, *Angew. Chem., Int. Ed.*, 1976, **15**, 630.
152 P. A. do Couto Jr., M. Nele and F. M. B. Coutinho, *Eur. Polym. J.*, 2002, **38**, 1471.
153 J. M. Newsam, T. Bein, J. Klein, W. F. Maier and W. Stichert, *Microporous Mesoporous Mater.*, 2001, **48**, 355.
154 P. P. Pescarmona and P. A. Jacobs, *Catal. Today*, 2008, **137**, 52.
155 F. Schüth, O. Busch, C. Hoffmann, T. Johann, C. Kiener, D. Demuth, J. Klein, S. A. Schunk, W. Strehlau and T. Zech, *Top. Catal.*, 2002, **21**, 55.

156 D. Demuth, K.-E. Finger, J. R. Hill, S. M. Levine, G. Löwenhauser, J. M. Newsam, W. Strehlau, J. Tucker and U. Vietze, *Combinatorial Material Development*, ed. R. Malhotra, American Chemical Society, Washington, D.C., 2002, **8**, 147.

157 S. A. Schunk, D. Demuth, A. Cross, O. Gerlach, A. Haas, J. Klein, J. M. Newsam, A. Sundermann, W. Stichert, W. Strehlau, U. Vietze and T. Zech, *High-Throughput Screening in Heterogeneous Catalysis*, ed. A. Hagemeyer, P. Strasser and A. F. Volpe Jr., WILEY-VCH Verlag GmbH & Co. KGaA, Weinheim, 2004, 19.

158 L. A. Baumes, P. Serna and A. Corma, *Appl. Catal., A*, 2010, **381**, 197.

159 H. Wang, Z. Liu, J. Shen and H. Liu, *Catal. Commun.*, 2004, **5**, 55.

160 N. Stock, *Microporous Mesoporous Mater.*, 2010, **129**, 287.

161 R. L. Ott and M. Longnecker, *An Introduction to Statistical Methods and Date Analysis*, ed. M. Taylor, Cengage Learning, Brooks/Cole, 2010, **6**, 35.

162 S. R. M. Pereira, F. Clerc, D. Farrusseng, J. C. van der Waal and T. Maschmeyer, *Comb. Chem. High Throughput Screening*, 2007, **10**, 149.

163 F. Schüth, *Stud. Surf. Sci.Catal.*, 2005, **157**, 161.

164 D. Farrusseng and C. Mirodatos, *High-Throughput Screening in Heterogeneous Catalysis*, ed. A. Hagemeyer, P. Strasser and A.F. Volpe Jr., WILEY-VCH Verlag GmbH & Co. KGaA, Weinheim, 2004, 239.

165 S. R. M. Pereira, F. Clerc, D. Farrusseng, J. C. van der Waal, T. Maschmeyer and C. Mirodatos, *QSAR Comb. Sci.*, 2005, **24**, 45.

166 D. C. Montgomery (ed.), *Design and Analysis of Experiments*, John Wiley & Sons, Inc., Hoboken, 2009, **4**, 121.

167 R. J. Martin, J. A. Eccleston and G. Jones, *J. Stat. Plan. Inference*, 1998, **73**, 91.

168 S. A. Schunk, C. Baltes and A. Sundermann, *Catal. Today*, 2006, **117**, 304.

169 S. Corthals, T. Witvrouwen, P. Jacobs and B. Sels, *Catal. Today*, 2011, **159**, 12.

170 S. A. Schunk, A. Sundermann and H. Hibst, *Catal. Today*, 2008, **137**, 36.

171 J. N. Cawse, G. Gazzolar and N. Packard, *Catal. Today*, 2011, **159**, 55.

172 T. A. Beltrán-Oviedo, I. Batyrshin and J. M. Domínguez, *Catal. Today*, 2009, **148**, 28.

173 D. G. Vlachos, A. B. Mhadeshwar and N. S. Kaisare, *Comput. Chem. Eng.*, 2006, **30**, 1712.

174 W. Yan, Y. Chen, Y. Yang and T. Chen, *Catal. Today*, 2011, **174**, 127.

175 R. Vijay and J. Lauterbach, *Stud. Surf. Sci. Catal.*, 2007, **171**, 325.

176 D. C. Montgomery (ed.), *Design and Analysis of Experiments*, John Wiley & Sons, Inc., Hoboken, 2009, **1**, 1.

177 D. E. Coleman and D. C. Montgomery, *Technometrics*, 1993, **35**, 1.

178 G. E. P. Box, W. G. Hunter and J. S. Hunter, *Statistics for Experimenters: Design, Innovation and Discovery*, John Wiley and Sons, Inc., Hoboken, 2005, 67.

179 V. L. Anderson and R. A. Mc Lean, *Design of Experiments: A Realistic Approach*, Marcel Dekker, Inc., New York, 1974, 123.

180 H. S. Bergh, *High-Throughput Screening in Heterogeneous Catalysis*, ed. A. Hagemeyer, P. Strasser and A. F. Volpe Jr., WILEY-VCH Verlag GmbH & Co. KGaA, Weinheim, 2004, 63.

181 Martin Holeňa, *High-Throughput Screening in Heterogeneous Catalysis*, ed. A. Hagemeyer, P. Strasser and A. F. Volpe Jr., WILEY-VCH Verlag GmbH & Co. KGaA, Weinheim, 2004, 153.

182 L. Végvári, A. Tompos, S. Göbölös and J. L. Margitfalvi, *Catal. Today*, 2003, **81**, 517.

183 A. Tompos, J.L Margitfalvi, E. Tfirst and L. Végvári, *Appl. Cat., A*, 2003, **254**, 161.

184 D. Wolf, O. V. Buyevskaya and M. Baerns, *Appl. Catal., A.*, 2000, **200**, 63.
185 M. Baerns and M. Holeňa, *Development of Solid Catalytic Materials*, ed. G. J. Hutchings, Imperial College Press, London, 2009, **7**, 13.
186 U. Rodemerck, D. Wolf, O. V. Buyevskaya, P. Claus, S. Senkan and M. Baerns, *Chem. Eng. J.*, 2001, **82**, 3.
187 O. V. Buyevskaya, A. Bruckner, E. V. Kondratenko, D. Wolf and M. Baerns, *Catal. Today*, 2001, **67**, 369.
188 U. Rodemerck, M. Baerns, M. Holeňa and D. Wolf, *Appl. Surf. Sci.*, 2004, **223**, 168.
189 I. Hahndorf, O. Buyevskaya, M. Langpape, G. Grubert, S. Kolf, E. Guillon and M. Baerns, *Chem. Eng. J.*, 2002, **89**, 119.
190 G. Grubert, E. Kondratenko, S. Kolf, M. Baerns, P. van Geem and R. Parton, *Catal. Today*, 2003, **81**, 337.
191 O. V. Buyevskaya, D. Wolf and M. Baerns, *Catal. Today*, 2000, **62**, 91.
192 A. Wolf and F. Schüth, *Appl. Catal., A.*, 2002, **226**, 1.
193 F. Clerc, S. R. M. Pereira, M. Lengliz, R. Rakotomalala, D. Farrusseng and C. Mirodatos, *Rev. of Sci. Instrum.*, 2005, **76**, 062208.
194 A. Corma, J. M. Serra, E. Argente, V. Botti and S. Valero, *ChemPhysChem.*, 2002, **3**, 939.
195 A. Corma, J. M. Serra, P. Serna, S. Valero, E. Argente and V. Botti, *J. Catal.*, 2005, **229**, 513.
196 S. Valero, E. Argente, V. Botti, J. M. Serra, P. Serna, M. Moliner and A. Corma, *Comput. Chem. Eng.*, 2009, **33**, 225.
197 T. R. Cundari, J. Deng and Y. Zhao, *Ind. Eng. Chem. Res.*, 2001, **40**, 5475.
198 K. Omata, T. Umegaki, Y. Watanabe, N. Nukui and M. Yamada, *Fuel Chem. Div. Preprints*, 2002, **47**, 574.
199 S. Nandi, P. Mukherjee, S. Tambe, R. Kumar and B. D. Kulkarni, *Ind. Eng. Chem. Res.*, 2002, **41**, 2159.
200 A. Groß, *Surf. Sci.*, 2002, **500**, 347.
201 S. Curtarolo, D. Morgan and G. Ceder, *Calphad*, 2005, **29**, 163.
202 J. R. Rostrup-Nielsen, J. Sehested and J. K. Nørskov, *Adv. Catal.*, 2002, **47**, 65.
203 H. S. Bengaard, J. K. Nørskov, J. Sehested, B. S. Clausen, L. P. Nielsen, A. M. Molenbroek and J. R. Rostrup-Nielsen, *J. Catal.*, 2002, **209**, 365.
204 G. Jones, F. Studt, F. Abild-Pedersen, J. K. Nørskov and T. Bligaard, *Chem. Eng. Sci.*, 2011, **66**, 6318.
205 X. Krokidis, J. W. Andzelm, N. Govind and V. Milman, *Appl. Cat., A*, 2005, **280**, 105.
206 L. J. Broadbelt and R. Q. Snurr, *Appl. Cat., A*, 2000, **200**, 23.
207 J. M. Ricart, F. Ample, A. Clotet, D. Curulla, J. W. Niemantsverdriet, J. F. Paul and J. Pérez-Ramírez, *J. Catal.*, 2005, **232**, 179.
208 M. C. Payne, M. Hytha, I. Štich, J. D. Gale and K. Terakura, *Microporous Mesoporous Mater.*, 2001, **48**, 375.
209 P. Raybaud, *Appl. Catal., A*, 2007, **322**, 76.
210 A. Hellman, K. Honkala, I. N. Remediakis, Á. Logadóttir, A. Carlsson, S. Dahl, C. H. Christensen and J. K. Nørskov, *Surf. Sci.*, 2006, **600**, 4264.
211 M. Head-Gordon, *J. Phys. Chem.*, 1996, **100**, 13213.
212 R. M. Dreizler and E. K. U. Gross, (eds.), *Density Functional Theory*, Springer-Verlag, Berlin, 1990, 1.
213 P. Hohenberg and W. Kohn, *Phys. Rev.*, 1964, **136**, B864.
214 F. Schüth, L. Baumes, F. Clerc, D. Demuth, D. Farrusseng, J. Llamas-Galilea, C. Klanner, J. Klein, A. Martinez-Joaristi, J. Procelewska, M. Saupe, S. A. Schunk, M. Schwickardi, W. Strehlau and T. Zech, *Catal. Today*, 2006, **117**, 284.

215 C. Klanner, D. Farrusseng, L. Baumes, C. Mirodatos and F. Schüth, *Angew. Chem., Int. Ed.*, 2004, **43**, 5347.
216 D. Farrusseng, C. Klanner, L. Baumes, M. Lengliz, C. Mirodatos and F. Schüth, *QSAR Comb. Sci.*, 2005, **24**, 78.
217 C. Klanner, D. Farrusseng, L. Baumes, C. Mirodatos and F. Schüth, *QSAR Comb. Sci.*, 2003, **22**, 729.
218 O. P. Sharma, V. Gupta, K. Sachdeva, N. K. Saini and H. Arya, *J. Nat. Conscientia*, 2011, **2**, 266.
219 A. Z. Dudek, T. Arodz and J. Galvez, *Comb. Chem. High Throughput Screening*, 2006, **9**, 213.
220 J. Procelewska, J. L. Galilea, F. Clerc, D. Farrusseng and F. Schüth, *Comb. Chem. High Throughput Screening*, 2007, **10**, 37.
221 J. Mon, M. Flury and J. B. Harsh, *J. Hydrol.*, 2006, **316**, 84.

Core-shell catalysts and bimodal catalysts for Fischer-Tropsch synthesis

Jun Bao[a] and Noritatsu Tsubaki*[b]
DOI: 10.1039/9781849727203-00216

In the past decades, numerous efforts have been made to develop efficient catalyst systems for Fischer-Tropsch synthesis (FTS) reaction. This chapter reviews two successful examples of catalyst design and development for FTS reaction: the core-shell catalyst and bimodal catalyst. Their preparation methods, structures and catalytic performances are introduced. The core-shell catalysts are designed by coating a zeolite membrane on the surface of pre-shaped catalyst pellet. The catalysts show excellent performance for the direct synthesis of isoparaffins based on the FTS reaction. The formation of heavy hydrocarbons is suppressed completely and the middle isoparaffins become the main products. This core-shell structure provides a tailor-made confined reaction environment that result in spatially confined effects and shape selectivity. The bimodal catalysts are prepared using oxide or active components as building blocks to produce small pores inside large pores of homo- or hetero-atom support. The large pore size of the bimodal catalysts can be successfully controlled. Both the experiment investigations and mathematical simulation have demonstrated that the bimodal pore structure can increase the FTS activity through higher diffusion efficiency in large pores and higher dispersion of active metal particles *via* increased surface area. Besides the spatial effect, new chemical effects might appear with the formation of hetero-atom structure. The concepts and preparation methods of core-shell catalysts and bimodal catalysts provide insight and inspiration for other catalytic reactions.

1 Introduction

The Fischer-Tropsch synthesis (FTS) reaction, discovered by Fischer and Tropsch in 1925,[1] is a catalyzed chemical reaction that converts syngas, *i.e.* a mixture of carbon monoxide and hydrogen, into a wide range of long-chain hydrocarbons and oxygenates. The advantages of FTS products over conventional petroleum-derived products are that they are free of sulphur, nitrogen, aromatic, and heavy metals. This makes the FTS products an ideal candidate for clean fuel production, as well as a starting material to other chemicals. In recent years, with the technological development of producing syngas from a great variety of sources, such as coal, natural gas, and biomass,[2] as well as the rise in the price of oil, the FTS has been a renewed and increasingly interesting area.

Although all group VIII metals display some activity in the hydrogenation of CO, the most active metals for FTS are ruthenium, nickel, iron and cobalt.[3–6] This is related to the capacity of the metals to dissociate CO. Ru-based catalysts have high activity but their utilization is limited because of the high price of ruthenium. Ni-based catalysts produce too much methane,

[a]*National Synchrotron Radiation Laboratory, University of Science and Technology of China, Hefei 230029, P.R. China*
[b]*Department of Applied Chemistry, School of Engineering, University of Toyama, Gofuku 3190, Toyama 930-8555, Japan. E-mail: tsubaki@eng.u-toyama.ac.jp*

and readily form volatile metal-carbonyls, limiting the reaction pressure and thus the FTS productivity. The relatively cheap Fe-based catalysts generally consist of precipitated iron, which is promoted with potassium and copper to improve the catalytic performance and with Al_2O_3 or SiO_2 as structural stabilizers. The high water-gas shift (WGS) activity of Fe-based catalyst allows them to utilize a large variety of feedstocks with variable H_2/CO ratio. The active phase of cobalt catalysts is metallic cobalt, which is usually supported on metal oxides due to the high price of cobalt and the better stability of supported catalysts. In general, the cobalt is poorly dispersed on the supports and Ru, Re, or Pt promoters are typically added to promote the reduction of cobalt oxides and increase active site densities. As far as preventing catalyst deactivation by carbon formation or oxidation.[6] Co-based catalysts are suitable for high hydrogen content syngas due to its low WGS activity.

The FT reaction yields predominately straight chain hydrocarbons and their distribution follows the Anderson-Schulz-Flory (ASF) law. The very long-chain hydrocarbons, *i.e.* the waxes, are widely used in the field of food, cosmetics and medicines. The C_9-C_{15} olefins are very suitable for the production of biodegradable detergents, whereas the paraffins are excellent lubricants. The hydrocracking or isomerization of waxes can directly produce high quality synthetic diesel fuel or gasoline. The productivity and product selectivity of the FT reaction, to a large extent, depends on the process conditions, as well as catalyst composition and structure. It still remains an important challenge to further increase the catalyst activity and control the product distribution for FTS. The past two decades of research have made significant progress in the field. For example, the utilization of supercritical fluid in the FTS has shown comprehensive advantages, including high conversion, high heavy hydrocarbons selectivity, efficient heat removal, quick diffusion of molecules and long catalyst life time.[7–10] Some new materials, such as nano-materials[11–13] and zeolites,[14–18] have been found to show unique performances for FTS. The advancements in characterization techniques also have led to a deeper insight into the reaction mechanism of FTS,[6,19,20] which is essential for the design and optimization of catalysts.

Solid catalysts, generally being more easily commercialized in the chemical industry and petroleum industry than liquid-phase homogeneous catalysts, are closely connected to nanoscience and nanotechnology. Catalyst active site densities depend on well-designed nano-scale assemblies to metals, metal oxides, organic species, and so on. Catalyst supports, which are not limited to inert supporting materials to disperse catalytic sites, have strong interactions with catalytically-active metals and provide nano-channels as mesoporous or microporous materials. All these nano-level chemical structures thoroughly control the mechanism, rate, direction and efficiency of the target catalytic reactions.

In FTS catalyst chemistry, the size and morphology of supported cobalt, iron carbide or ruthenium nanoparticles determine the reaction rate, hydrocarbon distribution, selectivity and catalyst stability.[21,22] The larger particle size of Co or Ru is more effective to produce heavier hydrocarbons, as wider metallic surface of each supported metallic particle provides enough continuous space for CH_2 chain propagation. But these larger particles will lower the overall reaction conversion as the supported metals are not well dispersed and total metallic surface areas are low. Conversely, if the particle size of Co or Ru is small, too much methane will form even though the reaction rate is enhanced to some extent. However, smaller metallic particles are not stable and tend to oxidize more easily. It is very important to design a suitable size of supported nano-particles of Co or Ru, to adjust the relationship between catalytic rate and product distribution. Furthermore, it is almost impossible to completely reduce supported Co or Ru nano-particles as catalyst supports such as silica, alumina or active carbon tend to make the very small metallic particles oxidize more easily or keep its original oxidized state in the precursor during reduction pre-treatment (*e.g.*, strongly interacting CoO or cobalt support compounds). Pressurized steam and CO_2 in the FTS reaction can further oxidize these metallic dots. Comprehensive design of supported metallic catalytic sites and support structures from the viewpoint of nanoscience is the first step to realize excellent catalysts. Generally, the structure of nanopores of FTS catalysts controls the product distribution from various aspects. Large pores favor heavier hydrocarbons and are efficient to increase mass transfer rates but their low BET surface areas will lower overall reaction rates. Diffusion rates of intermediates or products inside catalyst nanopores, where metallic particles are loaded, vary with these molecular sizes, nanopore sizes, acidic-basic, hydrophobic-hydrophilic properties of inner walls of nanopores, resulting in different product distributions and selectivities as the concentration gradients and elemental reaction step network are changed.

From the early 1990s, we have been engaged in the field of FTS. We are interested in this process because it plays an increasingly important role in energy development and environmental protection. In this chapter, we will introduce two successful examples in the development of FT catalysts: the core-shell catalysts and bimodal catalysts. Their preparation methods, structures and catalytic performances will be elaborated upon.

2 Core-shell catalysts for direct synthesis of isoparaffins from syngas: spatial-confinement effects and shape selectivity

In general, FTS products are almost normal aliphatic hydrocarbons and their distribution follows the ASF law. It is difficult to selectively obtain hydrocarbons with a narrow, specified carbon number distribution from the FTS reaction,[23,24] unless an additional procedure is employed, such as addition of 1-olefin,[25] or in supercritical phase reaction media.[7–10] The normal hydrocarbons are suitable as synthetic diesel fuel only. In order to be used as synthetic gasoline, they must be further hydrocracked and isomerized to branched, light hydrocarbons in a separate reaction. This means that for the direct synthesis of isoparaffins from syngas, the most efficient catalysts should contain both FTS sites and acidic sites.[26] The FTS sites convert the syngas into linear hydrocarbons, which then migrate to the acidic sites where they undergo further hydrocracking and isomerization to form branched hydrocarbons. The combination of a conventional FTS catalyst and a zeolite is an effective method to produce the isoparaffin in one step, since zeolite is the well-known hydrocracking/isomerization catalyst due to its acidic properties.[27] Several groups have tried to synthesize the isoparaffins by utilizing FTS catalyst of metal supported on zeolite.[28–30] However, these catalysts suffered from very low CO conversion for their extremely low reduction degree originating from the strong interaction between the metal and zeolite. Previous studies of Tsubaki *et al.* have found that a physical mixture of zeolite and cobalt catalyst can increase the yield of branched hydrocarbons in the FTS reaction.[31,32] Improving the dispersion of two kinds of active sites and rendering them close to each other is an efficient approach for enhancing catalytic performance.[33] However for this consecutive reaction, a crucial factor that affects the product selectivity is the ease of migration of the first reaction products to the active sites of the second reaction. For conventional bifunctional catalysts, the different active sites are randomly distributed on the surface, which provide an unrestricted, open reaction environment. The two coupled reactions happen independently and randomly, even though the distance between the two active sites is very short. This means that a lot of the products desorbed from the first reaction sites might leave the catalyst directly without further reaction on the other active sites.

To enhance the migration of intermediates, the normal hydrocarbons, to the acidic sites and improve the selectivity for isoparaffins synthesis, the Tsubaki group proposed a new concept of bifunctional catalysts having core-shell structure. This scheme for this type of catalyst is illustrated in Fig. 1. The core and shell components are independent catalysts for the different reactions, respectively. The core component is the conventional FTS catalyst. The shell that wraps around the core pellet is a layer of zeolite membrane with pore paths. The syngas firstly passes through the shell membrane to reach the inner core catalyst, forming the normal paraffins. Before leaving the catalyst, these hydrocarbons must enter the membrane channels and have a chance of being cracked at the acidic sites of the zeolite. Because of significant improvement of the collision probability between

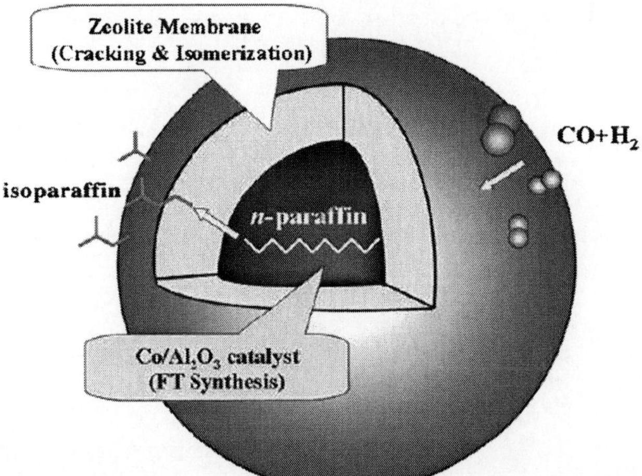

Fig. 1 Scheme of the core-shell catalysts. (Reproduced with permission from [39], copyright 2008 Wiley-VCH.)

intermediates and active sites, the core-shell structured bifunctional catalyst is expected to show high selectivity for isoparaffins. Furthermore, since the molecules with different steric hindrance have different diffusion diffusivities in the membrane channels, shape selectivity for the reactants or products may also be expected. Based on this concept, two kinds of core-shell catalysts for direct synthesis of isoparaffins from syngas have been developed.[34–39] Their preparation method, structures and catalytic performances will be discussed in the sections to follow.

2.1 [Co/SiO_2]/[H-ZSM-5] core-shell catalyst

This core-shell catalyst was prepared by coating the H-ZSM-5 membrane directly onto the surface of Co/SiO_2 pellets through a hydrothermal synthesis method.[34–36] TPAOH (tetrapropylammonium hydroxide) was used as the template, and $Al(NO_3)_3 \cdot 9H_2O$ and TEOS (tetraethyl orthosilicate) as the Al and Si sources, respectively. The effects of the crystallization time of zeolite and Co/SiO_2 pellet size on the structure and catalytic performance were investigated. Table 1 summarizes the structural and chemical properties of the core-shell catalysts. The pure H-ZSM-5 had larger specific surface area (432.5 m^2/g) and smaller pore volume (0.38 cm^3/g) than SiO_2, and these properties determined the increased surface area and the decreased pore volume of H-ZSM-5 coated catalysts. The weight of the H-ZSM-5 membrane increased with the crystallization time. Accordingly, the surface area increased while the pore volume decreased with the crystallization time. For smaller pellet size, the H-ZSM-5 membrane was coated more easily, as indicated in the weight increment of the Co/SiO_2(M)-Z-1 and Co/SiO_2(S)-Z-1 samples, even with only one day of synthesis. The amount of the surface acidic sites, determined by the NH_3-TPD, also increased with the crystallization time due to the incremental increase in zeolite loading. The Co_3O_4 particle size was similar for all catalysts, including the precursor Co/SiO_2 catalyst, as listed in Table 1. This indicates that the membrane coating process did not affect the Co cluster size on the SiO_2 surface.

Figure 2 shows the SEM images and the elemental distribution along the horizontal line of the section of the core-shell catalyst with different

Table 1 Physical and chemical properties of pure Co/SiO_2 and H-ZSM-5 zeolite coated Co/SiO_2 core-shell catalysts.[a] (Reproduced with permission from [36], copyright 2006 Wiley-VCH.)

Catalyst	Surface area/m^2 g^{-1}	Pore volume/ml g^{-1}	Zeolite coating/wt%	NH_3 uptake/ μmol g^{-1}	Co_3O_4 size/nm
Co/SiO_2	269.8	1.024	0	9.3	12
Co/SiO_2(L)-Z-1	275.3	0.529	11.5	47.0	11
Co/SiO_2(L)-Z-2	286.1	0.406	17.2	56.8	12
Co/SiO_2(L)-Z-7	310.9	0.282	24.3	76.4	13
Co/SiO_2(M)-Z-1	358.6	0.202	29.1	102.4	13
Co/SiO_2(S)-Z-1	372.6	0.178	29.6	105.2	13

[a] L, M and S mean that the core catalyst Co/SiO_2 with large (0.85–1.7 mm), middle (0.38-0.50 mm) and small (0.18–0.25 mm) pellet size; 1, 2, 7 represents the crystallization time in days in zeolite synthesis.

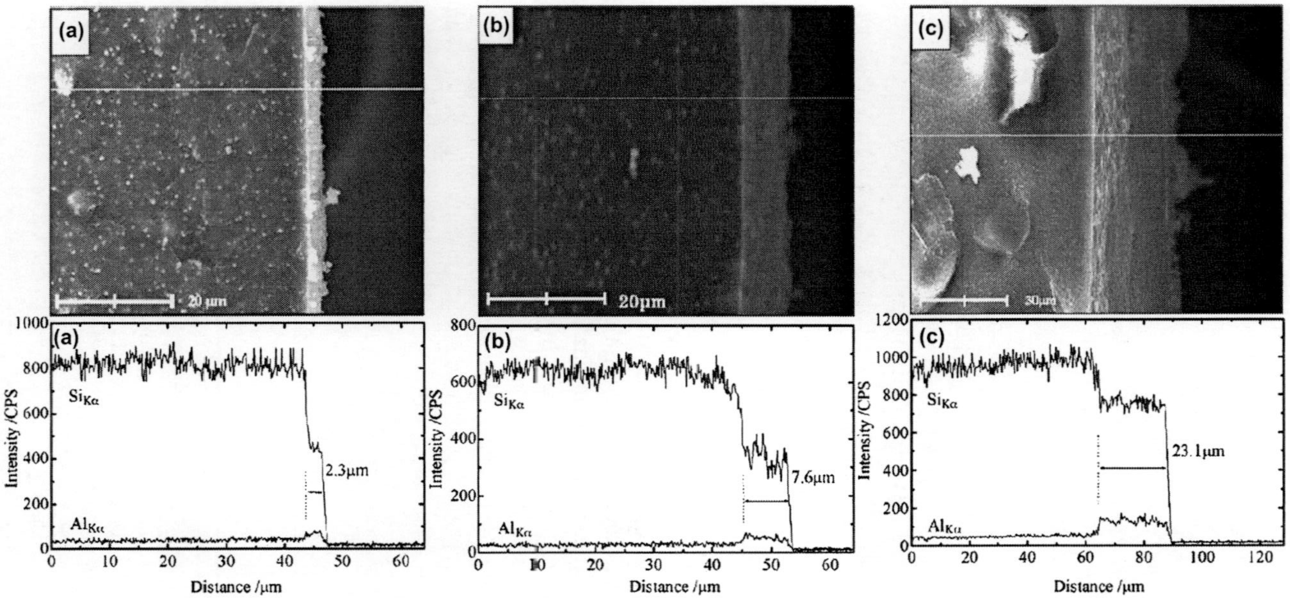

Fig. 2 Cross-sectional SEM images and the radial elemental distributicn of Al and Si by the EDX analysis. a) Co/SiO_2(L)-Z-1; b) Co/SiO_2(L)-Z-2; c) Co/SiO_2(L)-Z-7. (Reproduced with permission from [36], copyright 2006 Wiley-VCH.)

crystallization times. A compact H-ZSM-5 shell was observed crystallizing on the surface of Co/SiO_2 substrate. At the interface between the Co/SiO_2 substrate and the H-ZSM-5 membrane, the amount of Si dropped rapidly while that of Al increased slightly. Co/SiO_2(L)-Z-7 exhibited a flatter line for the Si signal intensity distribution than that of the other two catalysts, indicating that longer synthesis times during zeolite preparation led to a membrane with a more uniform structure. The membrane thickness of H-ZSM-5 shell can be controlled by the crystallization time.

The catalytic performances of the core-shell catalysts for direct synthesis of isoparaffins are presented in Table 2. The product distributions on these catalysts are shown in Fig. 3. For comparison, the pure Co/SiO_2, the physical mixture of H-ZSM-5 zeolite and Co/SiO_2 were also tested under the same conditions. From Fig. 3, the conventional Co/SiO_2 FTS catalyst exhibited a wide product distribution, in which the straight-chain hydrocarbons were the main products. When mixed with the H-ZSM-5 zeolite, the products distribution deviated from the ASF law, and the selectivity for isoparaffins and olefins increased, but there were still heavy paraffins up to C_{20}. The zeolite coated core-shell catalyst gave a completely different product distribution compared with the other two samples. Not only did the product distribution deviate from the ASF law, but also the heavy paraffins of C_{10+} were completely suppressed. The middle isoparaffins became the main products. The isoparaffins selectivity on the Co/SiO_2(L)-Z-7 was remarkably higher than that on the Co/SiO_2-Z-M, although the zeolite content in both catalysts was similar. This difference was caused by the unique core-shell structure of the zeolite coated catalyst. For the physical mixture of Co/SiO_2 and zeolite, there was no spatial restriction between the two reactions of FTS and hydrocracking/isomerization. It was a random process and the two reactions happened independently. The hydrocarbons desorbed from the FTS site might leave the catalyst directly without undergoing cracking on the zeolite sites. For the zeolite coated catalyst, its core-shell structure provided an integrated, confined reaction environment. Syngas firstly

Table 2 Effect of the zeolite crystallization time and pellet size of Co/SiO_2 on the reaction performance of core-shell catalyst. (Reproduced with permission from [36], copyright 2006 Wiley-VCH)

	Conversion/%	Selectivity/%			
Catalyst	CO	CH_4	CO_2	C_{iso}/C_n [a]	$C^=/C_n$ [b]
Co/SiO_2	98.4	15.7	10.6	0	0.28
Co/SiO_2-Z-M[c]	93.6	16.9	8.0	0.49	1.09
Co/SiO_2(L)-Z-1	83.6	22.7	10.0	0.37	0.14
Co/SiO_2(L)-Z-2	85.5	31.3	10.2	0.73	0.51
Co/SiO_2(L)-Z-7	86.1	37.4	7.0	1.88	1.35
Co/SiO_2(M)-Z-1	91.5	24.3	10.4	1.21	0.85
Co/SiO_2(S)-Z-1	89.1	22.4	6.9	0.74	0.63

Reaction conditions: 533 K, 1.0 MPa, $WCo/SiO_2/F = 10\,g\,h\,mol^{-1}$, $H_2/CO = 2$.

[a] C_{iso}/C_n is the ratio of isoparaffins to n-paraffin of C_{4+}.

[b] $C^=/C_n$ is the ratio of olefin to n-paraffin of C_{2+}.

[c] Co/SiO_2-Z-M represents the physical mixture of H-ZSM-5 and Co/SiO_2, with a weight ratio of (Co/SiO_2):H-ZSM-5= 4:1.

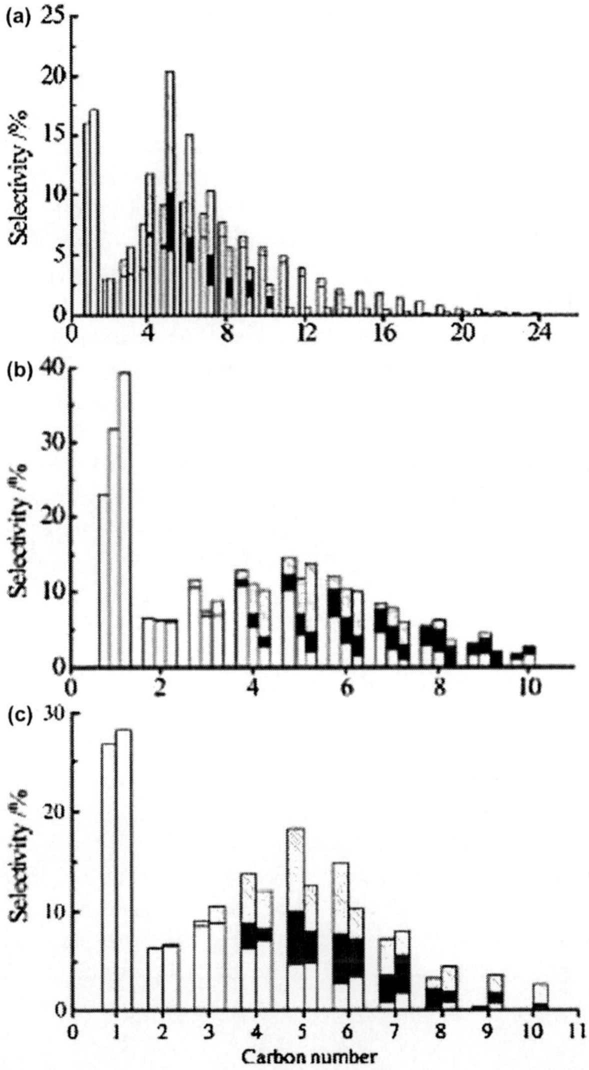

Fig. 3 (a) Product distributions on the pure Co/SiO_2 and the mechanically mixed Co/SiO_2-Z-X catalysts at 533 K. In each carbon number group, left: pure Co/SiO_2, right: Co/SiO_2-Z-X; (b) Product distributions on the core-shell catalysts with varied crystallization time. In every data group, left: 1 d, middle: 2 d, and right: 7 d; (c) Product distributions on the core-shell catalysts of 1 day crystallization with different pellet size of the Co/SiO_2. In every data group, left: 0.38–0.50 mm, and right: 0.18–0.25 mm. □: n-paraffin, ■: isoparaffin, \\\: olefin. (Reproduced with permission from [36], copyright 2006 Wiley-VCH.)

passed through the zeolite membrane channel to reach the core catalyst, forming normal paraffins *via* FTS. All the hydrocarbons must diffuse through the zeolite membrane before they left the catalyst, and all of them with straight-chain structure had a chance to be cracked at the acidic sites of zeolite. Since the hydrocarbon diffusion rate in the membrane depended on the chain length, the longer the chain length, the longer hydrocarbons reside inside the zeolite and the higher the chance that they will be cracked. Consequently, a high C_{iso}/C_n ratio and narrow product distribution were obtained. Furthermore, the membrane-coated catalyst has a much larger membrane area per unit reactor volume than conventional membrane reactors, which is another reason for the high selectivity for isoparaffins.

The core-shell catalysts with different crystallization times showed a similar CO conversion, which was lower than that of pure Co/SiO_2 or Co/SiO_2-Z-M. The slightly lower activity can be ascribed to coverage of some of the Co active sites by the zeolite coating. The activity did not decrease with the increased thickness of the zeolite coating, which indicates that the rate-controlling step was not the diffusion of the syngas through the zeolite membrane, but the FTS reaction itself on the Co sites under the present reaction conditions. Nevertheless, it is difficult to exclude the possibility that small zeolite pores might lower the diffusion rate of the syngas reaching the core FTS catalyst. The selectivity of isoparaffins, as well as olefins, increased with zeolite thickness. If the zeolite membrane is very thick, the diffusion rate of syngas will be lowered as the zeolite pore is not large. On the other hand, if the zeolite membrane is very thin, the amount of acidic sites is so limited that the rate of hydrocracking and isomerization will be lowered. The membrane thickness may be optimized, and depends on the rate balance between the FTS and the hydrocracking/isomerization, reaction temperature, zeolite type, and so on. The pellet size of the core catalyst had a strong effect on the core-shell catalyst property. Compared with the results of the Co/SiO_2(L)-Z-1 catalyst, which was also crystallized for one day, the membrane was easier to coat onto the smaller-sized pellet and resulted in a higher isoparaffin/n-paraffin ratio. The modification of either or both the membrane catalyst and the core catalyst, or optimizing the reaction condition might further enhance the catalytic activity and selectivity of the core-shell catalysts.

The 100 h reaction test is shown in Fig. 4, where the CO conversion, iso-to-normal ratio, olefin-to-paraffin ratio, the methane and CO_2 selectivity were stable during the entire experimental period. Another source of zeolite membrane stability may come from the metallic cobalt at the core FTS

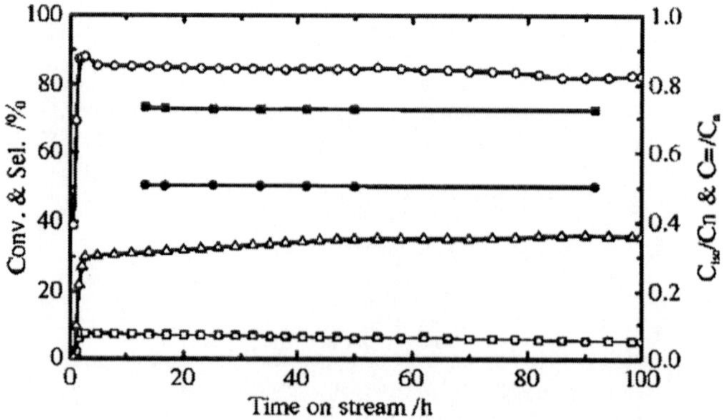

Fig. 4 Reaction performance of 100 h continuous FTS using Co/SiO_2(L)-Z-7. Standard reaction conditions; temperature: 533 K; O: CO conversion, ■: C_{iso}/C_n, ●: $C^=/C_n$, Δ: CH_4 selectivity, □: CO_2 selectivity. (Reproduced with permission from [36], copyright 2006 Wiley-VCH.)

catalyst, as the supported cobalt particles exist very closely to the zeolite membrane with intimate contact in the core-shell catalyst. Compared with the conventional gas-phase FTS process, the steady state is reached faster on the core-shell catalyst because the zeolite membrane can decompose the waxy products *in situ*. Also, temperature control in the conventional gas-phase FTS reaction is another issue which disturbs the steady state as the FTS is an exothermal reaction. Fortunately, on the core-shell catalyst the exothermal FTS and the endothermic hydrocracking of linear FTS hydrocarbons are integrated in a consecutive reactions mode. Hydrocracking reactions can absorb *in situ* FTS reaction heat, which leads to a quick arrival of the steady state.

2.2 [Co/Al_2O_3]/[H-beta] core-shell catalyst

In this case, the Co/Al_2O_3 catalyst and H-beta zeolite were chosen as the core and shell components for FTS and hydrocracking/isomerization reaction, respectively.[39] The fumed SiO_2 and $((CH_3)_2CHO)_3Al$ (aluminium isopropylate) were used as the Si and Al sources respectively, and TEAOH (tetraethylammonium hydroxide) as the template. The reactants should be selected carefully to avoid the deactivation of the obtained catalysts. For example, the frequently used template containing Cl^- or Br^-, as well as the conventional ion exchange method for preparing H-type zeolite from Na-type zeolite *via* NH_3-type zeolite could not be used here because even a small amount of Cl^-, Br^- or Na^+ will deactivate the FTS catalysts completely. During the preparation, the Co/Al_2O_3 pellet need to be refluxed in TEAOH and soaked in ethanol before being mixed with the hydrothermal synthesis precursor solution. Otherwise, no H-beta zeolite can be coated onto the FTS catalyst. This process activates the surface Al-OH groups of the pellet, with which the Al-OH and Si-OH groups in the zeolite can easily react to form Al–O–Al and Al–O–Si bonds between the substrate and the membrane. The surface Al-OH groups of the Co/Al_2O_3 pellets therefore anchor the zeolite membrane to the surface.

XRD analysis demonstrated the presence of H-beta zeolite on the Co/Al_2O_3 pellet after hydrothermal synthesis. The EDS surface elemental distributions of the Co/Al_2O_3 pellet before and after zeolite coating further confirmed the integrity of the zeolite membrane on the surface of the Co/Al_2O_3 pellet. No holes or crack was observed in the zeolite membrane. Figure 5 shows the SEM images and the elemental distribution along the horizontal line of the section of a coated Co/Al_2O_3 pellet. A compact H-beta zeolite shell can be observed crystallizing on the surface of the Co/Al_2O_3 pellet. Abrupt changes in Al and Si signal intensity happened at the interface between the Co/Al_2O_3 pellet and the zeolite layer, indicating a change from Al_2O_3 to H-beta zeolite. From the inner surface to the outer surface of H-beta membrane, the EDS signal of Si increased while that of Al correspondingly decreased. The gradient distribution of the SiO_2/Al_2O_3 ratio inside the zeolite shell was suggested to be related to the basicity of zeolite synthesis solution.[40] Because of the high pH of the synthesis solution, the surface Al of the core pellet was slightly dissolved and the Al species diffused away from the interface and established a concentration profile. During the synthesis, the Al species were incorporated into the zeolite and the

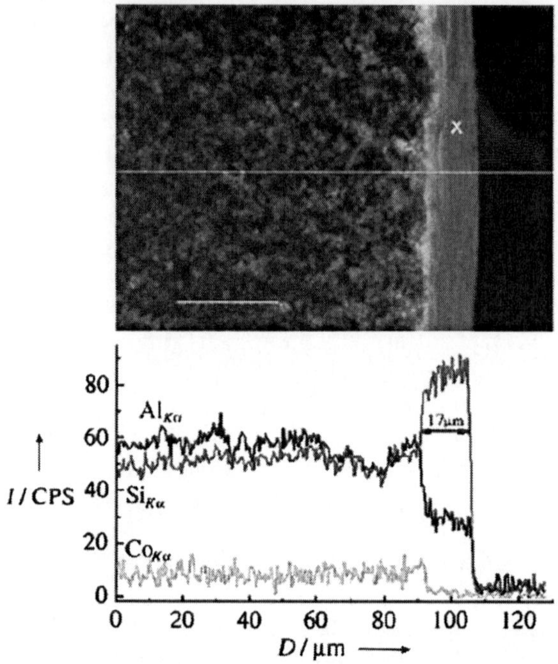

Fig. 5 The sectional SEM image of the coated Co/Al_2O_3 pellet, and the EDS line analysis results of this section. (Reproduced with permission from [39], copyright 2008 Wiley-VCH.)

Table 3 Catalytic performance comparison of the conventional Co/Al_2O_3, the physical mixture of Co/Al_2O_3 and beta-zeolite, and the beta-zeolite coated Co/Al_2O_3 catalyst. (Reproduced with permission from [39], copyright 2008 Wiley-VCH.)

Catalyst	Zeolite/%	Conv./%	Sel./%			
		CO	CH_4	CO_2	C_{iso}/C_n [b]	$C^{=}/C_n$ [c]
Co/Al_2O_3	—	84.04	21.89	4.77	0.19	0.17
Co/Al_2O_3-β-mix[a]	20.0	80.79	16.64	3.86	1.44	0.52
β/Co/Al_2O_3	20.7	74.34	13.58	2.65	2.34	0.46

Reaction conditions: 533 K, 1.0 MPa, $WCo/SiO_2/F = 10\,g\,h\,mol^{-1}$, $H_2/CO = 2$.
[a] Physical mixture of Co/Al_2O_3 and H-beta zeolite.
[b] C_{iso}/C_n is the ratio of isoparaffin to n-paraffin of C_{4+}.
[c] $C^{=}/C_n$ is the ratio of olefin to n-paraffin of C_{2+}.

concentration gradient in the solution led to a composition gradient inside the zeolite framework. The gradient distribution of the SiO_2/Al_2O_3 ratio reflected that the zeolite membrane was directly crystallized onto the pellet surface.

The catalytic performances of the H-beta zeolite coated Co/Al_2O_3, the pure Co/Al_2O_3 and the physical mixture of H-beta zeolite and Co/Al_2O_3 for direct synthesis of isoparaffins are shown in Table 3. Similarly with the H-ZSM-5 zeolite coated Co/SiO_2 catalyst, the H-beta zeolite coated Co/Al_2O_3 core-shell catalyst also shows much higher isoparaffins selectivity than that of the physically-mixed catalysts, although their zeolite content was similar. Furthermore, it is important to note that zeolite coated Co/Al_2O_3 catalyst showed desirable lower methane selectivity than that of the normal FTS catalyst, which might be related to the hydrophilicity of zeolite. It is well known that the hydrophilicity of zeolite increases when the Si/Al ratio decreases.[41] For the zeolite membrane coated catalysts, the part near the Al_2O_3 substrate had a relatively high Al content and hence high hydrophilicity. Because of the solubility rule (that is, like dissolves like), the presence of water can promote the accumulation of CO, leading to a higher CO/H_2 ratio in the interior part of the catalyst and a decrease of methane selectivity.[42]

3 Bimodal catalysts for Fischer-Tropch synthesis: chemical and spatial effects

Supported cobalt or iron catalysts are widely used in the FTS reaction. The activity of the catalysts depends on the degree of dispersion of the metallic atoms. The number of active sites is determined by the particle size, loading amount, and reduction degree.[43,44] Synthesis of highly dispersed supported catalysts requires strong interaction between the support and the precursor, but in turn such strong interactions generally lower the reducibility of such precursors.[45] The metal dispersion can be enhanced by increasing the surface area of the support. However, the support with large surface area usually contains small pore size, which results in poor intra-pellet diffusion efficiencies of molecules, especially in a multi-phase reactor. Slow transport of reactants to, and products from, active sites often controls the rate of primary and secondary reactions even on small catalyst pellets.[46] Meanwhile, the pore size of the catalyst also affects product selectivity, due to the spatial effect of the support. A bimodal structure of support with both large pores and small pores has excellent advantages for solving this contradiction because the large pores lead to a high-diffusion rate of reactants and products while the small pores provide a large active surface area, contributing to high diffusion efficiency and high dispersion of supported metal simultaneously, as theoretically expressed by Levenspiel.[47]

So far several preparation methods have been reported to form bimodal pore catalyst supports.[48,49] However, these methods used very corrosive reagents such as aqua regia, and the size of large pores of the obtained bimodal structure was as high as several hundred nanometers such that the effect of the bimodal structure was not obvious.[50] Furthermore, the reported methods are only effective for one specific oxide support, such as Al_2O_3 or SiO_2. Recently, the Tsubaki group developed a simple and general method to prepare the multi-functional bimodal support by building up nanoparticles from sol or solution of support precursor to form small pores inside the large pores of homo- or hetero-atom supports.[51–64] The formation scheme of the bimodal support is shown in Fig. 6. The nano-sol particles with small size were deposited onto inner walls of the large pore, *via* calcination and bond formation between surface silanol groups, to form small pores. The method can control the pore sizes precisely. Several kinds of bimodal support were developed by the method. Their performances for liquid-phase FTS reactions were investigated after the active component was supported.

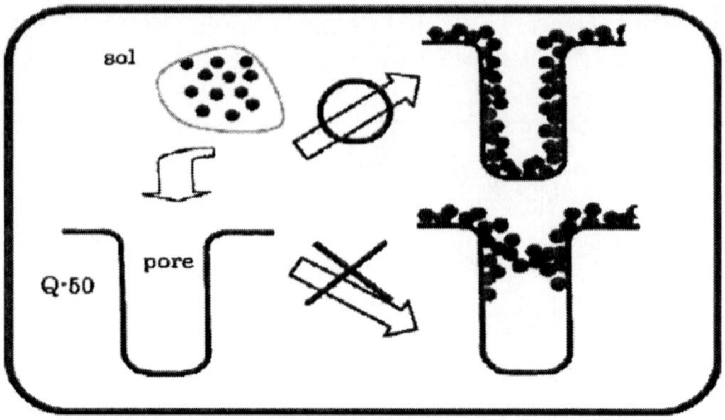

Fig. 6 Formation scheme of bimodal support. (Reproduced with permission from [56], copyright 2003 Springer.)

3.1 SiO_2–SiO_2 bimodal support

The bimodal support was prepared by incipient-wetness impregnation of commercially available silica gel (Cariact Q-50, Fuji Silysia, pore diameter: 50 nm) with the silica sol (Snowtex XS, Nissan Chemicals, SiO_2 content: 20 wt %, particle size: 5 nm) of different loadings.[51,60] Figure 7 shows the pore size distributions of silica gel Q-50, the bimodal support and Co supported bimodal catalyst. The silica gel Q-50 has a uniform distribution of large pores of 50 nm diameter, while the bimodal support exhibited two distinct types of pores having diameters of 6 and 45 nm. The small pores (6 nm diameter) were formed by the silica sol through dehydration of OH groups of the particles in the large pores of the silica gel in the calcination step. Therefore the large pore size of the bimodal support decreased from 50 nm of silica gel Q-50 to 45 nm. Table 4 shows the physical properties of various supports and catalysts. The BET surface area of the bimodal support significantly increased from 70 m^2/g of silica gel Q-50 to 106 m^2/g due to the contribution from the newly formed small pore. As a porous silica structure was formed inside the large pores of silica gel Q-50, the pore volume of bimodal support decreased from 1.2 ml/g of Q-50 to 0.46 ml/g. The increased BET surface area and the decreased pore volume proved that the particles of silica entered the large pores of Q-50, and deposited on the inner walls of Q-50 to form the small pores. If the silica sol did not enter the large pore, the observed pore volume would not change. If the silica particles were deposited at the entrance of large pores of Q-50, the large pores would be blocked, and the BET surface area of the bimodal support would decrease significantly. The BET surface and pore volume decreased and average pore size increased very slightly for the Co supported bimodal catalyst, since the supported cobalt crystallines blocked some smaller pores of the support, but the bimodal catalyst still kept a bimodal structure with two kinds of pore diameters of 8 and 47 nm.

The pore-size distribution of the bimodal support material can be controlled precisely by choosing a sol with a favorable particle size. Furthermore, if the diameter of a primary particle of the sol is D, the pore size, produced by these primary particles, is determined by diameter D and the

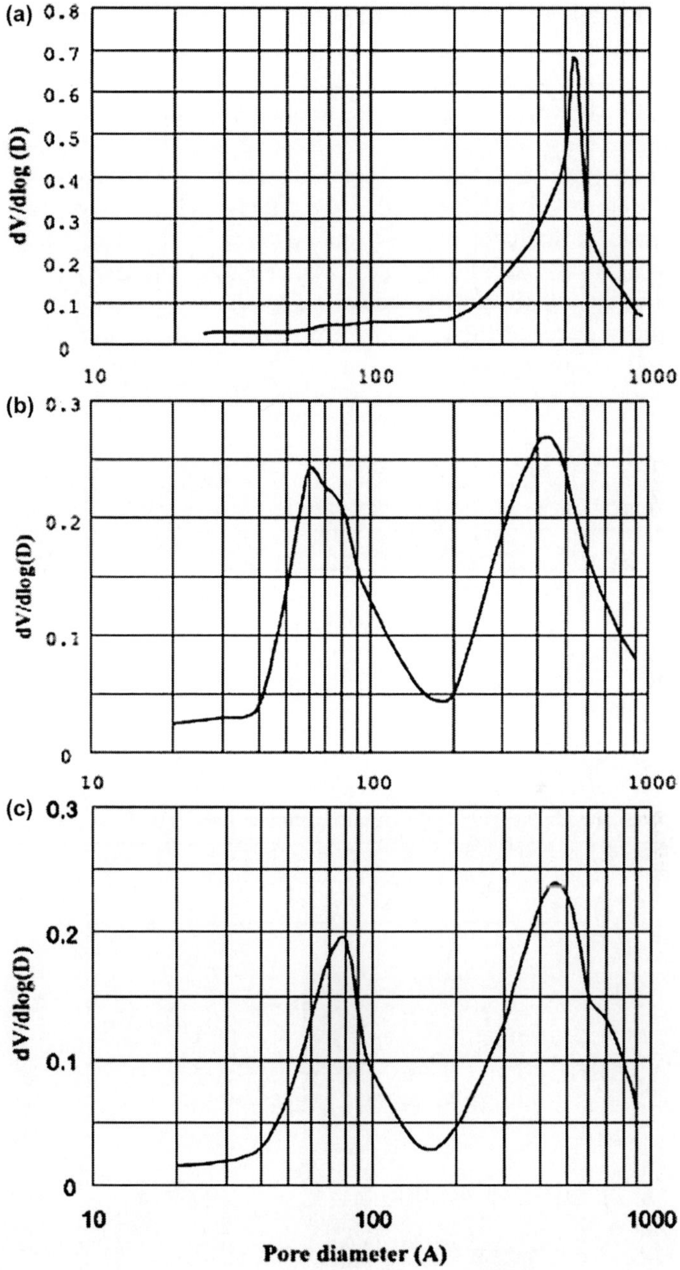

Fig. 7 The pore size distribution of (a) Q-50; (b) Silica bimodal support: the ratio of silica weight from silica sol to Q-50 silica weight was 11.2%; (c) Co supported bimodal catalyst. (Reproduced with permission from [51], copyright 2001 Elsevier.)

agglomeration of primary particles.[65] It will be 0.155D, 0.414D and D, when the primary particles aggregate into a closest packing, primitive cubic lattice, and diamond lattice, respectively. In this study, the primary particle size of silica sol was 5 nm, and it formed new small pores inside the large pores of silica gel Q-50. The most important point was that for the resulting silica bimodal supports, the newly formed small pore size was 6 nm, irrespective of the increased silica loading, as listed in Table 4.[56] These findings indicated that the primary particles of silica sol aggregated mainly into a diamond lattice inside the large pores of silica Q-50 in the calcination step. If the primary particles had aggregated mainly into a closest packing or primitive cubic lattice, the size of newly formed small pores would have been smaller than the diameter of the primary particle. If the primary particles would have formed secondary particles, the sizes of newly formed small pores would vary with the changed silica loading.

The performances of the catalysts prepared from the bimodal support or Q-50 and Q-3 support are compared in Fig. 8. The reaction rate of all of the catalysts reached a steady state after the reaction running for 1 h.

Table 4 The properties of the various supports and Co supported catalyst. (Reproduced with permission from [51] and [56], copyright 2001 Elsevier and copyright 2003 Springer.)

Sample	SiO_2 loading/wt%	Surface area/m^2/g	Pore volume/ml/g	Pore size/nm
Q-50 support	—	70	1.2	50
Q-3 support	—	546	0.30	3.0
Bimodal support-1	5.6	88	0.65	6.0, 45
Bimodal support-2	11.2	106	0.46	6.0, 45
Bimodal support-3	22.5	132	0.40	6.0, 45
Q-50 catalyst	—	52	0.82	54
Q-3 catalyst	—	513	0.16	3.7
Bimodal catalyst[a]	11.2	82	0.28	8.0, 47

[a] Derived from bimodal support-2, the cobalt loading is 10%.

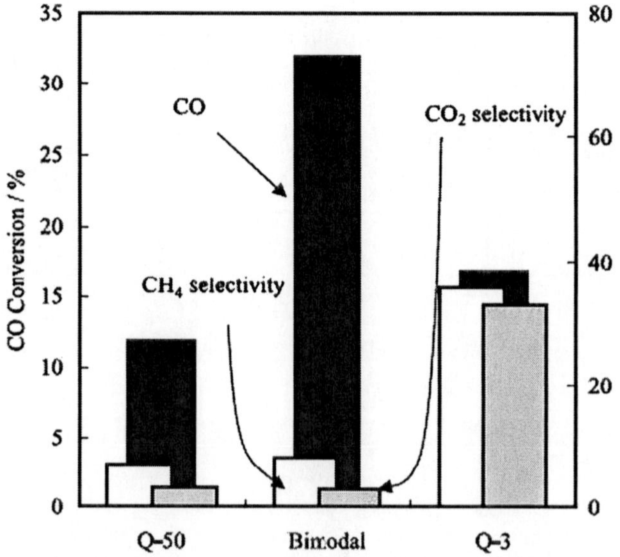

Fig. 8 The reaction performance of cobalt catalysts prepared from different support. Reaction conditions: 513 K, 1.0 MPa, W/F = 10 g h/mol, $H_2/CO = 2$, Cobalt loading: 10 wt%. (Reproduced with permission from [51], copyright 2001 Elsevier.)

Cariact Q-3 is similar to Q-50 but with a surface area of 546 m^2/g and an average pore size of 3 nm. Catalysts prepared from Q-3 support had the largest surface area but the smallest pore diameter, which exhibited a low catalytic activity and the highest CH_4 selectivity. For the catalysts prepared from Q-50 support, which had the lowest surface area and the largest pore size, the CO conversion, CH_4 and CO_2 selectivity were the lowest. The bimodal catalyst showed the best performances here. Its CO conversion was the highest, and meanwhile selectivities of CH_4 and CO_2 were as low as those of the catalyst prepared from Q-50. These differences are due to the bimodal structure. The small pores in the bimodal catalyst provided a large surface area, which led to a high dispersion of active metal particles and improved the catalytic activity, while the coexisting large pores promoted the mass transfer rate of syngas and 1-alkenes, and lowered the production of CH_4. Furthermore, as the byproduct H_2O could escape from large pores quickly, the formation rate of CO_2 was reduced since CO_2 arises from the water–gas–shift reaction ($CO + H_2O = CO_2 + H_2$).

3.2 ZrO_2–SiO_2 bimodal support

Several oxides, such as zirconia or alumina, have been used as promoters to synthesize highly active silica-based catalysts.[66,48] By introducing different sols other than SiO_2 sol into silica pellets, multifunctional bimodal support can be formed, where new chemical effects besides spatial effects of the bimodal structure might appear when the oxides react to build new phases of the material. The zirconia–silica bimodal support was formed from zirconia sol (particle size: 1.7–2.4 nm) and silica pellets Q-50.[52,53,56–58] The specific surface area increased from 70 m^2/g of the silica gel Q-50 to 201 m^2/g of the zirconia–silica bimodal support, and the pore volume decreased from 1.2 mL/g of Q-50 to 0.41 mL/g of zirconia bimodal support. The pores of the zirconia bimodal support exhibited maxima of 3.2 and 45 nm. After 10 wt% cobalt loading, the two pore sizes of the obtained zirconia bimodal catalyst were 5 nm and 47 nm, indicating some filling of narrower pores.

The performances of the catalysts prepared from the bimodal support and Q-50 or Q-3 for FTS are shown in Fig. 9. The performance of a cobalt-supported SiO_2–SiO_2 bimodal catalyst was also compared. The catalyst prepared from bimodal support showed higher CO conversion than that obtained with the catalysts derived from Q-50 and Q-3, and the selectivities of CH_4 and CO_2 were as low as those of the catalyst prepared from Q-50. For the zirconia bimodal catalyst, the CO conversion increased significantly as compared to the silica bimodal catalyst. The selectivity of CH_4 and CO_2 were almost the same. It was reported that on zirconia-modified cobalt silica catalysts, CO dissociation can be facilitated, leading to an increase in reaction rate.[67,68] ZrO_2 might form an active interface with cobalt, which is responsible to some extent for the enhancement in cobalt activity. The zirconia bimodal catalyst exhibited two promotional effects, chemical promotion and spatial promotion, to enhance the activity for liquid-phase FTS. One was the bimodal structure with both large and small pores in the supports, which improved the diffusion rate of syngas and products, and the dispersion of active metal particles, respectively. The other was the presence of ZrO_2 in the bimodal supports, which itself acted as a promoter for FTS.[69,70]

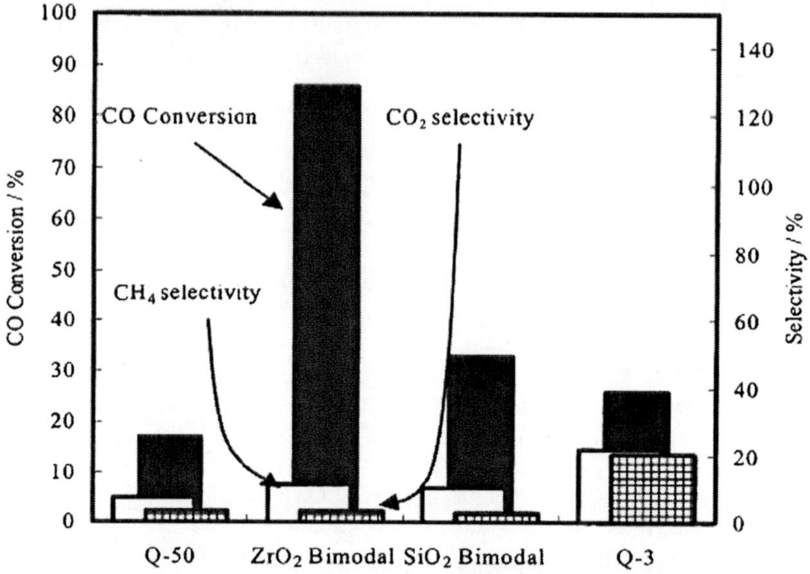

Fig. 9 The catalytic performance of cobalt catalysts prepared from different supports. Reaction conditions: 513 K, 1.0 MPa, W/F = 10 g h/mol, $H_2/CO = 2$, Cobalt loading: 10 wt%. SiO_2 bimodal: the ratio of silica weight from silica sol to Q-50 silica weight was 11.2%. Zirconia bimodal: the ratio of zirconia weight from zirconia sol to Q-50 silica weight was 16.8%. (Reproduced with permission from [56], copyright 2003 Springer.)

3.3 Al_2O_3–SiO_2 bimodal support

The preparation of the multifunctional Al_2O_3-SiO_2 bimodal support was a little different from that of the SiO_2-SiO_2 and ZrO_2–SiO_2 bimodal supports, which were obtained by building up nano-particles from silica or zirconia sol to form small pores inside tailor-made large silica pores. For the Al_2O_3-SiO_2 bimodal support, a polymer complex solution of aluminium was used instead of the sol to build up the small pores, because the polymer complex solution can form homogeneously distributed oxide particles and homogeneous packing of the particles. The polymer complex solution of aluminium was prepared by dissolving aluminum nitrate in a 0.3 mol/L polyethylene glycol (PEG, average molecular weight of 200) aqueous solution. Then, the polymer complex solution was introduced into the large pore of silica gel to form the bimodal pore structure by the incipient-witness method.[59,61]

The obtained bimodal alumina-silica support had two pore diameters of 4.5 and 47 nm, respectively. The intrinsic pore of the used silica Q-50 pellet decreased from 50 to 47 nm and a new 4.5 nm pore was formed by alumina. The BET surface area was enhanced from 70 m^2/g of Q-50 to 101 m^2/g of the bimodal one, with the main contribution from the newly-formed small pores. The pore volume of the obtained bimodal support decreased from 1.20 mL/g of Q-50 to 0.76 mL/g, indicating that alumina indeed formed the new small pores inside the uniformly distributed large pores of Q-50. Small alumina particles deposited onto inner walls of the large pore, *via* calcination and bond formation between surface Al-OH and Si-OH groups, to form small pores.

Table 5 Reaction performance and properties of various catalysts. (Reproduced with permission from [61], copyright 2005 Elsevier.)

Catalyst	Co particle size/nm		Reduction degree[b]/%	Co dispersion[c]/%	TOF[d]/s^{-1}	CO Conv./%	Selectivity/%		α
	XRD	H_2[a]					CH_4	CO_2	
Q-50	35	41	98.5	2.5	0.061	13.5	7.1	2.1	0.86
Q-3	4.5	3.2	60.3	38.1	0.014	9.6	18.1	16.8	0.82
SiO_2-SiO_2	20.6	30.3	87.6	3.3	0.072	17.8	8.9	2.2	0.86
Al_2O_3-SiO_2	15.3	23.2	90.2	4.3	0.121	38.9	10.2	1.3	0.87

Reaction conditions: P = 1.0 MPa, T = 513 K; W/F = 5 g · cat · mol^{-1}; $H_2/CO = 2$.
[a] Calculated from H_2 chemisorption at 373 K.
[b] Calculated by TPR from 373 to 1073 K.
[c] Determined by H_2 chemisorption.
[d] Based on H_2 total uptake.

The FTS reaction results are listed in Table 5, which showed that the alumina–silica bimodal catalyst had much higher catalytic activity than those of the silica–silica bimodal catalyst and Q-50 or Q-3 catalysts. The selectivity of CH_4 and CO_2 was almost the same as those of the Q-50 catalyst. Compared with silica–silica bimodal catalyst with similar pore structure, alumina–silica bimodal catalyst had larger pore volume, higher metal dispersion, higher reduction degree, and higher turnover frequency (TOF). Meanwhile, the alumina–silica bimodal catalyst was more reducible than the silica–silica bimodal catalyst. The *in-situ* diffuse reflectance infrared Fourier transform (DRIFT) spectra and temperature-programmed surface reaction (TPSR) experiment suggested that CO adsorbed on the alumina–silica bimodal catalysts was more reactive than when adsorbed on the others, resulting in the highest TOF. All of these factors are considered to be due to the hetero-atom structure in the alumina–silica bimodal structure, leading to its higher activity than that of the silica–silica bimodal analogy with a spatial effect only.

3.4 Fe-Cu-K/SiO_2 bimodal catalyst

In general, the bimodal catalysts have been prepared by a two-step process: the synthesis of bimodal pore support and subsequent impregnation of active components. Both the large and small pores of the final obtained catalysts were actually derived from the bimodal support and almost not affected by the introduction of the supported active components. However, the two-step preparation has the basic disadvantage that a considerable amount of active components has difficulty entering the small pores of the bimodal support during impregnation, which limits the metal dispersion on the surface of the support, and finally the catalytic activity. To further increase the performance and simplify the preparation of bimodal catalysts, the Tsubaki group proposed a new method that directly used the catalytically active components to generate small pores inside the large pores of the support. Based on the proposed method, supported iron bimodal catalysts containing copper and potassium promoters were prepared through a simple one-step impregnation.[71] The obtained Fe-Cu-K/Q-50 catalyst exhibited two distinct types of pores having diameters of 56.3 and 6.2 nm, respectively, reflecting the bimodal structure. Due to the formation of small pores, the BET surface area of the bimodal Fe-Cu-K/Q-50 catalyst increased to 94.8 m^2/g, and the pore volume of the catalyst decreased significantly from 1.26 mL/g to 0.47 mL/g, proving that these components entered the large pores of silica gel rather than having deposited on the outer surfaces. After calcination at 973 K, the small pores of the bimodal catalyst disappeared and the BET surface area decreased to 75.1 m^2/g due to the collapse of small pores and the agglomeration of active components. Figure 10a shows the SEM image of the bimodal catalyst, indicating that the inner surface of the large pore of the Fe-Cu-K/Q-50 had a rugged structure (arrow), displaying a bimodal pore structure as illustrated in Fig. 10b. These results revealed that the introduced active species indeed entered the large pores of Q-50 silica gel and acted as the "brick" to build the small pores inside large pores of the support, this was quite different from the previous bimodal catalyst, where the bimodal pore catalyst support was prepared first, followed by impregnation of the active component.

During the FTS reaction, the bimodal Fe-Cu-K/Q-50 catalyst showed a higher CO conversion than that of the one derived from Q-3 (Table 6), although its surface area was much lower. After calcination at 973 K, the CO conversion of the obtained catalyst decreased significantly. The selectivity of CH_4 and CO_2 did not change significantly. Furthermore, compared with the widely used co-precipitated iron catalysts, the supported bimodal Fe-Cu-K/Q-50 exhibited much higher FTS activity under the same reaction conditions. Its CO conversion ($W/F = 10\,g \cdot h/mol$) was close to 90%, about

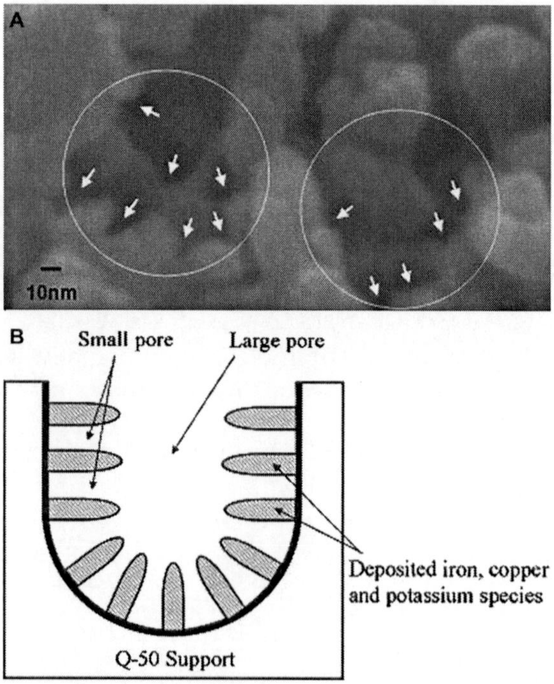

Fig. 10 (A): FE-SEM image of Fe-Cu-K/Q-50 catalyst; (B): Scheme of bimodal pore structure. (Reproduced with permission from [71], copyright 2009 Elsevier.)

Table 6 The reaction performance of various Fe-Cu-K catalysts.[a] (Reproduced with permission from [71], copyright 2009 Elsevier.)

Catalyst	W/F (g.h/mol)	CO conversion (%)	Selectivity (%)		α^b
			CH_4	CO_2	
Fe-Cu-K/Q-3	5	32.8	9.2	36.7	0.57
Fe-Cu-K/Q-50	5	53.7	10.8	37.1	0.65
Fe-Cu-K/Q-50[c]	5	45.5	9.6	34.1	0.61
Fe-Cu-K/Q-50	10	89.5	10.5	37	0.60
Fe-Cu-K/SiO_2[d]	10	13.5	12.7	14.6	0.61

[a] Reaction conditions: 553 K, 1.0 MPa.
[b] Chain growth probability.
[c] Calcination at 973 K.
[d] Prepared by co-precipitation.

7 times higher than that of the co-precipitated sample. The high CO_2 selectivity (~40%) indicated that the catalyst had a high WGS activity. These results suggest that the supported bimodal iron catalyst is a highly active catalytic material for FTS. Comparing the two unimodal catalysts, Fe-Cu-K/Q-3 and the high temperature treated Fe-Cu-K/Q-50, the latter with only the large pore showed a higher CO conversion although its specific surface area was much lower. This finding indicates that, in this case, the pore size of catalysts played a more important role than surface area in determing catalytic activity. The reaction rate in slurry phase here was mainly controlled by the diffusion of reactants and products.

3.5 Pore diffusion simulation model of bimodal catalyst

The experimental investigations have demonstrated that a catalyst with a bimodal pore structure can increase the catalytic activity through higher diffusion efficiency in large pores and higher dispersion of active metal particles *via* the increased surface area. On the basis of this, a mathematical model was developed to simulate reactants diffusion behavior inside the pores in different catalysts, especially in bimodal catalysts.[60]

Taking the Co/SiO_2 catalyst used for slurry FTS as an example, as shown in Fig. 11, supposing the Q-50 and Q-3 catalysts with single cylindrical pores with inner radius r and depth L, the reactants dissolved in the solvent moved in the bulk liquid and finally diffused into the pores and reached the inner surface of the catalyst particle. The relationship between the rate constants on different bases is given by

$$k_v \cdot \pi r^2 \cdot L = k_s \cdot 2\pi r \cdot L;\ k_v = \frac{2k_s}{r} \tag{1}$$

where k_V is the reaction rate constant based on pore volume, while k_S is based on surface area; r is the pore radius.

In general, the FT reaction kinetics follow a power rate law, and the rate equation is shown below:

$$v = k_v C_{H2}^{\lambda} C_{CO}^{y} = \frac{2k_s}{r} C_{H2}^{\lambda} C_{CO}^{y} \tag{2}$$

where x and y are the exponential dependencies of the rate on H_2 and CO concentration.

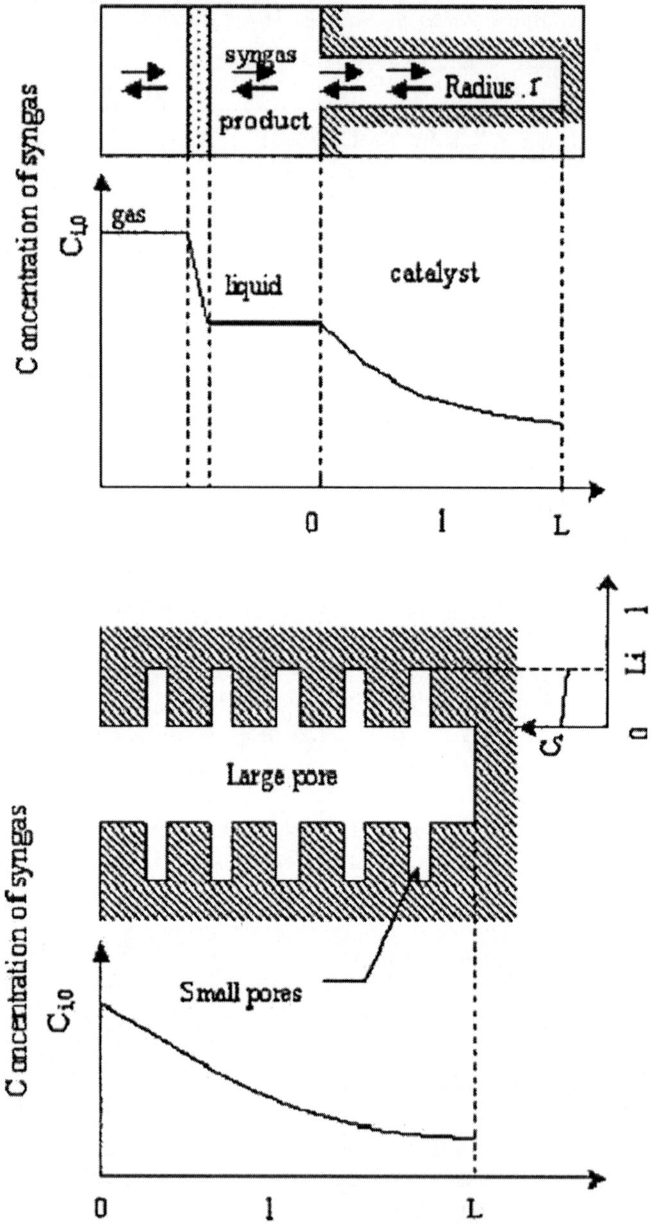

Fig. 11 Unimodal or bimodal pore structure with one or two size of pores as a model for different catalyst. (Reproduced with permission from [60], copyright 2005 Wiley-VCH.)

For bimodal catalysts, the diffusion effect in the small pores can be neglected because the small pore is not so deep, which will be proved in the next discussion. The diffusion in large pores is only considered here. The suppositional reaction constant based on large pore surface area (k^*_S) is introduced for simulation.

$$k_s \cdot S_{total} = k_s^* \cdot S_L; \; k_s^* = k_s \frac{S_{total}}{S_L} \tag{3}$$

where k_S is on the support total surface area, but k^*_S is on the large pore surface area; S_L is the surface area of the large pores, which can be calculated from the large pore volume V_L, and large pore radius r:

$$S_L = {}^{2V_L}\!/_{r} = 2 \times 0.186 \times 1000/47/2 = 15.83\ \mathrm{m}^2/\mathrm{g} \tag{4}$$

Similar to Equation 1, the suppositional reaction constant based on large pore volume (k^*_V) can be expressed as:

$$k^*_V = \frac{2k^*_S}{r} = \frac{2k_s \cdot S_{total}}{r \cdot S_L} \tag{5}$$

Thus, the reaction rate in bimodal catalyst can be expressed as:

$$v = k^*_V C^{\lambda}_{H2} C^{y}_{CO} = \frac{2k^*_S}{r} C^{\lambda}_{H2} C^{y}_{CO} = \frac{2k_s \cdot S_{total}}{r \cdot S_L} C^{\lambda}_{H2} C^{y}_{CO} \tag{6}$$

If the rate constant is based on the surface of the metal particles, Equation 2 should be converted to:

$$v = k_v C^{\lambda}_{H2} C^{y}_{CO} = \frac{2k_{Co} \cdot S_{Co}}{r \cdot S_{total}} C^{\lambda}_{H2} C^{y}_{CO} \tag{7}$$

where k_{Co} is the reaction rate constant based on the metal surface area; S_{Co} and S_{total} are the surface area of metal cobalt and the total BET surface area. This equation gives the relationship between reaction rate constant based on specific metal surface area (k_{Co}), and that based on pore volume (k_V) in a cylinder pore.

For a bimodal catalyst,

$$v = k^*_V C^{\lambda}_{H2} C^{y}_{CO} = \frac{2k_{Co} \cdot S_{total} \cdot S_{Co}}{r \cdot S_L \cdot S_{total}} C^{\lambda}_{H2} C^{y}_{CO} = \frac{2k_{Co} \cdot S_{Co}}{r \cdot S_L} C^{\lambda}_{H2} C^{y}_{CO} \tag{8}$$

In the slurry reaction, the CO and H_2 concentrations at the entrance of the pore are suggested to be the same as the concentrations in the bulk solution because the influence of reactant diffusion in the liquid laminar can be neglected due to rapid agitation. According to Fick's second Law:

$$D_{H2} \frac{d^2 C_{H2}}{dl^2} = v_{H2};\ D_{CO} \frac{d^2 C_{CO}}{dl^2} = v_{CO} \tag{9}$$

D_{H2} and D_{CO} are the effective diffusion coefficients for hydrogen and carbon monoxide in the pore, respectively; l is the position in the pore; v_{H2} and v_{CO} is the reaction rate per unit pore volume for H_2 or CO; and C_{H2} and C_{CO} are the concentrations of reactants.

The boundary conditions are:

$$C_{H2} = C_{H2,0},\ C_{CO} = C_{CO,0} \quad \mathrm{at}\, l = 0 \tag{10}$$

$$\frac{dC_{H2}}{dl} = \frac{dC_{CO}}{dl} = 0 \quad \mathrm{at}\, l = 1 \tag{11}$$

The reaction rate in Equation 2 and 6, can be expressed as

$$v = v_{CO} = \frac{1}{2} v_{H2} = k_v C^{\lambda}_{H2} C^{y}_{CO} \tag{12}$$

where k_V is the rate constant; x and y are the exponential dependencies on H_2 and CO concentration.

The rate equation for this system was determined experimentally. Finely crushed Q-50 catalyst with large pores only was tested in a slurry FT reaction to eliminate the most possible diffusion influence. The partial pressure of H_2 in the syngas was changed and the partial pressure of CO was fixed. N_2 was used as the balancing gas. It was found that the dependency on H_2 was first order. The following relationship was derived:

$$v = kC_{H2}f(C_{CO});\ v/C_{H2} = kf(C_{CO}) \tag{13}$$

where k is the apparent reaction constant; $f\,(C_{CO})$ is the function of CO concentration in the liquid.

Figure 12 shows the relationship between $V/(C_{H2})$ and C_{CO}. The expression of the function $kf(C_{CO})$ was determined using the best fitting equation

$$\frac{kf(C_{CO})}{\mathrm{mol\cdot s^{-1}\cdot g^{-1}}} = \frac{0.90}{\frac{C_{CO}}{\mathrm{mol\cdot m^{-3}}}+1.4} - 0.91\exp\left(\frac{-6.4C_{CO}}{\mathrm{mol\cdot m^{-3}}}\right) + 0.266 \tag{14}$$

From the figure, the reaction rate tended to be constant when CO concentration is high. This indicated that the dependency on CO concentration was zero-order under high CO concentration conditions. Consequently, the limitation of the function $f(C_{CO})$ should be 1. The apparent reaction constant was 0.266 $m^3\,g^{-1}\cdot s$. Thus, the expression of function $f(C_{CO})$ was:

$$f(C_{CO}) = \frac{3.38}{\frac{C_{CO}}{\mathrm{mol\cdot m^{-3}}}+1.4} - 3.42\exp\left(\frac{-6.4C_{CO}}{\mathrm{mol\cdot m^{-3}}}\right) + 1.00 \tag{15}$$

The catalyst effectiveness factor η was defined to measure how much the reaction rate is lowered because of the resistance to pore diffusion:

$$\eta = \frac{\textit{actual reaction rate within pore}}{\textit{rate if not slowed by pore diffusion}} = \frac{\bar{v}_{with\ diffusion}}{v_{without\ diffusion}} \tag{16}$$

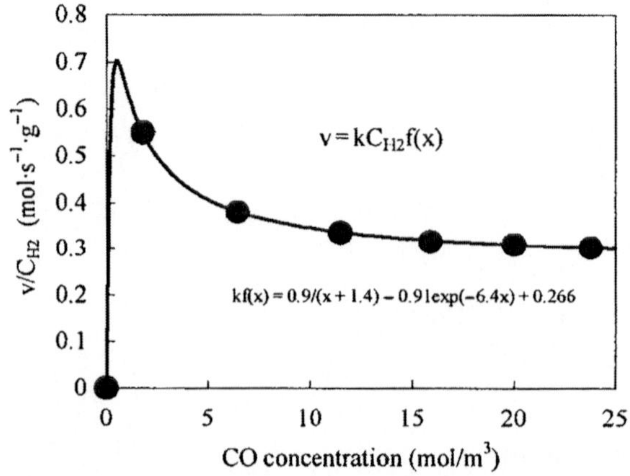

Fig. 12 Reaction rate as a function of CO concentration. (Reproduced with permission from [60], copyright 2005 Wiley-VCH.)

According to the CO or H_2 diffusion rate at the entrance of the pore and the ideal expression of the reaction rate, the effectiveness factor can be expressed as:

$$\eta = \frac{-D_{CO}\left(\frac{dC_{CO}}{dl}\right)_{l=0}\pi r^2}{k_v C_{H2,0} f(C_{CO,0})\pi r^2 L} = \frac{-D_{CO}\left(\frac{dC_{CO}}{dl}\right)_{l=0}}{k_v C_{H2,0} f(C_{CO,0})L} = \frac{-D_{H2}\left(\frac{dC_{H2}}{dl}\right)_{l=0}\pi r^2}{2k_v C_{H2,0} f(C_{CO,0})\pi r^2 L} = \frac{-D_{H2}\left(\frac{dC_{H2}}{dl}\right)_{l=0}}{2k_v C_{H2,0} f(C_{CO,0})L} \quad (17)$$

According to the CO consumption rate in pores $k_V \eta C_{H_{2,0}} f(Cco,0)$ and the specific pore volume V, the CO reaction rate per unit weight of catalyst can be expressed as $k_V \eta C_{H_{2,0}} f(Cco,0)V$. It is noticed that the units on CO consumption rate in pores are $mol \cdot m^{-3} \cdot s^{-1}$, but the unit of CO reaction rate per unit weight of catalyst $k_V \eta C_{H2,0} f(Cco,0)V$ is $mol \cdot g^{-1} \cdot s^{-1}$. On the other hand, CO flow rate per unit weight of catalyst is $(F/W)_{CO}$, so the CO conversion can be expressed as:

$$Conversion = \frac{k_v \eta C_{H2,0} f(C_{CO,0})V}{(F/W)_{CO}} \quad (18)$$

where V is the pore volume of the catalysts, and F/W is the reactant flow rate for the unit weight of the catalyst.

The diffusion coefficient was determined by the Stokes-Einstein method from hydrodynamics, as well as the experimental verification,

$$D = \frac{K_B T}{6\pi a f} \quad (19)$$

where D is the diffusion coefficient of a Brownian sphere of radius a in a continuum with viscosity f and temperature T, when no slip exists between the particle and the continuum, and K_B is the Boltzmann constant. Considering that the pores in catalysts are filled with high-boiling wax, the viscosity of FT wax at the experimental temperature (2.45 mPa · s, from the Shell Company) was used in the calculation.

The molecular diffusion coefficient obtained earlier does not consider the pore size effect. In gas-phase diffusion, there is a Wheeler equation to describe the diffusion coefficient in small pores:[72]

$$D = \frac{1}{3}\bar{u}\lambda(1 - e^{-2r/\lambda}) \quad (20)$$

where $\bar{u}$ is the mean speed of the gas molecule, λ is the mean free path, and r is the radius of the pore. In fact, the (1/3) $\bar{u}\lambda$ is the diffusion coefficient without a pore size effect in gas-phase diffusion,[73] so the $1 - e^{-2r/\lambda}$ term can be deemed to be the pore size effect factor. It is inferred that the hydrogen or carbon monoxide molecule in solvent follows Brownian motion, and that the hydrogen or carbon monoxide molecule is in a "viscous space". So the mean free path of hydrogen can be expressed as follow:[73]

$$\lambda = \frac{1}{\sqrt{2}\pi d^2 (P/K_B T)} = \frac{1}{\sqrt{2}\pi d^2 N_A C_{H2,0}} \quad (21)$$

where d is the molecule diameter, P the pressure, K_B the Boltzman constant, T the temperature, N_A the Avogadro constant, and $C_{H2,0}$ the concentration of hydrogen in solvent. As a result, the real diffusion coefficient in pores can be derived as:

$$D_e = D(1 - e^{-2r/\lambda}) \tag{22}$$

Assuming that the solubilities of H_2 and CO in the solvent follow Henry's Law, with the Henry coefficients being obtained from the literature,[74] the solubilities of H_2 and CO in the solvent (cetane) can be calculated to be 46.6 and 23.8 mol/m^3, respectively,under the reaction conditions.

The proper measurement of pore depth of a porous catalyst pellet can be expressed as follows:[75]

$$L = \frac{volume\ of\ pellet}{exterior\ surface\ available\ for\ reactant\ penetration\ and\ diffusion} \tag{23}$$

For a spherical pellet, $L = 3/r$, where r is the catalyst pellet radius.

In this experiment, the catalysts were sieved at 80–100 mesh, and the pellet radius was about 75 μm. The depth of the pore was approximately 25 μm. All the parameters for the simulation are shown in Table 7.

According to Equations 17 and 18, the dC_{CO}/dL and dC_{H2}/dL at $l = 0$ can be calculated.

$$\left(\frac{dC_{CO}}{dl}\right)_{l=0} = \frac{-k_V C_{H2,0} f(C_{CO,0}) L}{D_{CO}} = -\frac{conv. \times (F/W)_{CO} L}{D_{CO} \times V} \tag{24}$$

$$\left(\frac{dC_{H2}}{dl}\right)_{l=0} = \frac{-2k_V C_{H2,0} f(C_{CO,0}) L}{D_{H2}} = -\frac{2 \times conv. \times (F/W)_{CO} L}{D_{H2} \times V} \tag{25}$$

They and the reactants concentrations ($C_{CO,0}$ and $C_{H2,0}$) were set as initial values in the simulating Equations 26 and 27 using computational software (EQUATRAN®-G).

$$D_{H2} \frac{d^2 C_{H2}}{dl^2} = 2k_V C_{H2} f(C_{CO}) \tag{26}$$

$$D_{CO} \frac{d^2 C_{CO}}{dl^2} = k_V C_{H2} f(C_{CO}) \tag{27}$$

Table 7 The parameters for simulation and some simulation results. (Reproduced with permission from [60], copyright 2005 Wiley-VCH.)

Catalyst	Q-50	Q-3	Bimodal
Pore size (nm)	54	3.7	8.0; 47
Co particle size (nm)	37	1.4	22.6
Conv. (%)	11.8	16.9	32
D_{H2} (10^{-10} m^2/s)	4.74	0.434	4.28
D_{CO} (10^{-10} m^2/s)	3.38	0.308	3.05
$C_{H2,0}$ (mol/m^3)	46.6	46.6	46.6
$C_{CO,0}$ (mol/m^3)	23.8	23.8	23.8
L (10^{-5} m)	2.5	2.5	2.5
k_V (k^*_V for bimodal catalyst) (s^{-1})	0.0258	1.094	0.403
k_{Co} (10^{-8} m/s)	1.05	74.1	3.81
Catalyst Effective Factor η	0.98	0.17	0.75
ΔLn(conv.)/Δ(1/T)(K)	13409.4	7237.7	9387.8
Activation energy (kJ/mol)	111.5	60.2	78.0

In the simulation, the pore depth L was divided into 1,000 segments. On a given k_V or k^*_V, the program gave a series of C_{CO}, C_{H2}, dC_{CO}/dL and dC_{H2}/dL. Different k_V or k^*_V were tested until the simulated dC_{CO}/dL and dC_{H2}/dL values fit the boundary conditions (Equation 11). Consequently, the metal surface kinetic constant k_{Co} was calculated by Equations 7 and 8. The simulated reactant concentration profiles in different pores are shown in Fig. 13. For Q-50 and bimodal catalysts, the concentration profiles of reactants inside the pores were changed slightly, while in the case of the Q-3 catalyst, it decreased quickly. The results suggest that the pore diffusion resistance in the bimodal catalyst is not strong. From the reactant concentration distribution profiles, the CO concentration in Q-3 catalyst pores will decrease to 0 when $l/L = 0.22$, which may be another reason for the high CH_4 selectivity.[76]

Some other simulation results are compared in Table 7. First, the explanations for neglecting the inner pore diffusion in small pores of bimodal catalysts will be given. The large pore size of bimodal catalyst (47 nm) is 7 nm smaller than the Q-50 catalyst (54 nm), due to the fact that nanoparticles formed the small pores on the walls of the original Q-50 pores. So the mean depth of small pores is about 3.5 nm. According to Equations 19–22, the H_2 and CO diffusion coefficients in small pores are calculated to be about $6.68 \times 10^{-11}\,m^2 \cdot s^{-1}$ and $3.33 \times 10^{-11}\,m^2 \cdot s^{-1}$, respectively. The metal surface reaction kinetic constant k_{Co} in bimodal catalysts is $3.81 \times 10^{-8}\,m \cdot s^{-1}$. From the relationship of the kinetic constant based on surface area, and the kinetic constant based on pore volume in cylinder pores (Equation 7), the reaction kinetic constant k_V in small pores is 0.521 s^{-1}. The simulation results of the diffusion of reactants in small pores of the bimodal catalyst showed that the catalyst effectiveness factor η is very close to 1 (0.999996) in these small pores. From Fig. 13d, the reactant concentration distribution profile inside the small pores suggests that the assumption of neglecting the inner-pore diffusion is reasonable.

The reaction rate based on specific metal surface k_{Co} increased with decreasing metal particle size. For bimodal catalysts, there are some small cobalt particles in small pores and some large particles in large pores. The reaction rate of the small cobalt particles is as high as that of the Q-3 catalyst, while that of the large cobalt particles is comparable to the Q-50 catalyst. These lead to a moderate average reaction rate for the bimodal catalyst. The catalyst effectiveness factors of Q-50, Q-3, and bimodal catalysts are 0.98, 0.17 and 0.75, respectively, which means that there is almost no pore diffusion resistance in the Q-50 catalyst, while a strong pore diffusion resistance exists for the Q-3 catalyst. Furthermore, the pore diffusion resistance in bimodal catalysts is not important. The apparent activation energies estimated from the CO conversions between 513 K and 533 K also show an apparent activation energy of Q-50 without pore diffusion resistance is 111.5 KJ/mol. This value is about twice the activation energy of Q-3 with a strong pore diffusion resistance. Also, the moderate activation energy of bimodal catalyst suggests that the pore diffusion resistance is not very strong. Conclusions drawn from the simulations are very consistent with the experimental results.

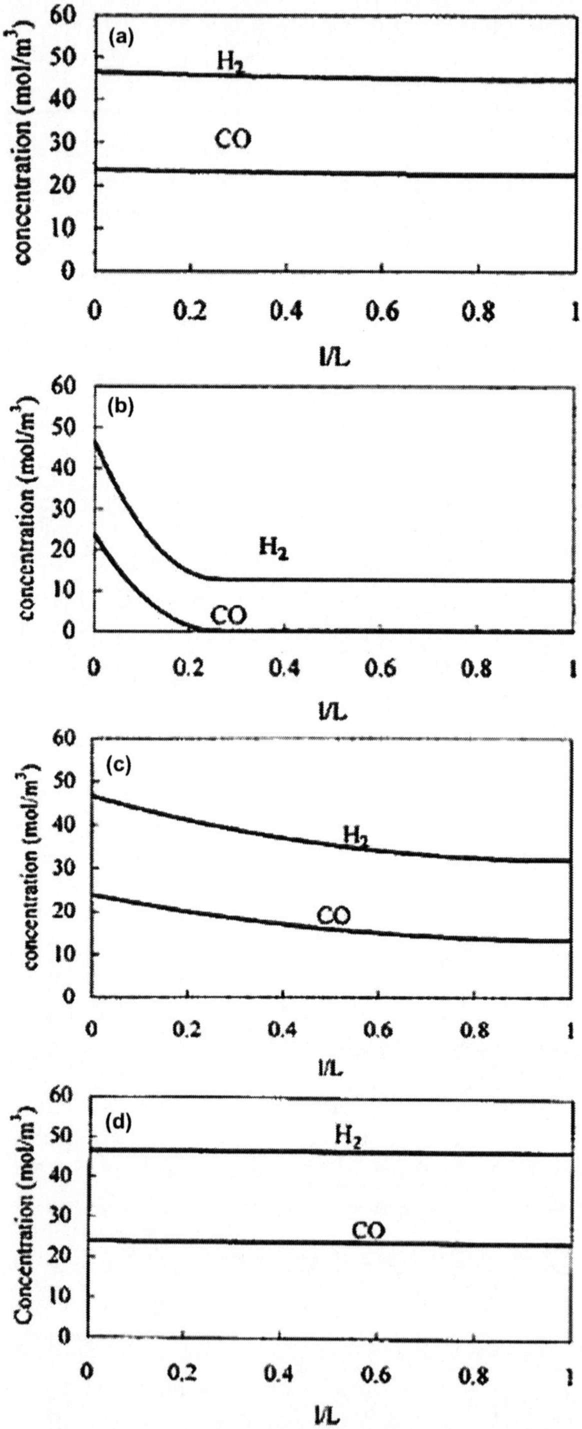

Fig. 13 Simulation of the effect of catalyst pore size on the reactant concentration profiles inside the pores. (a) Q-50 catalyst; (b) Q-3 catalyst; (c) bimodal catalyst; (d) small pore in bimodal catalyst. (Reproduced with permission from [60], copyright 2005 Wiley-VCH.)

4 Conclusions

Although it was developed over 80 years ago, FTS is still an important technology for the production of clean transportation fuels and chemicals. Over the past decades, substantial efforts have been devoted to increasing the catalyst activity, especially improving the product distribution, which still remains an important challenge. This chapter has elaborated on two kinds of novel FTS catalysts: the core-shell catalysts and bimodal catalysts. They are successful examples of catalyst design and development for FT reaction.

The core-shell catalysts are designed by coating a zeolite membrane on the surface of a pre-shaped catalyst pellet. This core-shell structure provides a custom-tailored confined reaction environment, which produces a spatially confined effect and shape selectivity. Its successful application in the direct synthesis of isoparaffins from syngas provides insight and inspiration for other consecutive reactions, such as one-step synthesis of DME, as the shell and core components are independent catalysts for different reactions.

For bimodal catalysts, both the experimental investigations and mathematical simulations have demonstrated that the bimodal pore structure can increase the FTS activity through higher diffusion efficiency in large pores and higher dispersion of active metal particles *via* increased surface area. The method described in this chapter has been exploited to prepare various supports and catalysts with distinct bimodal pore structures by using oxide or active components as building blocks to produce small pores inside large pores of homo- or hetero-atom structures. Furthermore, the large pore size of the bimodal structure can be successfully controlled. The method can be easily extended to prepare various bimodal catalysts with different chemical compositions. Besides the spatial effect, new chemical effects might appear with the formation of hetero-atom structure.

Acknowledgements

We thank Drs. Yoshiharu Yoneyama and Dr. Yi Zhang, University of Toyama, Japan, for helpful discussions.

References

1 G. Henrici-Olive and S. Olive, *The Chemistry of the Catalyzed Hydrogenation of Carbon Monoxide*, Springer, Tokyo, 1984.
2 H. L. Chum and R. P. Overend, *Fuel Process. Technol.*, 2001, **71**, 187.
3 A. A. Adesina, *Appl. Catal. A*, 1996, **138**, 345.
4 J. P. Hindermann, G. J. Hutchings and A. Kiennemann, *Cat. Rev. - Sci. Eng.*, 1993, **35**, 1.
5 H. Schulz, *Appl. Catal. A*, 1999, **186**, 3.
6 A. J. Henricus and D. van, *The Fischer-Tropsch Synthesis: A Mechanistic Study Using Transient Isotopic Tracing*, [Ph. D Thesis of Tech], Eindhoven University of Technology, 2001.
7 X. Huang and C. B. Roberts, *Fuel Process. Technol.*, 2003, **83**, 81.
8 L. Fan, K. Yokata and K. Fujimoto, *AIChE. Journal*, 1992, **38**, 1639.
9 N. Tsubaki, K. Yoshii and K. Fujimoto, *J. Catal.*, 2002, **207**, 371.
10 K. Fujimoto, L. Fan and K Yoshii, *Top. Catal.*, 1995, **2**, 259.

11 Y. K. Kee, H. S. Roh, Y. T. Seo, D. J. Seo, W. L. Yoon and S. B. Parka, *Int. J. Hydrogen Energy*, 2008, **33**, 2036.
12 D. Mahajan, P. Gutlich and U. Stumm, *Catal. Commun.*, 2003, **4**, 101.
13 A. M. Saib, A. Borgna, J. van de Loosdrecht, P. J. van Berge, J. W. Geus and J. W. Niemantsverdriet, *J. Catal.*, 2006, **239**, 326.
14 A. N. Pour, Y. Zamani, A. Tavasoli, S. M. K. Shahri and S. A. Taheri, *Fuel*, 2008, **87**, 2004.
15 A. Martínez, J. Rollán, M. A. Arribas, H. S. Cerqueira, A. F. Costa and E. F. S. Aguiar, *J. Catal.*, 2007, **249**, 162.
16 K. S. Triantafyllidis, V. G. Komvokis, M. C. Papapetrou, I. A. Vasalos and A. A. Lappas, *Stud. Surf. Sci. Catal.*, 2007, **170**, 1344.
17 M. P. Rohde, D. Unruh and G. Schaub, *Catal. Today*, 2005, **106**, 143.
18 A. Martínez and C. López, *Appl. Catal. A*, 2005, **294**, 251.
19 N. Lohitharn and G. Goodwin Jr. J., *J. Catal.*, 2008, **257**, 142.
20 G. Jacobs, Y. Y. Ji, B. H. Davis, D. Cronauer, A. J. Kropf and C. L. Marshall, *Appl. Catal. A*, 2007, **333**, 177.
21 G. Prieto, A. Martínez, P. Concepción and R. Moreno-Tost, *J. Catal.*, 2009, **266**, 129.
22 Z. J. Wang, S. Skiles, F. Yang, Z. Yan and D. W. Goodman, *Catal. Today*, 2012, **181**, 75.
23 R. B. Anderson, *The Fischer-Tropsch Synthesis*, Academic Press, New York, 1984.
24 J. Patzlaff, Y. Liu, C. Graffmann and J. Gaube, *Appl. Catal. A*, 1999, **186**, 109.
25 N. Tsubaki and K. Fujimoto, *Fuel Process. Technol.*, 2000, **62**, 173.
26 Z. W. Liu, X. H. Li, K. Asami and K. Fujimoto, *Appl. Catal. A*, 2006, **300**, 162.
27 (a) F. G. Botes and W. Bohringer, *Appl. Catal. A*, 2000, **267**, 217; (b) A. Feller, A. Guzman, I. Zuazo and J. A. Lercher, *J. Catal.*, 2004, **224**, 80.
28 Y. W. Chen, H. T. Tang and J. G. Goodwin Jr., *J. Catal.*, 1983, **83**, 415.
29 H. H. Nijs and P. A. Jacobs, *J. Catal.*, 1980, **66**, 401.
30 B. Imelik, Y. Naccache, J. C. Vedrine, G. Coudurier and H. Praliaud, *Catalysis by Zeolite*, Elesevier, Amsterdam, 1980.
31 X. Li, K. Asami, M. Luo, K. Michiki, N. Tsubaki and K. Fujimoto, *Catal. Today*, 2003, **84**, 59.
32 T. Zhao, Y. Yoneyama and N. Tsubaki, *Ind. Eng. Chem. Res.*, 2005, **44**, 755.
33 Q. J. Ge, Y. M. Huang, F. Y. Qiu and S. B. Li, *Appl. Catal. A*, 1998, **167**, 23.
34 J. He, Y. Yoneyama, B. Xu, N. Nishiyama and N. Tsubaki, *Langmuir*, 2005, **21**, 699.
35 J. He, B. Xu, Y. Yoneyama, N. Nishiyama and N. Tsubaki, *Chem. Lett.*, 2005, **34**, 148.
36 J. He, Z. Liu, Y. Yoneyama, N. Nishiyama and N. Tsubaki, *Chem. Eur. J.*, 2006, **12**, 8296.
37 G. Yang, J. He, Y. Yoneyama and N. Tsubaki, *Stud. Surf. Sci. Catal.*, 2007, **167**, 73.
38 G. Yang, J. He, Y. Yoneyama, Y. S. Tan, Y. Z. Han and N. Tsubaki, *Appl. Catal. A*, 2007, **329**, 99.
39 J. Bao, J. He, Y. Zhang, Y. Yoneyama and N. Tsubaki, *Angew. Chem. Int. Ed.*, 2008, **47**, 353.
40 (a) D. E. Beving, A. M. P. McDonnell, W. S. Yang and Y. S. Yan, *J. Electrochem. Soc.*, 2006, **153**(8), B325; (b) Y. S. Yan, M. E. Davis and G. R. Gavalas, *Ind. Eng. Chem. Res.*, 1995, **34**, 1652.
41 M. Noack, P. Kolsch, J. Caro, M. Schneider, P. Toussaint and I. Sieber, *Micro. Meso. Mater.*, 2000, **35–36**, 253.
42 G. Jacobs, T. K. Das, J. L. Li, M. S. Luo, P. M. Patterson and B. H. Davis, *Stud. Surf. Sci. Catal.*, 2007, **163**, 217.

43 E. Iglesia, S. L. Soled and R. A. Fiato, *J. Catal.*, 1992, **137**, 212.
44 B. G. Johnson, C. H. Bartholomew and D. W. Goodman, *J. Catal.*, 1991, **128**, 231.
45 S. L. Soled, J. E. Baumgartner, S. C. Reyes and E. Iglesia, *Proc. Mater. Res. Soc. Symp.*, 1995, **368**, 113.
46 E. Iglesia, S. C. Reyes and R. J. Madon, *J. Catal.*, 1991, **129**, 238.
47 O. Levenspiel, *Chemical Reaction Engineering*, Wiley, New York, 1972.
48 T. Inui, M. Funabiki, M. Suehiro and T. Sezume, *J. Chem. Soc. Faraday Trans.*, 1979, **175**, 787.
49 M. Inoue, K. Kitamura and T. Inui, *J. Chem. Technol. Biotechnol.*, 1989, **46**, 233.
50 R. Takahashi, S. Sato, T. Sodesawa and M. Yabuki, *J. Catal.*, 2001, **200**, 197.
51 N. Tsubaki, Y. Zhang, Y. Yoneyama, S. Sun, H. Mori, X. Li and K. Fujimoto, *Catal. Commun.*, 2001, **2**, 311.
52 Y. Zhang, Y. Yoneyama and N. Tsubaki, *Chem. Commun.*, 2002, 1216.
53 Y. Zhang, Y. Yoneyama, K. Fujimoto and N. Tsubaki, *J. Chem. Eng. Jpn.*, 2003, **36**, 874.
54 Y. Zhang, Y. Yoneyama and N. Tsubaki, *J. Jpn Petro. Inst.*, 2003, **46**, 335.
55 S. Sun, K. Fujimoto, Y. Zhang and N. Tsubaki, *Catal. Commun.*, 2003, **4**, 361.
56 Y. Zhang, Y. Yoneyama, K. Fujimoto and N. Tsubaki, *Top. Catal.*, 2003, **26**, 129.
57 M. Shinoda, Y. Zhang, Y. Yoneyama, K. Hasegawa and N. Tsubaki, *Fuel Process. Technol.*, 2004, **86**, 73.
58 Y. Zhang, M. Shinoda and N. Tsubaki, *Catal. Today*, 2004, **93**, 55.
59 Y. Zhang, M. Koike and N. Tsubaki, *Catal. Lett.*, 2005, **91**, 193.
60 B. Xu, Y. Fan, N, Y. Zhang and N. Tsubaki, *AIChE J.*, 2005, **51**, 2068.
61 Y. Zhang, M. Koike, R. Yang, H. Sukamon, V. Tharapong and N. Tsubaki, *Appl. Catal. A*, 2005, **292**, 252.
62 Y. Zhang and N. Tsubaki, *Stud. Surf. Sci. Catal.*, 2006, **172**, 233.
63 Y. Zhang, Y. Liu, G. Yang, S. Sun and N. Tsubaki, *Appl. Catal. A*, 2007, **321**, 79.
64 Y. Zhang, S. Nagamori, M. Yamaguchi, S. Rengakuji and N. Tsubaki, *Catal. Commun.*, 2008, **9**, 902.
65 M. F. L. Johnson and J. Mooi, *J. Catal.*, 1968, **10**, 342.
66 H. P. Withers, K. F. Eliezer Jr and J. W. Mitchell, *Ind. Eng. Chem. Res.*, 1990, **29**, 1807.
67 M. Ichikawa and T. Fukushima, *J. Phys. Chem.*, 1985, **89**, 1564.
68 A. L. Borer, C. Bronnimann and R. Prins, *J. Catal.*, 1994, **145**, 516.
69 A. Hoek, A. H. Joustra, J. K. Minderhoudt and M. F. Post, UK Patent Application GB 2 125062 A, 1983.
70 F. Rohr, O. A. Lindvag, A. Holmen and Z. A. Blekkan, *Catal. Today*, 2000, **58**, 247.
71 Y. Zhang, J. Bao, S. Nagamori and N. Tsubaki, *Appl. Catal. A*, 2009, **352**, 277.
72 A. Wheeler, *Adv. Catal.*, 1951, **3**, 249.
73 P. W. Atkins; *Physical Chemistry*, 3rd ed, Oxford: Oxford University Press, 1986.
74 B. B. Breman, A. A. C. M. Beenackers, E. W. J. Rietjens and R. J. H. Stege, *J. Chem. Eng. Data*, 1994, **39**, 647.
75 O. Levenspiel, *Chemical Reaction Engineering*, Wiley, New York, 1972.
76 R. J. Madon and E. Iglesia, *J. Catal.*, 1994, **149**, 428.